The Evolution of Cranes at Fairfield shipyard.

More than any other structure, towering high above the ships they built and often visible for miles around, cranes defined a shipyard.

Fitting-out cranes.

Fitting-out cranes lifted heavy loads into hulls once launched. These loads were typically boilers, main machinery as well as auxiliary and gun mountings and armour plates. At Fairfield, several sheerlegs were used for this purpose culminating in the 120 ton sheerlegs shown here designed by Andrew Laing and built in the shipyard's own shops. The increasing size of ships and their components, particularly the main armament mountings for battleships, required a much larger capacity crane. In 1911, the giant cantilever crane shown here was installed and tested to 250-tons with a normal maximum working load of 220 tons.

Berth cranes.

In the riveted era, berth cranes, either fixed or travelling, lifted frames, beams, sections and plates into position on the building berth. In the prefabrication era, berth cranes lifted into position. Berth cranes gradually evolved from pole derricks lifting just a few tons structures capable of lifting up to 80 tons and more in the prefabrication era. From c2 onwards, construction at the Govan yard no longer required berth cranes.

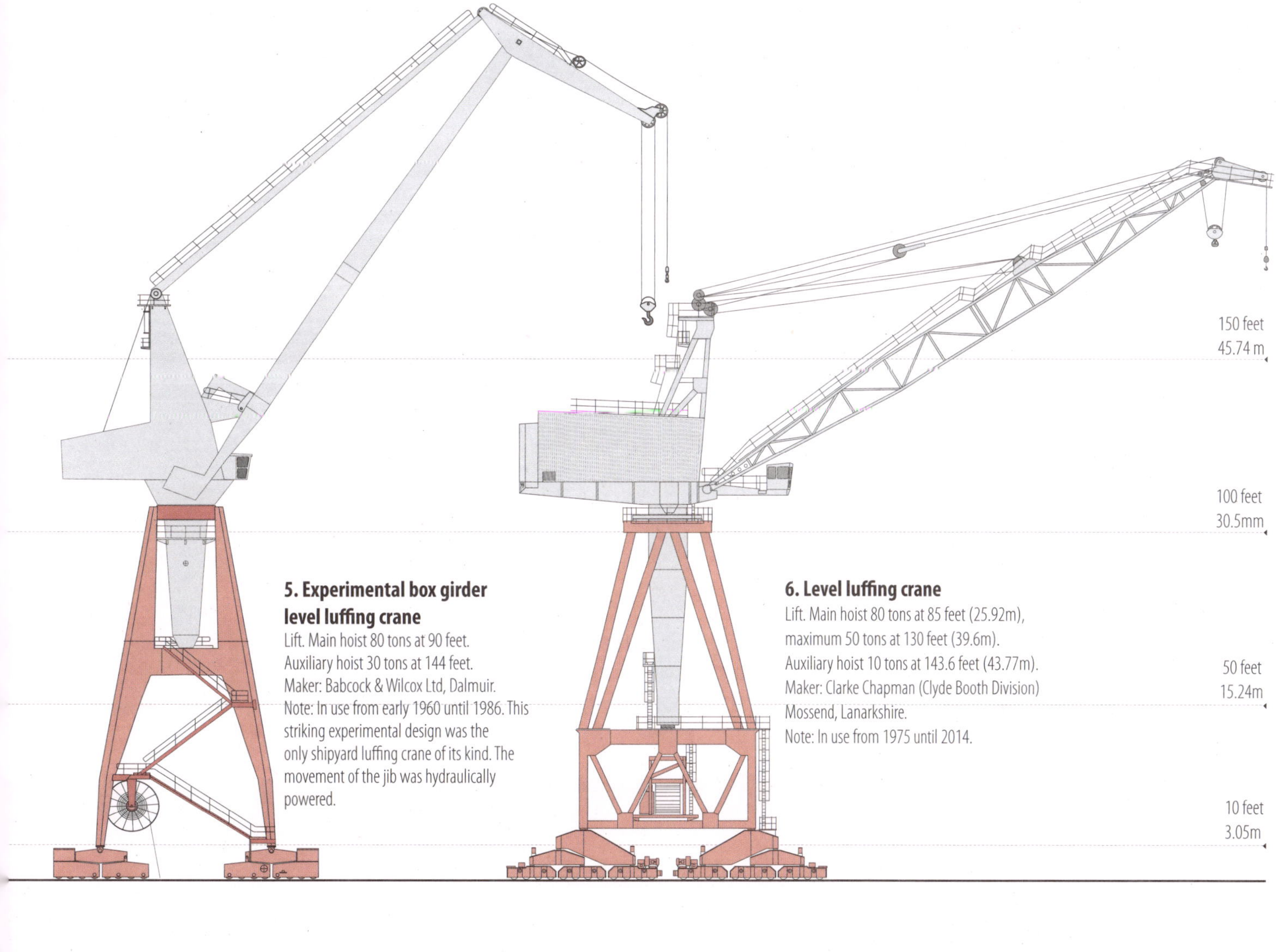

5. Experimental box girder level luffing crane

Lift. Main hoist 80 tons at 90 feet.
Auxiliary hoist 30 tons at 144 feet.
Maker: Babcock & Wilcox Ltd, Dalmuir.
Note: In use from early 1960 until 1986. This striking experimental design was the only shipyard luffing crane of its kind. The movement of the jib was hydraulically powered.

6. Level luffing crane

Lift. Main hoist 80 tons at 85 feet (25.92m), maximum 50 tons at 130 feet (39.6m).
Auxiliary hoist 10 tons at 143.6 feet (43.77m).
Maker: Clarke Chapman (Clyde Booth Division) Mossend, Lanarkshire.
Note: In use from 1975 until 2014.

FAIRFIELD

FAIRFIELD

A Shipyard Success Story 1834–2024

IAN JOHNSTON

Seaforth
PUBLISHING

Frontispiece: A view from the giant cantilever crane showing the building berths of the main yard on a misty day in 1918. Nearest the camera on No 1 berth is the battlecruiser HMS *Rodney*, sister ship to HMS *Hood*. Suspended in March 1917, work never resumed on what would have been Fairfield's largest warship. The existing structure was broken up on the slip in 1919. (*Imperial War Museum*)

Half title page: The Cunard Atlantic liner *Campania* under construction in 1892.

First published in Great Britain in 2024 by Seaforth Publishing
An imprint of Pen & Sword Books Ltd George House, Beevor Street,
Barnsley S71 1HN

www.seaforthpublishing.com
Email info@seaforthpublishing.com

British Library Cataloguing in Publication Data
A CIP data record for this book is available from the British Library

ISBN 978-1-3990-8966-1 (Hardback)
ISBN 978-1-3990-8967-8 (ePub)
ISBN 978-1-3990-8969-2 (PDF)

Pen & Sword Books Limited incorporates the imprints of Atlas, Archaeology, Aviation, Discovery, Family History, Fiction, History, Maritime, Military, Military Classics, Politics, Select, Transport, True Crime, Air World, Frontline Publishing, Leo Cooper, Remember When, Seaforth Publishing, The Praetorian Press, Wharncliffe Local History, Wharncliffe Transport, Wharncliffe True Crime and White Owl.

Typeset and designed by Mousemat Design Limited
Printed and bound in India by Replika Press Pvt. Ltd.

FSC
www.fsc.org
MIX
Paper | Supporting responsible forestry
FSC™ C016779

CONTENTS

ACKNOWLEDGEMENTS

Writing the history of such a significant company as the Fairfield shipyard at Govan was a major challenge. This was not so much the case for early sections where the events that shaped the Company were well documented and archival sources mostly plentiful, but for the sections within living memory, where those directly involved in events, sometimes controversial, were able to comment, correct and advise.

In particular, I would like to thank the following individuals for giving freely of their experience: John Bowen, Jim Craig, Steinar Draegebo, Paul Friedberg, Bill Gallagher, Jamie Gilchrist, Bobby Gordon, Sven Hanssen, David Millar, Ian Morrison, Eileen Reid, Paul Sweeney and Jim Thomson.

Special thanks to Abigail Morris for general assistance and good humour in finding facts, people and sources as well as help with the Ship List.

I would like to thank the following organisations: BAE Systems for their support in funding images for this book. Glasgow City Archives based at the Mitchell Library for assistance and help over many visits and in particular to Dr Irene O'Brien, Michael Gallagher, Barbara McLean and Nerys Tunnicliffe. Pat Cassidy of Govan Workspace Ltd for support and access to Fairfield Heritage and the museum dedicated to the history of the Fairfield shipyard. Thanks also to Paul Bevand and Michael Conway

For reading over and commenting on the manuscript: Professor Hugh Murphy; Paul Sweeney MSP.

To John R Hume a very special thanks for reading over the manuscript and writing the Foreword as well as making helpful suggestions. These include, in particular, his insight into the relationship between Charles Randolph, David Elder and John Elder and the early years of the business that would become Fairfield which I have endeavoured to include. John must also be recognised and thanked for the pivotal role that he played along with his colleague, Michael Moss, in ensuring that the records from which this history has been written survived into preservation following the collapse of Upper Clyde Shipbuilders in 1971.

As a young man, I walked past this shipyard many times, heard the Fairfield hooter go several times a day, heard the distant sound of caulkers' hammers through the night and saw thousands of men stream out onto Govan Road. Little did I realise then the struggles that this yard faced for survival. In October 1971, when shop stewards stood at the main gate at the height of the Upper Clyde Shipbuilders work-in, I entered the yard for the first time. I was awe-struck at the sheer scale and purposefulness of what I saw, played out against the drama of the dynamic soundscape of machinery, instructions being shouted and the direct action of metal against metal. In those moments I thought, finally, as a Glaswegian, so this is what it's all about – it was another world.

Finally, I would like to express my thanks to Robert Gardiner for his support in publishing my books over the past 13 years.

As ever, I would like to thank my wife Linda for tolerating my absence in the attic writing this and other books. Linda is also responsible for the liberal application of commas throughout the text for which I thank her.

FOREWORD

When I first learned that Ian Johnston was going to write a book about the Fairfield yard, I was delighted, for the Fairfield/Govan yard has been part of my life for many years, and I am a considerable admirer of Ian's writing, and of his extraordinary skill as a graphic artist.

It was a great pleasure to hear that Ian wanted me to write this Foreword. I have thought long and hard about what form this should take, and have started several drafts, none of which, in retrospect, seems to be worthy of a decent introduction to a masterly study.

For this is what Ian has created, making sense of what at times is a bewildering story. When I started studying and teaching industrial history in the late 1950s and early 1960s, I assumed that the history of such a well-known and indeed pivotal industry as Clyde shipbuilding would be well known. How wrong I was. Ian's study reveals just how difficult it is to get to grips not with the industry as it existed at that time but with the impact of societal and technical change in the 60 or so years I have to a greater or lesser extent known shipbuilding on the Clyde and some of the key players in the changing fortunes of that industry. And here I know I relapse into the difficulty of making sense of an industry in which change is always happening. Ian's graphic depiction of change in the types of ships built in the Govan yard is one of the great strengths of his most thoughtful and creative analysis of Fairfield/Govan – the ability to change, to adapt to alteration in actual and potential markets. and to developing techniques. But though such changes are bewildering, there are aspects of shipbuilding that do not change in principle: the design of vessels to suit different purposes, to ensure that such vessels are able to withstand the forces that affect a ship at sea, and the development of production techniques which allow such vessels to be constructed as efficiently as possible but with the underlying necessity of insuring structural integrity. The need to reconcile these often conflicting essentials poses challenges to the management and the workforce for which they are responsible. To these basic essentials should be added others, more obvious in a way – the development and maintenance of good relationships with customers, based on effective delivery of agreed outcomes, but also on reliability on service. But that is not enough. It is said of a good building that it must afford 'Firmness, Commodity and Delight' and this is just as true of a ship. Fairfield/Govan, as Ian shows in his study, is a yard which has over the years, despite major changes in ownership and types of vessels produced, consistently built not only soundly designed and constructed vessels, but ones which by and large have that indefinable quality of 'grace'.

This can, I believe, be traced back to Charles Randolph and before him to a tradition in Govan of belief in a loving and caring God. In the specific instance of Randolph, my growing understanding rooted in Ian Johnston's study, is that after Robert Napier had sacked John Elder and his father, David, had also left the Napier business (in whose success he had been critical), Randolph saw in John, the son he never had, someone who could carry on his business with enthusiasm and transform it by developing the concept of applying 'compounding' to marine propulsion. The magnificent, even awe-inspiring, engine works the new partnership constructed in Tradeston was, I believe, intended to show Glasgow and its business community that the new firm was going to challenge Robert Napier's firm, and the purchase of the Fairfield Estate and the development of the largest integrated iron shipbuilding yard and engine works in the world was further evidence of a desire to take on Napier. But, as far as I am in a position to judge, these moves were motivated not just by sound business thinking but also by a real act of kindness and caring by Randolph. I leave it to readers of Ian's study to think about this reasoning and to speculate if other key figures in the yard such as Isabella Elder, Francis Elgar, Sir James Lithgow, Iain Stewart and Eric Mackie may also have been motivated by kindness as well as business sense. Govan Workspace, under Pat Cassidy and Ian himself are also open to such an interpretation.

But let me return to the beginning of this Foreword, to say that I found Ian's study enthralling, full of insights and based on the soundest of research. This is a major contribution not only to the history of the Fairfield/Govan yard but of shipbuilding on the Clyde in general. I am deeply moved by being asked to contribute this.

John R Hume
February 2024

INTRODUCTION

More books have been written about the Fairfield shipyard in Govan than about any other British shipyard. This is yet another one and it will almost certainly not be the last. The shipyard's attraction stems from its newsworthiness. This came about firstly through its great technical achievements, record breaking, and the prestige and national interest generated in its early years. Much later, its prolonged and ultimately successful fight for survival within a rapidly diminishing industrial landscape that characterised shipbuilding in Britain from the 1960s onwards continued this interest.

This book is an attempt at combining that long history in recognition that a British shipyard has survived into the twenty-first century despite the odds. Inevitably, in such a lengthy history extending for close on 190 years, only a fraction of what actually happened can be told. In recording the broad sweep of events, this book drops into occasional detail and thus is rather episodic.

The name Fairfield stems from the site chosen to build the shipyard on, previously Fairfield Farm, part of Fairfield Estate, in Govan, at that time a separate township to the immediate west of Glasgow. Although the yard began operation under the title Randolph Elder & Co in 1864 and then John Elder & Co from 1871 until 1885, it was then renamed the Fairfield Shipbuilding & Engineering Co. However, the yard was always known by shipyard workers and locals from the outset as Fairfield. Even today, in 2024, the name is synonymous with the shipyard.

Randolph Elder & Co were relative late comers in the developing story of Clyde shipbuilding with yards such as Robert Napier's and Tod & McGregor's having been instrumental in beginning the process where Clydeside became the most important river in the UK for shipbuilding and marine engineering. The first man who would take the Fairfield yard to prominence was John Elder whose mechanical inventiveness and, in particular, his pursuit of fuel efficiency, then as now of great importance, resulted in a practical compound steam marine engine that was significantly more fuel efficient than the marine steam engines then in use. Success in building compound engines led Elder with his partner Charles Randolph, to decide that building the hulls to put their engines in made commercial as well as technical sense and thus the Fairfield shipyard came into being. What Elder might have gone on to do was cut short by his early death at the age of 45.

The second person instrumental in the rise of Fairfield was William Pearce, an Admiralty-trained naval architect from Chatham. Like many of his contemporaries, he moved to the Clyde where some of the most interesting developments in marine engineering and shipbuilding were taking place. In 1870, following John Elder's demise, Pearce began an association with Fairfield which led to his outright ownership and control of the business. Through his energy and a degree of showmanship, Pearce secured a number of major contracts which resulted in the fastest Atlantic liners of the day. By the time of his death, unexpectedly at the age of 55, Fairfield had become the leading Clyde shipyard both in terms of scale of operations and the prestige of the contracts secured. This period ends in the construction of many famous ships, most notably a pair of record-breaking Cunarders, *Campania* and *Lucania*.

If the first half of the twentieth century, punctuated by two world wars with a prolonged depression in between, can be seen as a period when Fairfield settled into its position as a leading British shipbuilder, the events that followed raised the yard to one of national importance but for entirely different reasons.

On its 100th anniversary in 1964, Fairfield, through a major scheme of investment, was considered to be among the most modern shipyards in Europe. The reality, however, was that Fairfield, along with much of the UK shipbuilding industry, was heading into financial difficulties in part because of low productivity and outmoded labour relations.

In 1965, the greatest shock to the prestige of Clyde shipbuilding was delivered when Fairfield went into administration despite a full order book and 5000 employees. The ensuing crisis led to one of the most remarkable periods in the life of this shipyard and industrial Britain. Fairfield became a national proving ground, an 'experiment' into new methods, improved productivity, industrial relations and management practices aimed at pointing the way to profitability not just at Fairfield but across the industry as a whole. The Fairfield Experiment as it became known, remains one of the most intriguing episodes in the long history of this yard. The experiment, which had its detractors, would prove to be short lived, with Fairfield absorbed into another Government-funded project, much greater in scale, the amalgamation of shipyards into regional groups making them theoretically leaner, more productive and thus competitive.

The grouping of which Fairfield became a part was the ill-fated Upper Clyde Shipbuilders Ltd, (UCS) comprising four other yards on the upper Clyde including Fairfield's great rival, John Brown & Co at Clydebank. With a large order book but insufficient working capital to finance its contracts, after three and a half years and a series of crises requiring cash bail-outs, UCS was forced into liquidation. The collapse of UCS led to a prolonged period of acrimony, high drama and eventual resolution. This was characterised by the shop stewards and the action they took in the form of the 'Work-in'. This very successful campaign promoting the 'right to work' caught the public imagination and wrong-footed the Government, resulting in the yards being retained.

At Fairfield, a further Government-backed company was formed in 1972 taking the name Govan Shipbuilders Ltd. Thus, the Fairfield name, so long associated with technical excellence and craftsmanship, was consigned to history. Five years later, the clamour to nationalise the shipbuilding industry resulted in the formation of British Shipbuilders. During this period, the industry continued to decline until, in 1983, the decision was made by Government to divest itself of endemic loss-making by privatising the remaining yards, including Govan Shipbuilders.

As a persistent loss-maker, there seemed to be little hope for the Govan yard until, in 1987, a Norwegian industrial conglomerate, Kvaerner, acquired the yard. Assisted by the UK Government, Kvaerner made significant investment in plant and facilities, to enable the construction of sophisticated gas and chemical carriers profitably. At the same time, fundamental improvements to productivity agreements and industrial relations were made. Altogether, this resulted in a remarkable turn-around, repositioning the Govan yard as one of the most efficient in Europe on the cusp of profitability. However, in 1999, Kvaerner made the decision to retreat from shipbuilding, not just on Clydeside, but across all of its worldwide shipbuilding interests.

Yet again, the future of the Govan yard looked bleak. However, as commercial shipbuilding in Britain was reduced to statistical irrelevance, the reorganisation of the defence industries made Govan an acquisition target for newly formed BAE Systems plc. With the former Yarrow yard at Scotstoun already part of this new company, the Govan and Scotstoun yards were integrated, becoming the largest builder of surface warships in the UK.

At the time of writing in 2023, BAE Systems have embarked upon a major investment programme at its Clyde shipbuilding facilities including a new shipbuilding hall at Govan. With contracts for sophisticated warships and a state-of-the-art shipyard, shipbuilding is set to continue at Fairfield for decades to come.

Ian Johnston
March 2024

ERA ONE 1830–1884

'The Great Country of the Steam Engine[1]

By the time the shipyard at Fairfield Farm became operational at Govan in 1864, the Clyde was already well on its way to becoming the most important marine engineering and shipbuilding river in the UK. The subsequent rise of this shipyard to the top of the shipbuilding industry by the 1890s was down to two individuals, John Elder, a pioneering marine engineer and William Pearce, a naval architect and entrepreneur.

Long before John Elder rose to prominence, the development of marine engineering and later, shipbuilding on the Clyde had already been marked by a succession of gifted engineers, not least his father, David Elder. To set the historical context in which the Fairfield yard would flourish and draw a line from the early pioneers of the marine steam engine to John Elder and the breakthroughs he would make, some exploration and consideration of his predecessors' achievements is necessary.

Various individuals had carried out pioneering work in adapting the steam engine, as developed by James Watt, to propel vessels. Most notably in the UK, this was achieved by Patrick Miller and William Symington in the 1780s and later, Symington and the steam boat *Charlotte Dundas* in 1803. Although these experiments were a practical success, they did not lead to commercial reward. This was not achieved until Henry Bell's *Comet* sailed down the River Clyde in January 1812, becoming the first commercially successful steamship in Europe.

After the success of *Comet* in 1812, a series of steam boats were ordered for service on the river and by 1817 there were 24 such boats sailing on the Clyde. All of these vessels were built either in Port Glasgow or Greenock, as the Clyde at Glasgow was not yet navigable. What Glasgow did have, as a city of 120,000 in 1816, was a developing manufacturing infrastructure which included the skills necessary to keep industry and its machinery working. Skilled people included millwrights and iron founders, sometimes referred to as engineers. Steam engines were the high technology of the day and two millwrights, James Cook and James Robertson, both with premises in Glasgow at Tradeston, and Duncan McArthur with works in Glasgow at Camlachie, were responsible for many of the early steam engines.[2]

Steam Pioneers

David Napier 1790–1869

However, Henry Bell and John Robertson, the Glasgow engineer who had manufactured *Comet*'s steam engine, were not the only people interested in making marine steam engines in Glasgow.[3] David Napier, an engineer who trained in his father's works in Dumbarton and later at the Camlachie Foundry[4] in Glasgow, emerged as a talented engineer who had made the boiler for *Comet* in 1812. He would go on to construct his first marine engine for the vessel *Marion*, named after his wife, in 1816. Napier designed and manufactured a large number of steam engines including for the paddle steamer *Rob Roy* in 1818. Although not able to build hulls, Napier nevertheless conducted a series of trials to establish the hull form for this vessel using models on a nearby stream known as Camlachie Burn which was adjacent to his Camlachie Foundry. The *Rob Roy*, with its hull built by William Denny, the Dumbarton shipbuilder, became the first vessel to operate a service on the open sea and later, in French ownership, as the *Henri Quatre*, it became the first French steamboat. In 1821, Napier purchased ground at Lancefield on the north bank of the Clyde at Glasgow where he erected works which he called the Lancefield Foundry. At the same time, he was able to excavate a small dock into the river wall, by his works, an early version of the fitting-out basin. To this basin he could bring hulls built elsewhere to have their engines placed aboard using large sheer legs. David Napier continued to build marine engines until 1832. In 1836, Napier left Glasgow for London where he established a shipyard on the Isle of Dogs.

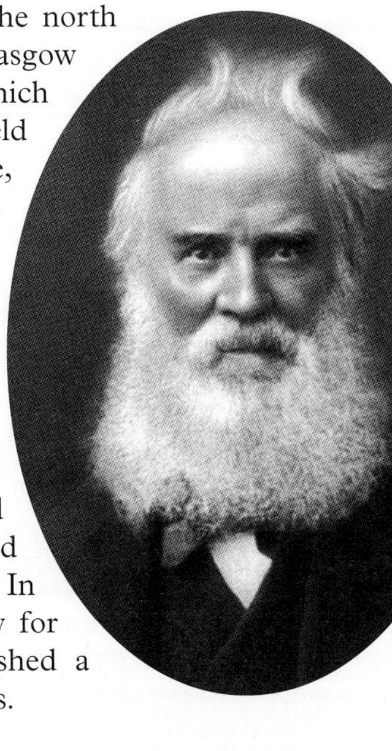

David Napier, 1790–1869. Responsible for some of the earliest marine engineering achievements, Napier designed and manufactured the boiler for *Comet* in 1812 and thereafter the machinery for a series of ocean going steam ships.

Robert Napier 1791–1876

David Napier was followed into marine engineering by his cousin and brother-in-law Robert Napier. When David moved his business from Camlachie to Lancefield, he leased the Camlachie Foundry to Robert, although it was not until 1823 that Robert was awarded the contract for his first steam engine, for the paddle steamer *Leven*. Robert Napier had the good fortune to employ as his works manager David Elder, probably the finest of the many marine engineers by now working in Glasgow. Robert took over the Camlachie Foundry from his cousin David in 1821.

Robert knew how to seize the moment and this arrived in 1839 when Canadian, Samuel Cunard, travelled to London to negotiate a contract with the Admiralty for a regular mail service between the UK and North America. Cunard met with the secretary of the East India Company, James Cosmo Melvill who, impressed with Robert Napier and the steamship *Berenice*, advised Cunard that he should talk to Napier in Glasgow.

It is worth pausing in the year 1840, for, while it is a fact that the Clyde would go on to become the most important marine engineering and shipbuilding river in the UK, this might not necessarily have been the case. The Thames was at this time the most important shipbuilding river in the UK, with well-established shipbuilders such as the Thames Iron Works, Samuda Brothers and the Millwall Ironworks where John Scott Russell, along with Isambard Brunel, would build the colossal *Great Eastern* in 1859. Proximity to the seat of Government, the Admiralty and the city, that was driving a massive overseas empire, had conferred many advantages for local industries. Notable Thames marine engineers included John Penn & Sons, Maudslay Son & Field and the Rennie brothers. Meanwhile on the Clyde, despite a clutch of talented, inventive marine engineers in and around Glasgow, as yet there was very little shipbuilding in the upper reaches, although Scotts at Greenock on the lower Clyde, had been established as far back as 1711.

David Elder 1785–1866

David Elder was an outstanding engineer whose mechanical genius contributed greatly to the reputation that Robert Napier would later acquire as the 'Father of Clyde Shipbuilding'. Elder arrived in Glasgow in 1817 from his native Kinross to train as a mechanical engineer, becoming the manager of Robert Napier's works in 1821. Elder designed and oversaw the manufacture of engines that Napier contracted for, designed and manufactured many of the machine tools used in their construction and, through apprenticeships at Napier's works, instructed many of the leading marine engineers that would follow him.

In a memoir about David Elder written in 1886, the following quote gives some indication of his abilities and the esteem in which he was held:

> The early celebrity of the Clyde district as the cradle of steam navigation is largely due to Mr David Elder, who may be regarded as the father of marine engineering upon the Clyde. He knew his business thoroughly, both theoretically and practically; and, as we have already seen . . . he despised everything superficial, and it was a cause of great pride and gratification to him that no engine ever put out of his hands had failed from insufficient strength or imperfect workmanship.[5]

Robert Napier, 1791–1876 is considered to be the 'father of Clyde shipbuilding' through the contributions he made to marine engineering and shipbuilding, not least of which was the contracts he signed with Samuel Cunard in 1839 resulting in the subsequent building of the machinery and many of its ships on the River Clyde.

David Elder, 1785–1866. A man for whom exactness was everything. David Elder was responsible for the design and manufacture of the steam engines that propelled the ships built by Robert Napier.

John Elder 1824–1869

John Elder was born on 8 March 1824, to David and Grace Elder, one of three sons, all of whom would become engineers.[6] He was educated at Glasgow High School, excelling at mathematics and technical drawing. After a brief time at Glasgow University studying civil engineering, Elder began an apprenticeship at Robert Napier's works where Elder's father was manager. At the end of his apprenticeship, John Elder gained experience at the engineering works of B Hick & Sons at Bolton. He followed this with a period as a draughtsman at Grimsby Docks. In 1848, Elder returned to Glasgow where he took charge of the drawing office at Napier's works, again working with his father. At this time, as was the case during his apprenticeship, Elder would see the steam machinery for many of Cunard's Atlantic ships being built.[7]

John Elder, 1824–1869, with Charles Randolph, patented the compound marine steam engine that brought great fuel efficiency to marine transportation. Elder went on to develop and make famous his shipyard at Fairfield, although he would not live to see its full realisation.

Charles Randolph 1809–1878

Charles Randolph was born in 1809 in Stirling. He attended school at Glasgow High and studied Classics at Glasgow University before following an interest in engineering which took him to Anderson's College in Glasgow, one of the early constituents of the present day University of Strathclyde. From there, he began an apprenticeship at Robert Napier's Camlachie works which was managed by David Elder. On completion of his apprenticeship, Randolph worked in Manchester as a millwright and, while there, met John Elliott.

Charles Randolph, 1809–1878, one of many notable engineers apprenticed at Robert Napier's works on Clydeside. Randolph established a general engineering business in 1834 which became the starting point for the Fairfield story.

Randolph & Co

In 1834 Charles Randolph returned to Glasgow and, in October of that year, set up a small millwrights business in Centre Street, Tradeston, by building a workshop of 60 x 40 feet. Randolph was 25 years of age. Soon after, he was joined by his cousin Richard Stedman Cunliff, referred to at the time as a 'yard merchant', who took charge of the financial and administrative side of the business. As a general engineering business, Randolph & Co designed and manufactured machinery such as saw milling and thrashing machinery for factory owners. This included machinery for Alexander's thread factory in Duke Street, Glasgow as well as for factories as far away as New South Wales.

Randolph Elliott & Co

In 1839, John Elliott joined Charles Randolph as a partner, the Company then being styled Randolph, Elliott & Company. Through Elliott's contacts, many in Northern Ireland, the business grew, requiring various extension to the workshops in Centre Street. The general engineering work undertaken by the company included 'fitting up appliances for directing and conveying power in spinning mills, weaving mills, printworks, paper mills, gunpowder mills and almost every sort of establishment in which steam or water power was required.'[8] In 1842, John Elliott died, but by this time the success of the Company was assured and it became one of the leading general engineering firms in Scotland.

The Clyde and Cunard

While these events were unfolding and the business of engineering expanding at various points on the Clyde, the reputation of the River began to grow through its association with the shipping lines of the day and the ships built for them. A prime example of this was Cunard, that great shipping enterprise, for a long time considered to be Britain's national line. The Cunard connection goes back long before John Elder contemplated building ships or even thought about the compound engine. The driving force behind Cunard was the burgeoning transatlantic route from Britain to North America. While sailing ships had been crossing the Atlantic for centuries, the first steam ships to make the crossing were *Savannah* in 1819 using a steam engine to supplement sail, and *Royal William* using steam throughout in 1833. These vessels were respectively of American and Canadian origin. It was not until 1838 that the Great Western Steamship Company, a division of the Great Western Railway Co, launched the first British steam ship designed specifically for the North Atlantic route. This ship was the *Great Western*, designed by Isambard Brunel and built in Bristol with steam engines by Maudslay Sons & Field on the Thames.

As mentioned earlier, in February 1840, Samuel Cunard, a wealthy merchant from Halifax, Nova Scotia, arrived in London to present his bid to provide a regular steam ship service between Liverpool and Halifax. The requirement for this had been issued by the Admiralty a few months earlier, specifying a service to carry the mail across the Atlantic. Until this point, there had been little involvement of the fledgling marine industries on the Clyde in the transatlantic trade, although the *Sirius* of 1837 had been built by Robert Menzies & Son in Leith with engines manufactured in Glasgow by Thomas Wingate & Co at Springfield. In 1839, *British Queen* was built by Curling & Young on the Thames with engines by Robert Napier.

As for the Clyde, while shipbuilding in Port Glasgow and Greenock was well established, so far Glasgow had little to offer in the late 1830s. What it did have, however, was a number of companies involved in the design and manufacture of marine steam engines and none more so than Robert Napier's. When Cunard made his pitch to the Admiralty to run the mails to North America, he had three ships in mind but no shipbuilder to construct them nor sufficient capital to do so. Cunard was advised by James Cosmo Melvill, secretary to the East India Company, for whom Napier had built their first steamer *Berenice*, that Robert Napier in Glasgow was the person to approach. The construction of the first Cunarders on the Clyde did much to further enhance the reputation of the river as an important centre for marine engineering and shipbuilding.

Randolph Elder & Co

It was against this background, in September 1852, that John Elder, then 28 years of age, was inducted into Randolph's engineering business as a partner, whereupon the name was changed to Randolph Elder & Company. Both partners had a common background in marine engineering at Napier's engine works. Coupled to the obvious capability of Randolph's works, and the steadily growing demand for marine steam engines, it is hardly surprising that they should turn to the design and manufacture of steam machinery.

The first marine engine built by Randolph & Elder was for a vessel constructed by Alexander Stephen & Sons, at Kelvinhaugh shipyard on the Clyde. They had been contracted to build the 684-ton passenger cargo paddle steamer *William McCormick* for the Londonderry Steamboat Co. The machinery driving the paddle wheels was of the well-established side lever type. On 24 January 1853, however, Randolph Elder & Co's patent 'for an arrangement of compound engines adapted for the driving of the screw propeller' was filed.[9] Thirteen patents would follow between the years 1853 and 1867, five attributed to Charles Randolph and John Elder together and nine to John Elder by himself. The next marine engine built by Randolph Elder, following the traditional engine for *William McCormick*, was an important milestone in the development of the marine steam engine. The machinery

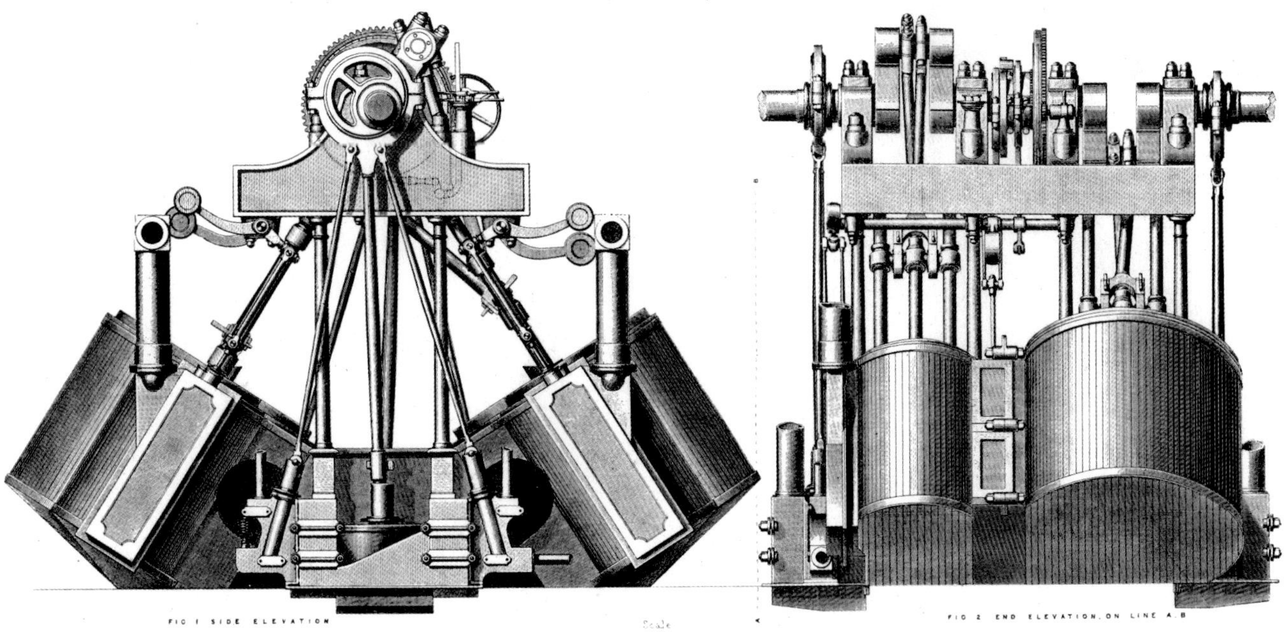

Cross sections of the compound engine as patented by Charles Randolph and John Elder on 15 March 1856 and installed on the Pacific Steam Navigation's *Valparaiso*. This was the third patent filed and the second by Charles Randolph and John Elder. A balanced arrangement of two compound engines each of three cylinders (one high pressure and two low pressure).

for the screw steamer *Brandon* in 1854, under construction at Barclay Curle's Kelvinhaugh slip dock, was a complete departure, being the first of a radically new type of marine steam engine. In 1854 Randolph Elder & Co also began the manufacture of boilers at works in Dale Street, close to their Centre Street engine works.

The Compound Engine

Brandon was constructed for the London & Limerick Steam Navigation Co Ltd. On completion in July 1854, her trial trip showed a coal consumption of 3.25lbs per horsepower per hour compared to the typical consumption figure of 4 to 4.5lbs for steamers then in service. Succeeding vessels employing this compound engine, *Inca* and *Valparaiso* of 1856, constructed for the Pacific Steam Navigation Company, further demonstrated remarkable fuel efficiency with consumption figures of 2.5lbs and 3lbs. Two years later, the steamer *Callao* was able to achieve a figure of 2.7lbs.[10] By the end of the 1850s, even this figure was improved upon when *Bogota* and *Lima*, built by Robert Napier in the early 1850s, both of which were re-fitted in 1859 by Randolph Elder & Co with compound engines, showed a further increase in fuel efficiency. Newspaper reports of the time give some idea of how significant the impact of these engines was:

> The *Bogotá* before conversion was fitted with side lever engines of 400 hp and, on her voyage to Madeira, ran 1417 miles in six days on the consumption of 274.5 tons of coal, giving an average of 9.75 knots on the consumption of 38cwts of coal per hour. The same ship, when fitted with Randolph Elder and Co's engine, ran a distance of 2470 miles in nine days, 21 hours on the consumption of 232 tons of coal thus presenting an average speed of 10.42 knots per hour on a consumption of 19cwts per hour, the average indicated horsepower being 950, with average consumption of 2.25 pounds of coal per hp per hour.[11]

While the Pacific Steam Navigation Co took advantage of the fuel efficiency offered by the compound engine, other shipping lines were less quick to do so. It was not until April 1860 that the Admiralty, presiding over a vast range of warships, ordered a set of compound engines from Randolph Elder to test for themselves the fuel efficiency of the type.

HMS *Constance*

It was not until 1862 that the Admiralty converted the wooden sailing frigates, *Arethusa*, *Octavia* and *Constance*, to screw propulsion. *Arethusa* and *Octavia* were fitted with regular (non-compound) steam engines, constructed by the celebrated Thames engine builders Penn & Sons and Maudslay Sons & Field. The third frigate, *Constance*, was fitted with Randolph Elder's six-cylinder compound engines, making it the first Admiralty warship to be fitted with compound steam machinery. On 30 September 1865, all three frigates left Plymouth bound for Madeira. *Constance* arrived in Madeira on 6 October, *Octavia* on the 8th and *Arethusa* on 10th. The last two were 130 and 200 miles respectively behind *Constance*, both completing the journey under sail. *Arethusa* required 3.64lbs of coal per horsepower per hour, *Octavia* 3.17 and *Constance* 2.51. This was a complete validation of the efficacy of the compound engine.

Ship	Distance behind	Consumption	Mean ihp (indicated horse power)
Constance	–	2.51	1747
Octavia	130	3.17	1399
Arethusa	200	3.64	1052

Generally attributed to the engineering genius of John Elder alone, it is likely that Charles Randolph also played an important role in the development of the compound engine. In a newspaper article reviewing the life of Charles Randolph, who died on 11 November 1878, it noted that:

> Mr Randolph had given much attention to the question of economy of steam, and followed it up until the idea took very definite shape that the steam might be used in a type of engine not previously in vogue among marine engineers – namely one in which it should be used first in a high pressure cylinder, and then be re-expanded in a low pressure cylinder – such engines being termed compound . . . It is not necessary that we should attempt to apportion the merit due to each of the inventors, but this we may say, namely, that while the conception originated with Mr Randolph, the complete working out of the idea of the compound engine to its full practical issue was in great measure due to the persistence and perseverance of Mr Elder, whose faith in its ultimate success never faltered.'[12]

Randolph Elder & Co, Shipbuilders

Between 1854 and 1859, Randolph Elder & Co, with engine works at Centre Street and boiler works at Dale Street, became marine engineers, fitting or re-fitting 25 ships with their fuel efficient compound steam machinery. Four of the ships that the Company re-engined belonged to the Pacific Steam Navigation Co, a shipping line that

Foremen at Randolph Elder & Co's Centre Street Works *c*1852.

would derive great commercial advantage by employing vessels fitted with these engines in the long routes from Liverpool to ports on the west coast of South America.

With business as marine engineers occupying the majority of the Company's time, at some point towards the end of that decade, they decided to become shipbuilders, taking the first step towards this in November 1860 by purchasing a ten year lease of the unoccupied shipyard, Govan Old, on the east side of Water Row in Govan. This yard had been formerly occupied by Robert Napier. On 9 September 1861, their first ship to be launched, the steamer *Talca,* built for the Pacific Steam Navigation Co, entered the Clyde. Of just over 700 tons, *Talca* was 190 feet long by 30 feet, powered by compound engines of 260 ihp (indicated horse power). Although *Talca* was the first ship launched, the first ship ordered was *MacGregor Laird* for the African Steamship Co, named after that shipping line's founder and explorer. John Elder's youngest brother, Alexander, was Superintendent Engineer of the African Steamship Co which ensured that a number of ship contracts were placed at his brother's yard. One of these was *Calabar*, a sister ship to *MacGregor Laird.*

With orders for ships plentiful, Randolph Elder & Co also took the contracts for two floating docks. These were designed by the Scottish engineer and inventor Robert

MacGregor Laird, the first ship built by Randolph Elder & Co, completed at the Govan Old yard in 1862. This yard had previously been occupied by Robert Napier.

A line engraving of the floating dock at Callao 1866, from *The London Illustrated News.*

William Thomson. Constructed in sections at Govan, these were transported to their final destinations to be assembled, completed and put into service. This was a significant achievement for the time and a tribute to Thomson's design and the precision of the shipyard's construction. The first dock was erected at Surabaya, Indonesia, and floated in November 1862, followed by the second, contracted for by the French Government, for use at its naval base in Saigon. This dock was completed in sections at Govan in January 1863 and subsequently transported to Saigon. A third floating dock, also designed by Robert William Thomson, was built in 1866 for the Pacific Steam Navigation Co and destined for service at Callao, Peru.

A New Shipyard and Engine Works

In total, just twelve ships and two floating docks would be constructed at the Govan Old yard. At some point towards the end of 1862 or early in 1863, Randolph Elder & Co, made the decision to expand their shipbuilding activities by laying out a new yard and began the search for a suitable location.

They did not have to look far, as less than half a mile to the west of the Govan Old yard, lay the grounds of the Fairfield Estate, much of which was for sale. In October 1863, newspaper reports state that Randolph Elder & Co, had acquired ground at Fairfield with a river frontage of 1200 feet extending 500 feet back from the river. Moreover, these reports indicated that the yard would be operational within weeks.[13] This suggests that work constructing the various buildings and workshops had begun earlier in 1863. In March 1864, shipbuilding operations at the new Fairfield yard began and, in February, the lease, of the Govan Old yard, which did not expire until 1870, was let to Dobbie Hedderwick & McGaw, who took possession in March 1864.

The choice of site for the new yard was part of the general westward, down-stream, relocation of shipbuilders, forced in some cases to do so by the Clyde Navigation Trust and the expansion of their port facilities at Glasgow harbour. This was not necessarily a bad thing, as the ever increasing size of ships required larger premises for their construction.

It is clear from newspaper reports that by 1870/71, Randolph Elder & Co, had secured a total of 70 acres of ground at Fairfield, all of which had been owned by the

Shipyards at Govan c1855/60s

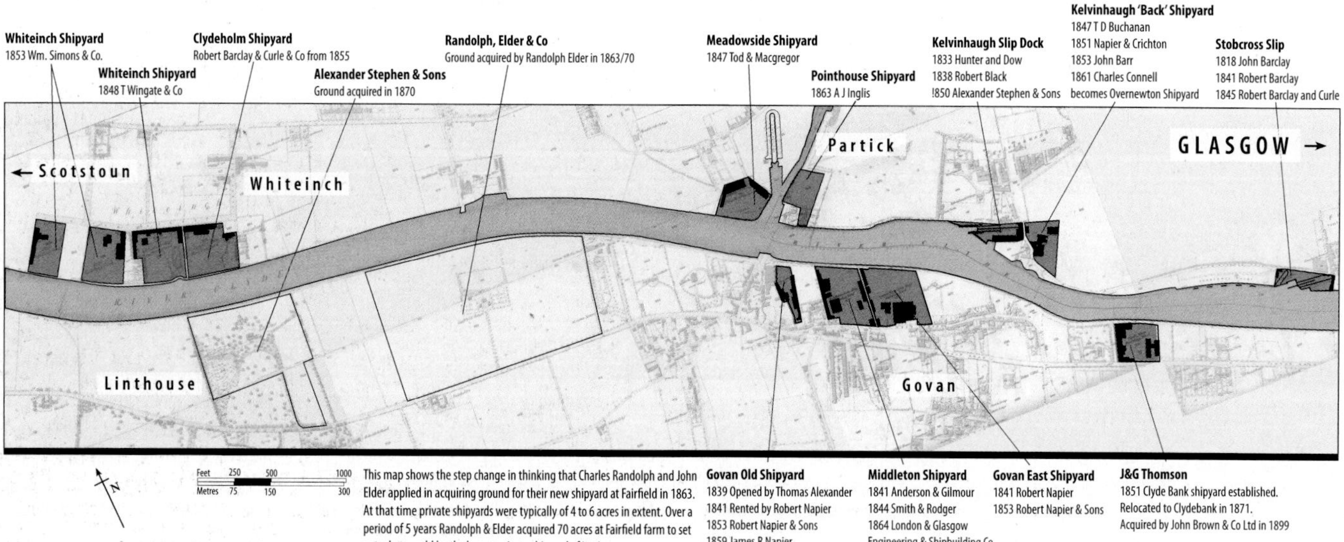

This map shows the step change in thinking that Charles Randolph and John Elder applied in acquiring ground for their new shipyard at Fairfield in 1863. At that time private shipyards were typically of 4 to 6 acres in extent. Over a period of 5 years Randolph & Elder acquired 70 acres at Fairfield farm to set out what would be the largest private shipyard of its time.
The map also shows that shipbuilding was already well established on the Clyde at Govan and elsewhere on the river long before marine engineers Randolph, Elder & Co began shipbuilding.

Trustees of Dr J White, a noted chemical manufacturer at Rutherglen. This land, however, was not acquired all at once and a major shipyard was not initially built. The 70 acres were bounded to the east by Morris Pollok's Govan Silk Works and his gardens and mansion house, the Govan–Renfrew turnpike to the south, arable land to the west, including Linthouse Estate, and the River Clyde to the north.

Looking at available records, it seems that about 15 acres were purchased in the first instance, with additional plots thereafter. The shipyard, as first laid out, occupying 15 acres, was adjacent to Fairfield House and gardens which lay at the very centre of the eventual 70 acres. It was not until closer to 1870/71 that all of this ground had been acquired. Even then, in 1885, Fairfield House and gardens, although owned by John Elder & Co, successor to Randolph Elder & Co, was occupied by a variety of tenants. Fairfield House, in the middle of the shipyard, was not demolished until 1896.

Randolph Elder's plan was to build a private shipyard and engine works combined[14] which would be larger than anything so far seen on the Clyde or elsewhere. This would enable Randolph Elder & Co's engine works at Centre Street to be transferred and integrated into the new shipyard. For this purpose, a new engine works would need to be constructed at Fairfield. The last major element in the plan for the new yard was the construction of a fitting-out basin or dock at a right angle to the river, where newly launched hulls could be taken to have machinery shipped and to complete the lengthy process of fitting-out, without obstructing traffic in the river channel. This basic concept was the logical development of shipyard layout on a narrow river such as the Clyde.

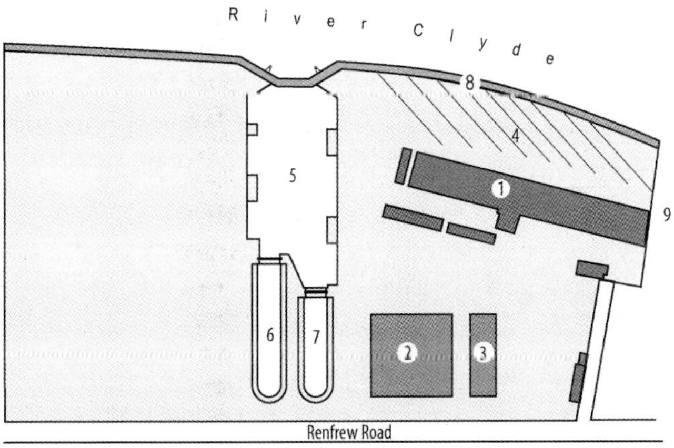

1871 Randolph, Elder & Co, shipyard and engine works showing proposed docks

1. Iron workers sheds
2. Engine shops
3. Smithy
4. Building slips
5. Proposed fitting-out basin
6. Proposed graving dock 500 feet long
7. Proposed graving dock 380 feet long
8. Footpath carried by a timber platform over launching ways
9. Ground occupied by Govan Silk Works

The Clyde Navigation Trust, from whom building approval was required, frustrated John Elder and his successors in their desire to construct a fitting-out basin with dry docks at the new Fairfield works. Approval would prove to be a difficult, protracted and acrimonious process. This proposal dating from 1871, shows the original intention for the site. The basin was eventually constructed without the dry docks as the need for them had diminished after the Clyde Navigation Trust built their first dry dock upstream from Fairfield in 1869.

By the time *Payta* was launched at the Govan Old yard in July 1864, the fifth vessel launched there in five weeks, three vessels had been laid down at Fairfield. By the beginning of 1865, the new yard at Fairfield was booming, with 12 vessels on order or under construction. Additionally, two floating docks had been contracted for, each over 3000 tons deadweight. That year, output statistics for the Clyde continued to rise, recording a total of 268 ships launched with a tonnage of 158,300. A further 180 were on the stocks, amounting to an additional 130,000 tons.[15]

The Fairfield Blockade Runners

The Fairfield yard had become operational towards the end of the American Civil War and, perhaps it is not surprising given the number of vessels already built on the Clyde to beat the blockade imposed by the Union, that the first vessels to be built at Fairfield were designed and built to run the blockade of the Confederate States.

Ordered by Donald MacGregor, on 4 February 1864, these ships, four in number, with a fifth added soon after, and referred to as the *Falcon* class, were in reality for Alexander Collie & Co, London based agents for the Confederacy. Like most of the blockade runners built on the Clyde, the vessels had to be fast and capable of carrying a minimum 1000 bales of cotton. When the first of this class, *Falcon*, began her maiden voyage across the Atlantic to Bermuda, she called at Cobh, then known as Queenstown, to take on coal. This earned the unflattering description published in *The Liverpool Albion*:

> … she is a queer looking craft, having three funnels and being polacca rigged, that is, with large lateen sails, or enormous lugsails, and is furnished with three boilers and three furnaces. She is very fast, and on a recent occasion, steamed from Ardrossan to Belfast in 4.5 hours.[16]

According to a document held in the Fairfield archives in Glasgow, '*Falcon* brought a cargo of 1080 bales of cotton out of Wilmington, the largest ever brought, was chased and got away easy, allowed to be the fastest boat in the blockade.'[17]

While *Falcon* reached Bermuda and subsequently carried out several successful blockade running voyages, her sister vessel, *Condor*, was less fortunate. On her first blockade running voyage from Halifax to Wilmington, *Condor* ran aground and sank on 1 October 1864 at Fort Fisher near Wilmington, North Carolina.[18]

The New Shipyard at Fairfield

In January 1865, one of the first descriptions of the new shipyard was published, describing firstly the yard's contract for two floating docks, work on hand and, finally, the physical description of the yard:

> Taking the combined tonnage of these two [floating] docks at 8000, which is no more than enough, and adding that to the 14,000 tons of shipping we have referred to, we find that this one company have in hand no less than 22,000 tonnage in craft of one kind or another, large and small. Apart altogether from the composite, brass, the immense quantities of timber, and other material used, there are being employed in the fitting up of these vessels somewhere about 17,000 tons of iron. Keeping these extraordinary facts in view, looking around upon the great extent and wonderful completeness of the general works of this firm alone, and remembering that in connection with them there is constantly employed a body of 3000 and more workmen, we have surely proof strong and full enough to show the popularity, enterprise, and superiority of our Clyde shipbuilders . . . We may state that the yard of the firm is about 1200 feet long by 530 feet wide, enclosing some 15 acres of ground. One of the buildings in it is alone 600 feet long by 130 feet wide. The machinery and apparatus in connection with the joiners, engineers, carpenters, blacksmiths, boatbuilders, riggers, sawyers, and all the other various departments is of the newest and most approved kind. Rails have been laid all round the works for wagons, trucks, and small steam engines, and two steam travelling cranes, capable of going the whole length of the yard, are placed in different parts of the work, so that the whole of the surface ground is made available.[19]

Liverpool to Valparaiso

The fuel efficiency of the compound engine was of great commercial significance to the Pacific Steam Navigation Company, whose trade routes to South America were among the longest undertaken at the time. This was well illustrated in an article published in February 1864 describing the voyage of the paddle steamer *Chile* from Liverpool to Valparaiso. Built at the Govan Old yard, *Chile* was a paddle steamer 275 feet in length with engines of 400nhp (nominal horse power) which gave her a speed of 10 knots. The steamer left Liverpool on 11 November 1863 with a full cargo and a large number of passengers. She arrived at Madeira after steaming for 4 days 22 hours, in the fastest time so far recorded between those locations. After some goods and passengers disembarked, the ship sailed for St Vincent in the Caribbean, reaching there after 8 days and 23 hours from Liverpool. After spending two days there, *Chile* set off for Cape Horn, which she rounded, eventually arriving at Valparaiso. The total time of steaming was under 31 days and the distance covered 9000 miles. The coal consumption was 1050 tons.[20]

An Obstructive River Authority (part 1)

By late 1863, work had begun preparing the ground at Fairfield for the first phase of the new shipyard. The proposed works represented a step change in the private shipbuilding facilities hitherto constructed on the Clyde (or anywhere else) at that time and represented something in the region of a four-fold increase in size. The plan was to construct a shipyard capable of building the largest vessels then foreseen, meaning vessels of up to 550 feet in length. Here a large engine works would be built, capable of machining and assembling most aspects of a marine engine with the exception of large castings and forgings. Finally, a dock was to be constructed at a right angle to the river, where ships could be taken after being launched to be completed. This included the installation of the boilers and engines from the nearby engine works. Altogether, the proposed works would represent the largest integrated private shipbuilding and marine engineering works of their kind anywhere. These works, however, would take time to complete and, quite apart from removing the existing Centre Street and Dale Street works to new workshops at Fairfield, it would require the permission of the Clyde Navigation Trust for the alterations that would be made to the river bank and wall. This would prove to be a serious and protracted issue, compounded by the public right of way that ran along the river bank.

While erecting the iron working and other shipyard buildings, permission was sought from the Clyde Navigation Trust for the creation of what was termed the launching slope, where the building berths would be laid out. In December 1863, John Elder wrote to the General Manager of the Trust, George Reith,[21] asking for approval to proceed with their proposal to lay out launching slips of a 'permanent nature'. The river's edge had previously been formed into a bank on top of which was a public right of way. Elder expressed urgency in approving his plans, as they had a number of vessels under construction which would require to be launched within three months. The primary function of the Clyde Navigation Trust, which had legal rights to the river and the river wall, was to ensure the free passage of navigation at all times as determined by various Acts of Parliament and most recently the 1857 Navigation Act.

By early 1864, a draft agreement had been reached between both parties which enabled the shipbuilders to lay out slips for the first vessels to be laid down in the new yard. The next stage was the construction of berths for larger vessels and the cutting of the fitting-out basin into the river wall. On 21 January 1865, Reith, having consulted the Trustees, said that, with reference to new building slips, Elder's plan was approved as per existing slips but applied four conditions to this approval, the second of which read:

> '. . . you will undertake to remove the same [the permanent launching ways] and restore the whole river ground and dyke to its original state whenever required by the Trustees to do so.'[22]

This stringent condition, potentially ruinous to the shipbuilding business should it have been applied, appears to have been ignored by Randolph Elder & Co, as they failed to sign and return the conditions sent to them as part of the draft agreement but, nevertheless, instructed civil engineering work to proceed in forming the launching slips. While the Trustees expressed their desire to support the shipbuilders, recognising the importance of this industry, at the same time they felt obliged to uphold their statutory duties to maintain the river and navigation thereon by imposing these conditions. While negotiations with the Trustees were underway, the Clyde was affected by a major stoppage of work.

The Lockout of 1866

In 1866, against a general downturn in the shipbuilding industry, the demand by shipyard workers for a shorter working week came to a head. This demand brought much of the shipbuilding and engineering industries on Clydeside to a halt, as the respective sides adopted a siege mentality in what would become an all too familiar means of dealing with disputes. The issue of a shorter working week took hold in the early months of 1866 and, by 16 April, resulted in a delegation of men from the engineering and shipbuilding trades meeting with the employers to press for a reduction in hours worked per week from 60 to 57. When the employers agreed to concede the hours but with wages reduced accordingly, men in some of the establishments went on strike. On 13 May, ship joiners at Randolph Elder & Co also went on strike. While they returned to work soon after, the Company's engineers at Centre Street did not go back. Various trades in other yards followed suit, as the strike gained momentum.

In response to the strike, employers took the initiative, by forming themselves into the Clyde Shipbuilders' and Engineers' Association and, thus formed, imposed a three month long lockout across the Clyde, starting on 28 May. Although there was no good time to do this, from the employer's perspective, the downturn in trade then being experienced made it acceptable. It also enabled employers to use the lockout as a means of breaking the unions. Men who were on strike and part of a union or society, were more able to resist the employer's tactic through benefit pay-outs while non-union labour would have to get by as best they could, underlining a key reason to become a member of a trade union.

In June, representatives from a number of societies,[23] formed a Workers' Committee which was empowered to take 'such measures as may be deemed advisable towards a settlement'. When the Employers' Association was approached and asked if they would receive a delegation of the Workers' Committee with a view to the settlement of the present dispute, the reply from a spokesperson said:

> I am directed to say that it appears to the employers unnecessary to have a conference with the delegates or members of the trades' unions, and that they have passed a resolution at today's meeting, by which those willing to resume work on the terms already offered, may do so. It will thus be seen that the various works are to be opened today for the reception of such non-society men as have shown a disposition to come to terms.[24]

While this was not the first strike to take place on Clydeside, it illustrates the nature of the standoff between employers and employees that would be played out again and again. In this instance, there was a slow return to work on the employers' terms, although some trades held out until August of that year.

On 14 April 1868, the three partners in Randolph Elder & Co, Charles Randolph, Richard Cunliff and John Elder, agreed that, on the natural termination of the partnership on 30 June 1868, John Elder would take over the business on the following day, 1 July. On the date of termination, Charles Randolph and Richard Cunliff resigned, leaving John Elder in sole control of the business. Elder agreed to pay Randolph £10k and Cunliff £5k for their shares of capital and thereafter, annually, until the balance of capital was all paid up.[25]

An Obstructive River Authority (part 2)

Throughout 1867/8, discussions between the shipbuilders and the Clyde Navigation Trust took place, principally between John Elder and George Reith, in consultation with the Trustees' engineer, Andrew Duncan.[26] Elder submitted plans which showed a removable bridge over the mouth of the proposed basin to maintain the public footpath.

On 13 October 1868, Elder wrote to the Trust, referring to negotiations that had been going on for over a year between himself and the Trust's engineer regarding the basin and permanent launching slips, the latter to be for the 'largest class of vessels and ironclads'. This lengthy letter detailed the various points that Elder had taken into account in line with Duncan's observations and recommendations. The entrance to the basin would now be set back about 90 feet from the present river wall, whereby ships launched could be taken into the basin at an angle and not at a right angle to avoid disruption to traffic on the river channel. Elder said that the largest ship so far launched in the river, the 380 foot long ironclad *Black Prince,* launched at Napier's yard in 1861, had required piling to take place into the river channel for almost two months to create the building slip. Elder's plan would eliminate the need for this, by forming his building slips at an acute angle thus minimising the impact of slipways and vessels launched into the river. To illustrate these points, Elder went to Reith's office with a plan and cut outs of ships of various sizes up to 350 feet in length, to literally animate, on Reith's desk, how this would work (see photo of cut out ships).

In response to Elder's letter and plan and visit with cut out ships, the Trustees were generally approving, but, in a letter to the shipbuilder, applied no fewer than eleven conditions this time, including point 9, a repeat of the wholly unreasonable condition referred to earlier:

> Messrs Randolph Elder shall be bound at any time within six months after being required to do so by the trustees, to restore the whole river wall removed by them together with the public footpath of twenty feet in width to the same condition in all respects as the same were in at the time of being removed . . . in case of failure so to restore the river wall, bank and footpath, Messrs Randolph Elder & Co shall pay to the Trustees the sum of eight thousand pounds sterling as the cost and value of restoring the same.

Additionally, the Trustees noted that the shipbuilder might use his basin to load goods not connected with his business and were insistent that vessels entering his basin apart from newly launched ships, would have to pay berth dues to the Trust.

By March 1869, Reith wrote that he had only just received a reply from Elder in which several important

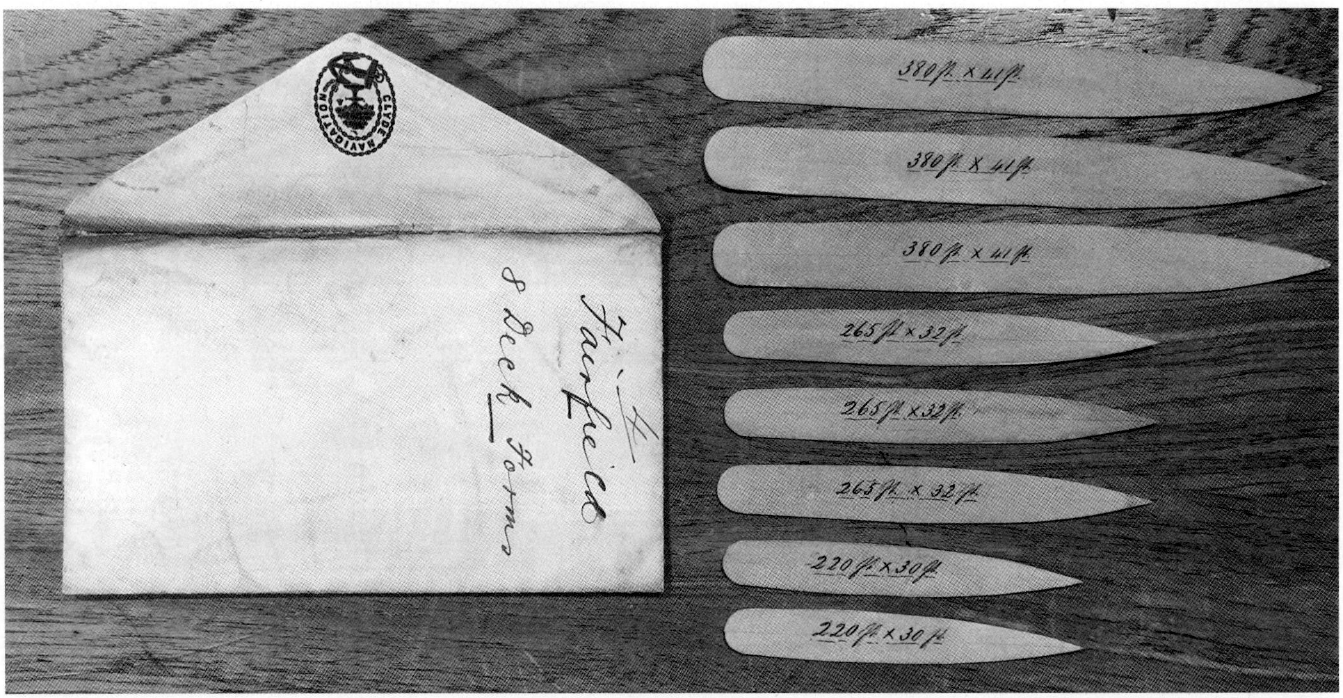

The cut-out ship hulls that John Elder used to demonstrate proposed access to the fitting-out basin to George Reith of the Clyde Navigation Trust. He laid a plan of the yard out on a desk top and used the cut-outs to show that by setting the entrance to the proposed fitting-out basin further back from the river, ships turning into the basin would do so at a more gradual angle and not block the river channel.

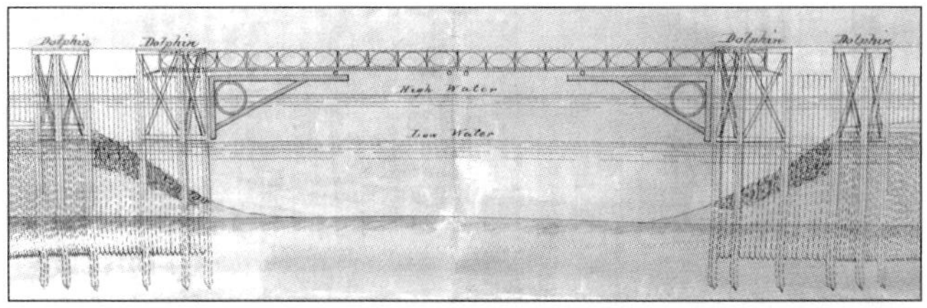

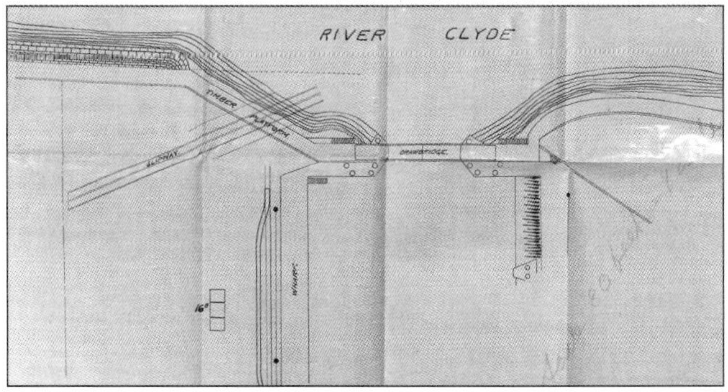

Maintaining the public footpath and a method of continuing it across the entrance to the fitting-out basin was a major problem for the shipyard until an Act of Parliament in 1914 allowed the footpath to be removed. The drawing at top, from November 1871, is an elevation for a bridge with a removable centre span. The bottom drawing from September 1890, is a plan for a drawbridqe. This plan shows the proposed shipbuilding berth at the west side of the basin which was excavated but never completed, being used as a timber pond instead. The last means of conveying the public over the mouth of the basin was a small ferry not much larger than a rowing boat.

alterations had been made to the conditions. While he could agree to some of the alterations, Elder's alteration to Condition 9, that it be replaced with 'simply maintain the works' cannot be accepted. Reith then referred the matter to the Law Committee of the Trust for an opinion.

In April 1869, Reith wrote to Elder, asking that he complete the agreement sent to him. Moreover, Reith also requested that the unsigned agreement dating from 1865 be signed and returned. The stumbling block preventing Elder doing so was, of course, the condition which would require Elder to restore the river wall across his building berths and block off his fitting-out basin. The impasse resulted in both parties seeking legal opinion to determine a way forward.

On 22 April, Reith wrote to Elder saying that the Law Committee of the Trustees could not agree to his alterations and, in reaching this decision, they took the opinion of one of the 'first professional men in Glasgow', Anderson Kirkwood, as to the legal rights of the parties. In response, John Elder consulted J H Lloyd, a London based barrister, who, in reviewing the ensuing Trustees report, considered that:

> The conditions to which the Trustees attach their consent are founded on an erroneous assumption as to their rights and powers. No express consent on their part to the proposed works is either asked or required. Their business is merely to take care that the navigation is not obstructed and that the works as carried out shall be satisfactory to their engineer or a neutral engineer in case of difference and the attempt to prescribe terms such as those contained in the report of the sub-committee on the suggestion of the Manager is wholly beyond their competence and an arbitrary stretch of authority.

The Trustees were unimpressed and not minded to change their position. This they expressed to John Elder. He, however, had gone to London to seek medical assistance for disease of the liver and possibly, at the same time, to consult with his barrister. By late August 1869, George Reith was anxious to reach some form of accommodation with Elder and wrote to Elder's solicitor in Glasgow asking 'is there any word from Mr Elder yet, or is anything further to be done with a view to an arrangement between him and the Trustees?' Just a few weeks later, on 17 September, John Elder died and there the issue remained until taken up by his brother in law, John Ure, Isabella Elder's brother and formerly the Clyde Navigation Trust's principal engineer, who would shortly take up position as a partner in the Elder shipyard.

Returning to events prior to John Elder's death, in April 1869 Randolph Elder & Co held a soirée for its Centre Street employees in the Prince of Wales Halls [Sauchiehall Street]. J L K Jamieson, the Works Manager acted as Chairman. He noted that the last time the Company had held such a soirée, trade was depressed and only 1500 persons were employed across all departments. Now, with trade much revived, the company employed close upon 8000.

The sail plan for the *Congo* and her two sister ships, *Bonny* and *Roquelle*. All three were completed in 1869 for the British & African Steam Navigation Co.

An Untimely Death

Although already mentioned in connection with the Clyde Navigation Trust, the passing of John Elder and its consequences deserves a little more attention. In the early summer of 1868, Elder was an exceedingly busy man. In addition to winning new contracts, running two sites (the shipyard at Fairfield and the engine works at Centre Street and Dale Street), Elder was writing a detailed paper concerning circular ships to be given in London in May of that year. He was also overseeing the early stages in the design of the large engine works building which would enable the old works to be relocated to the shipyard site. On top of that, he was dealing and probably very frustrated with the Clyde Navigation Trust over the saga of the river wall, the building slips and his proposed fitting-out basin.

John Elder's Circular Ship

On 25 May 1869, John Elder delivered his paper entitled 'Circular Ships of War with Immersed Motive Power', to the prestigious Royal United Service Institution in London.[27] Based on experiments he had conducted with a 5 foot diameter model, Elder proposed, in considerable detail, a circular coastal or harbour defence warship, 280 feet in diameter. Such a design, he contended, would provide a very stable gun platform and present a very small target to enemy shell fire. His design, unusual enough on its own, also showed novel forms of propulsion, ranging from conventional screws to a water jet propeller, driven by a centrifugal pump. Although supported by Sir Edward Reed, the Admiralty Chief Constructor, nothing further came of Elder's paper at that time.

With the exception of the ongoing dispute with the Clyde Navigation Trust, everything else that Elder had planned regarding the shipyard and its future development was falling into place when he died of cirrhosis of the liver, on 17 September 1869, while seeking medical attention in London. John Elder was only 45; his death came as a great shock to all who knew him, and, not least, the workforce at Fairfield and the people of Govan and beyond, such was his fame and reputation.

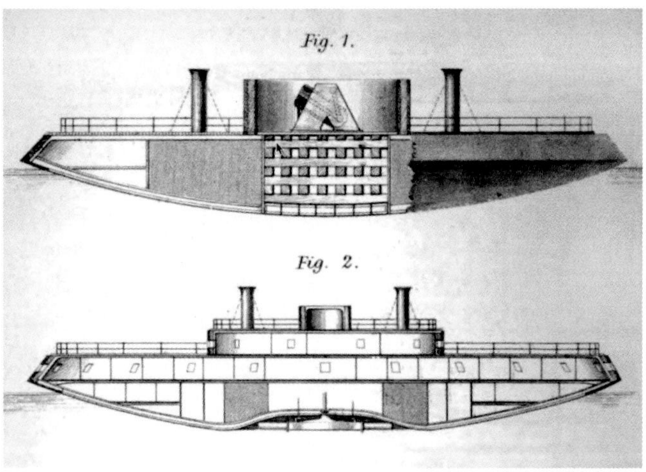

Two of several drawings of a 300ft diameter circular ship that John Elder proposed to the Royal United Service Institution, London, in 1869.

The eminent engineer and physicist, William Macquorn Rankine, who knew Elder well, wrote a lengthy tribute to him and his achievements entitled, 'A Memoir of John Elder, Engineer and Shipbuilder', published in 1870.[28] This memoir gives a full account of Elder's background and engineering achievements, as well as his religious beliefs. In the early part of his memoir, Rankine says of Elder, the student of engineering:

> To those who knew him well, and enjoyed the advantage of personal communication with him, it was manifest that his eminence was not due so much to teaching by others as to the fact that John Elder was that rare character – a man of genius; and therefore in a great measure independent of that external control and guidance which are necessary for the training of ordinary students.

Later Rankine quotes one of Elder's 'oldest and most trustworthy men' who told him:

> I never saw anyone like him, nor expect to see his like again! He was so just, so true, so kind to everyone. Every man trusted him and knew that he would do all that was possible to benefit them in every respect. He had many plans for their good, which, alas! he was not spared to carry out . . . His funeral was one of the most impressive sights I ever witnessed. The busy works south of the Clyde were shut, forge and hammer at rest, and silent as the grave. The forest of masts along the river were draped in flags, lowered half-mast in sign of mourning. A very army of workmen, dressed like gentlemen, followed his body – column after column.

John Elder died a wealthy man, leaving an estate of £188,000 equivalent to almost £23m at 2020 rates. He may have had an indication that he was dying or at least knew there was a strong possibility of it as, in his will, he left an annuity of £200 per annum to both his brothers David and Alexander, rising to £300 on the death of their mother. The bulk of his estate was, however, left to his wife Isabella. At the time of his death, Elder had seen much of his vision for the shipyard completed, while the new engine shops, designed by Angus Kennedy to Elder's brief, were incomplete and the fitting-out basin yet to be started, owing to ongoing difficulties with the Clyde Navigation Trust. Nevertheless, after five years of operation, the yard had launched over 40 ships, but was still some way short of the fully integrated works he had envisioned. It was not until the end of 1871, that the engine shops were completed. At that time, this building, along with the shipyard itself, comprised the largest private works of their kind anywhere.

The steamship *Sin Nanzing* completed in 1870. Although powered by steam she was fully rigged for sail, typical of all steam ships during the early steam era.

John Elder & Company

Isabella Elder, John Ure, John Jamieson and William Pearce

Under the terms of her husband's will, Isabella Elder became sole proprietor of the business as well as executor of his estate. She undertook to manage the business for a period of nine months, during which time she arranged for her brother, John Francis Ure, to take over as General Manager. John Jamieson continued as Engine Works Manager while William Pearce, Shipbuilding Manager at Napier's shipyard, was offered the position of Shipbuilding Manager at Fairfield, which he accepted. On 1 July 1870, this arrangement was formalised, with the assets, property and good will of the business transferred from Isabella Elder to a new co-partnership of John Ure, John Jamieson and William Pearce. Under the terms of the agreement, each partner had to contribute £5k to the capital of the company and was able to draw an annual salary of £1,200. Profits and losses were to be shared in the proportions: Ure 15/28ths, Jamieson 7/28ths and Pearce 6/28ths. The business was renamed John Elder & Company in tribute to the highly regarded, deceased engineer and shipbuilder.

Isabella Elder, 1828–1905. When her husband died in 1869, Isabella was left a wealthy woman. She used her wealth to provide medical and recreational facilities for the people of Govan as well as funding the John Elder Chair in Naval Architecture and supporting medical education for women at the University of Glasgow.

John Francis Ure, a civil engineer, was formerly the Principal Engineer for the Clyde Navigation Trust. In 1858, he had relocated to Newcastle-upon-Tyne, where he occupied the post of Engineer to the Tyne Commissioners. In 1869, in response to the death of his brother-in-law, Ure resigned this post to return to Glasgow, where he took charge of the shipyard while still acting in a reduced role as Consulting Engineer on the River Tyne.

William Pearce was born in 1833 in Brompton, Kent, and trained as a naval architect at Chatham Royal Dockyard. In 1861, when the Admiralty decided that iron was to be used for the first time in a Royal Dockyard in the construction of the hull in one of its warships, HMS *Achilles*, Pearce, with colleagues, was made responsible for her construction. In 1863, he left the dockyard to become a surveyor for Lloyds Register, a post that took him to Glasgow. In 1864, he was appointed Shipyard Manager at Robert Napier's Govan shipyard, where he designed and built the ships *Pereire* and *Ville de Paris* for the Compagnie Générale Transatlantique. This was a significant demonstration of his abilities as a naval architect.

John Lennox Kincaid Jamieson was a Glasgow-born marine engineer who, in 1856, became the Superintendent Engineer of the Pacific Steam Navigation Company for whom Randolph Elder and later, John Elder & Co, would build many ships. In 1866, he was appointed head of the engineering department at Randolph Elder's Centre Street works, a position he kept until 1879.

The immediate task facing the partnership of Ure, Jamieson and Pearce was the completion of the Fairfield Works. On Elder's death in September 1869, the shipyard buildings had been completed, work on the engine works had begun but the main building berths and fitting out basin were yet to be cut and formed. Because this required removal of the river wall which had been built by the Clyde Navigation Trust, their permission to start work on the berths and basin had to be sought.

Difficulties with the Clyde Navigation Trust (part 3)

When Ure, Jamieson and Pearce took over the management of the Fairfield Works, they were set on completing the yard as John Elder had intended. However, as before, the issue with the Clyde Navigation Trust over the entrance to the proposed dock, the dock and its use, remained a serious obstacle. On 5 January 1870, Reith wrote to John Ure, with whom he was already well familiar, that the whole subject had again been considered and the Trustees who '… cannot legally grant the unconditional privilege which your friends at Fairfield are desirous of obtaining.'

With the Trust unwavering, the issue was finally brought to an end in 1872 when the Company approached the Sheriff resulting in permission to proceed being granted. What amounted to the last shot in this protracted disagreement was fired on 6 February 1873, in a letter to the editor of the *North British Daily Mail.* The author, simply signed as John Elder & Co, but possibly, John Ure, was replying to an article in the same newspaper printed the day before concerning the minutes of the Clyde Trust's Committee on Law Business and in which a defence of the Trust's position as interpreted by Bailie Bain, a City of Glasgow Councillor,[29] was offered regarding the dock at Elder's shipyard.

> Sir, we have seen in your paper of yesterday Bailie Bain's remarks, at Tuesday's meeting of the Clyde Trustees, respecting the intersection of the river wall for access to our proposed basin and graving dock works at Fairfield. The Bailie must have overlooked, no doubt inadvertently, the facts of the case when he made his observations. The Trustees and their officials have throughout interposed every objection in their power to the execution of these works, and they only yielded after they found that neither their legal advisors nor their engineers could support them in their opposition. We have completely succeeded in carrying our original views in all their integrity. The matter might have been settled long

Constructed for Dutch owners, *Willem III* was intended for passage to the Dutch East Indies via the newly opened Suez Canal. Just two days into her maiden voyage from Nieuwediep to Batavia in May 1871, fire broke out while the vessel was in the English Channel. Although badly damaged, the ship was towed into Portsmouth and later repaired. She continued in service under various names and shipping lines until scrapped in 1903.

ago had we submitted to the following conditions proposed to us (in writing) by the Clyde Trustees as essential conditions to the withdrawal of their opposition.

1, that we should whenever required by the trustees, close up the entrance to our dock thus rendering a large proportion of our works entirely useless after an actual and proposed expenditure . . . of upwards of £250,000.

2, that we should pay to the Trustees the same dues for our dock, constructed on our own property under our own cost as are levied within the harbour of Glasgow, constructed at the cost of the Trustees.
3, that so far as these works are concerned we should for ever abandon our *locus standi* to object in Parliament to any measure which the trustees might introduce in relation to the River Clyde and the works thereof. To these conditions Mr Elder, like us, his successors, would on no account accede. We venture to add, that the proposed works, besides furnishing employment to a great number of people, will contribute much to increase the trade of the river, an object which the Trustees profess to have at heart, although certainly they do occasionally take strange ways of manifesting their zeal.
I am, etc. John Elder & Co.

The Pacific Steam Navigation Co was not the only company to return repeatedly to John Elder & Co for ships. At a time when shipbuilding in the UK was far in advance of its European counterparts, many steamship lines that would later favour their home yards came to the Clyde. One of these was Stoomvaart Maatschappij Nederland (Netherland Steamship Company) of Amsterdam, formed in 1870 to establish regular trade between the Netherlands and the Dutch East Indies via the Suez Canal. From the outset, this company entered into discussions with John Elder for vessels fitted with compound engines. The very first to be built was the *Willem III*, delivered in May 1871. After a few days into her maiden voyage, the vessel caught fire and was burnt out not far from the Isle of Wight. Despite this, the Netherland Steamship Company would go on to contract a further 22 vessels from John Elder and Fairfield.

The Fairfield Works as initially completed were described in *The Scotsman* newspaper of 6 July 1872 and reproduced here.

In the present article we propose to deal with the firm which, for the name it has achieved, the enterprise it has exhibited, its rapid development, and the extent of its resources, is probably unrivalled by any other private shipbuilding firm in the world. We refer to Messrs John Elder & Company. From the outset the ambition and aim of the firm was to be in the van. With this object in view, the late Mr Elder planned the new premises on a scale of great magnitude ...

The new Engine shop, which is one of the largest of its kind in the world, measures 296 feet from east to west and 293' 10" from north to south. Its height from the ground to the top of the water gutter is 50' 11". To provide accommodation for and keep separate the different departments, there are three galleries on either side of the main entrance (each 30 feet wide by the entire length of the shop) which are approached by spacious staircases inside the building. The galleries are supported on strong iron columns and girders, and in the open space between the galleries there are two very heavy travelling cranes made by Sir William Armstrong. Each crane is capable of raising a dead weight of 40 tons. Attached to the galleries there are projecting platforms on two of which heavy castings can be lifted by the travelling cranes with the utmost ease. The cranes cover a span of 51 feet. Abundance of light and ventilation is provided for the building and there are ample facilities by spacious doorways on every side for egress and ingress. All the machinery is new and of the most powerful kind. At the east end of the building the boilers are made. It is quite a singular arrangement to have both the boiler and engine making departments under the same roof but in practice it has been found to answer very well. Motive power is supplied to the machinery by three pairs of small diagonal high-pressure engines. The engine shop we may add, is built from designs furnished by Mr Angus Kennedy, civil engineer and architect, Bath Street, Glasgow.

As for the smiths' shop which immediately adjoins the engineers' department, its dimensions are in keeping with the rest of the works. It measures 293' 10" in length by 102 feet in width. The height to the top of the wall is 23 feet; but, as the roof has a very steep slope, it is nearly as high as the engine shop, taking its height to the top of the ventilator. Altogether the smiddy contains 100 fires and the blast is supplied by two pairs of Roots patent blowers … Several patent furnaces have been erected in the smiths' shop for the purpose of heating the iron … One other feature of the smiths' shop worth noticing is the iron rack, which is capable of holding 500 tons of finished iron, or considerably more than the capacity of any other smiths' rack in this country.

In the older part of the Fairfield shipbuilding yard the buildings are not remarkable for any specially exceptional or interesting features if we except their length, which is 1200 feet from east to west, by an average width of 50 feet. This, as the merest novice may perceive, represents an enormous area under cover. Practically, it is one building from end to end over the whole 1200 feet but of course it is divided and subdivided into different departments, the smiths occupying one part, the angle ironworkers another and so on. The smiths' department is at the east end of the range and contains upwards of 72 hearths. Then follow a number of angle iron furnaces each 40 feet in length and so on until the premises occupied by the artificers in wood – the sawmillers, joiners, and spar makers – is reached at the other end. This sawmill is 150 feet square and the riggers' loft above is of corresponding dimensions. Attached to the joiners' shop there is a spacious shed for drying the wood, which is taken into the yard by a travelling crane fitted up at the edge of the river.

Large stores adjoin the various departments for the preservation of tools, rivets etc. As an indication of the extent to which rivets are used, we may mention the fact that at the present time fully 100 tons of rivets per month are being made for and by the firm. Of the boatbuilding (the firm build all or nearly all their own boats) and other finishing departments … they are on a scale of magnitude and completeness commensurate with the rest of the vast establishment. It would be unfair however, to pass over in silence the thoughtful arrangements made by the late Mr Elder for the relief of his workmen in case of accident. Almost at the entrance gate there is a wooden building erected for the purposes of an hospital, containing every appliance necessary to treat the injuries to which workmen are liable. Another 'institution' of the yard is the water closet system, which is so arranged that no man can spend more than 10 minutes without the knowledge of the man in charge, and when this limit is exceeded, half an hour's pay is deducted. We have already pointed out that the shipbuilding yard is 70 acres in extent. Not more than 50 acres however are actually used at present. The next largest yard on the Clyde, that of Messers J & G Thompson at Dalmuir, only covers about 30 acres so that in point of size the yard of Messrs John Elder and Co leaves all others very far in the rear. It is probable that all, or nearly all, the 70 acres will come to be occupied when the firm have completed their arrangements for the construction of a graving dock at the western boundary of the yard. The idea of having a graving dock formed part of the original scheme of Mr Elder when the Fairfield works were commenced but owing to some disagreement or misunderstanding with the Trustees of the Clyde Navigation, the work was abandoned after the foundations had been dug, and progress has been completely suspended during the last two years. We understand however that the firm contemplate being in a position to proceed with the dock, which would undoubtedly form a valuable adjunct to the works at an early date.

At the present time Messrs John Elder & Co, have 16 vessels on hand or contracted for, their aggregate tonnage being upwards of 36,000 tons; six of them are about or above 4000 tons each but there are none of them calling for special notice. At the Fairfield works alone the firm now employ upwards of 4000 hands, and at their Centre Street engine works and boiler works they employ nearly if not fully 2000 more.

In reference to John Elder's death, the article noted that 'Mr Elder's plan has now been carried out to the extent of providing accommodation for all the departments at Fairfield, but not unfortunately, before he himself had gone over to the Great Majority.'

The part of Elder's plan that was incomplete at the time of the above article, was the fitting-out basin. Quite apart from the obstructions, deliberations, negotiations and ultimate resolution with the Clyde Navigation Trust, it seems that nature also intervened to frustrate completion of the scheme. When the ground behind the river wall was being excavated to form the dock, the combination of a high tide and severe gale one evening was sufficient to breach the river wall and flood the partially excavated dock.[30] Only the east wharf was initially made usable, with a 60-ton capacity sheerlegs constructed to install machinery from the adjacent engine works. The remainder of the dock, the head and west side, was left almost in a state of nature before the dock was properly completed about the turn of the twentieth century.

In July 1875, with the transfer of machines and manufacturing from the old works to the Fairfield Works complete, the Centre Street Engine Works and Dale Street Boiler Works were put up for sale.

In the mid 1870s, Elder received two major contracts

The steamship *John Elder*, 1870. So highly did the Pacific Steam Navigation Co think of John Elder and the important role his compound engine had played in their routes from the UK to South America, they named this ship after him following his death.

from the Admiralty. The first was the machinery for the ironclad battleship HMS *Inflexible*. The second was for the hull and machinery for the armoured cruiser HMS *Nelson*.[31] These were significant contracts from a highly significant source and the first for the hull as well as the machinery. The only orders received from the Admiralty previously had been for machinery only: for the frigate HMS *Constance* ten years earlier and coastal defence ship HMS *Hydra* in 1871.

A very poor reproduction of a photograph but the earliest one of the yard available. Taken about 1871, it shows barges under construction for service in China. The tall wooden fence at right separated the public footpath running along the banks of the river from the shipyard.

Launch of HMS *Nelson*

On 4 November 1876, the armoured cruiser HMS *Nelson* was launched in front of distinguished guests and thousands of onlookers in the yard and on the other side of the river. At 2pm that afternoon, while the shores (wooden props) were being removed from the hull, William Pearce stood on the launching platform, signalling the start of the ceremony. A trumpeter sounded a note for silence, whereupon the Reverend John McLeod of Govan Parish Church read out the service written specifically for ships of the Royal Navy. Following this, Pearce gave the order 'Stand By', whereupon Miss Helen Shaw Stewart[32] came forward to break the bottle of wine that was fixed to the ship's bow. Pearce then asked if all was ready, handing Miss Stewart a mallet with which she struck a knife, severing the rope and causing the dog shores (the timber props holding the hull in place) to fall away, thus enabling the hull, weighing 5000 tons, to begin the short journey down the slipway into the river. As this failed to happen, Pearce commanded 'Pump Away'. whereupon three hydraulic pumps began to exert pressure on the hull, which slowly began to slide down the launching ways. Once in the water, the hull was brought to a stop by four heavy chains, two on either side at the stern and two anchors at the bow. These were dragged on the ground to slow the hull's progress.

The launch was played out to the tune of Rule Britannia as performed by the band of the 25th Lanarkshire Rifle Volunteers. The hull was taken upriver by tugs to Clydebank Quay, where her engines would later be fitted, as the fitting-out basin at Fairfield was still not operational.

At the luncheon held in the drawing offices after the launch, William Pearce said that the Admiralty knew that they could get ships in the Clyde at least equal to any they could build themselves. He continued 'It must be gratifying also to know that the Clyde alone could renew the British Navy in two years. There were at present 40,000 men engaged in shipbuilding on the Clyde and, in case of war, there would be very little difficulty in replacing the navy ship, or in producing a fleet that could compete with the united navies of the world in two years.'[33]

In March 1877, Andrew Laing, aged 21, joined John Elder & Co as a draughtsman in the engine drawing office at Centre Street. This was the first step in what would be a brilliant career. At the time of Laing's arrival at Centre Street, Pearce had begun construction of the *Arizona* for the Guion Line. The machinery for this ship had been designed by Bryce Douglas, the Engine Works Manager. *Arizona* was a record breaker, as were the two Guion Line ships that followed from Fairfield, *Alaska* and *Oregon*. Laing's ability was such that in 1881, he was appointed by Bryce Douglas as Chief Draughtsmen. In 1885, Laing became Assistant Manager and two years later, Manager, when Bryce Douglas left Fairfield to become Head of Engineering at the Barrow shipbuilding works and later, Managing Director.

On 1 July 1878, both John Ure and John Jamieson retired from the business, leaving William Pearce as the sole proprietor. Pearce was about to begin his career as the head of the Fairfield Works having secured a remarkable contract for a very unusual vessel, the yacht *Livadia*.

Livadia

In October 1878, the Russian Imperial Yacht, *Livadia*, in service with the House of Romanov, was wrecked on the Black Sea. The design of a replacement vessel was quickly put in hand, resulting in an unusual experimental vessel, which was also to be given the name *Livadia*. In the following year, just as the Guion liner *Arizona* was about to leave Fairfield and set up a new Atlantic speed record, William Pearce entered into discussions with Russian authorities. These culminated in the award of the contract to build the yacht at Fairfield. The new *Livadia* was no ordinary vessel but one to the design of Andrei Alexandrovich Popov, a Russian Admiral who had recently designed and built two circular monitors for the Imperial Russian Navy. His design for the Imperial Yacht followed this unusual circular design concept, which had been previously proposed by none other than John Elder.

Admiral Popoff, undoubtedly aware of Elder's theoretical designs, had begun experimenting with circular vessels. In 1870, this had led to the construction of two monitors, *Novgorod* and *Vitse-admiral Popov,* built at Mykolaiv for service on the Black Sea. Completed in 1874 and 1876, these vessels were not entirely successful, one of the main difficulties being how to steer a circular hull driven

At left, the fourth son of Czar Alexander II, Grand Duke Alexei who attended the launch of *Livadia* at Govan in 1880. He is seen here later in life as General Admiral of Russia, 1883. At right, is Grand Duke Konstantin with whom William Pearce together with Admiral Popov discussed the design and contract for *Livadia* at Biarritz.

Bow view of *Livadia* in the fitting-out basin in 1880 showing the great beam of the ship with three funnels abreast. Sponsons rising from the lower hull support an outrigged area of the upper deck housing the ships' boats. The original 90-ton sheerlegs are in view while the bridge at the mouth of the basin to carry the right of way appears to be under construction.

by propellers. However, these designs were sufficiently interesting for Popov to be given the task of designing the new Imperial Yacht. In developing its hull form, and to assist steering, Popov effectively added a bow and stern to the circular plan of what would become *Livadia* in what was termed a 'turbot-shaped' hull. To test the hydrodynamics of this hull, a scale model was trialled at the ship test tank in Amsterdam Royal Dockyard, under the supervision of naval architect Dr Tidemann, a contemporary of William Froude.

The choice of John Elder & Co as shipbuilder appears to have been encouraged by Popoff, possibly in part because of Elder's paper but most likely because the Fairfield yard was at this time a world leader. Before William Pearce signed the contract to build *Livadia* with a guaranteed speed of at least 14 knots, he knew that Tidemann's experiments had made it an absolute certainty that, with proper machinery, a speed of at least 15 knots would be achieved.[34] Pearce, having given that assurance to the Czar's brother, the Grand Duke Konstantin, the plans were approved by the Czar and the contract placed. Pearce also conducted a few experiments of his own with a one-tenth scale steel model on Loch Lomond. *Livadia*'s machinery, comprising three compound engines driving three shafts, was designed by Bryce Douglas, while the interiors were by Glasgow architect William Leiper. These were in the style of Louis XVI at Fontainbleau Palace, while the drawing room was in the Crimean-Tartar style.

Livadia was far from the only contract being worked on at Fairfield at this time. The very rapid progress made on the yacht was in parallel with four vessels for the Compagnie Générale Transatlantique, two for the British and African Steam Navigation Co, and three for the Admiralty.

The day of *Livadia*'s launch, Wednesday 7 July 1880, began with the arrival of the Czar's son Grand Duke Alexei, at St Enoch's station in Glasgow. The Grand Duke had travelled through the night by train from London. He was greeted on the platform by a guard of honour formed by Russian sailors, 200 of whom had travelled to Glasgow from Hull. Grand Duke Alexei and his staff arrived at the shipyard at 11.30am to be met with a great cast of dignitaries including the Duke and Duchess of Hamilton, the Russian Ambassador, the Third Sea Lord Rear Admiral Houston Stewart, Admiral Popoff and Captain Goulaeff who had superintended the construction of the ship. Thousands of people filled the shipyard and watched from the other side of the river as the ceremony took place. This took the form of a Russian Orthodox choral service, made possible by bringing a choir from a Greek church in London. The Duchess of Hamilton then smashed a bottle of wine on *Livadia's* bow and the ship slid down the launching ways to the tune of the Russian national anthem played by the 74th Highlanders.

With the ship safely afloat and moored in the fitting-out basin, guests retired to a luncheon held in a marquee set up in the yard. Following a series of toasts, Grand Duke Alexei said

> Mr Chairman, ladies and gentlemen, I ask to be excused in having to speak publicly. I would wish simply to say two words and no more – to propose to you the health of the Duchess of Hamilton and to propose to you, prosperity to Mr Pearce and to Glasgow, which is the centre of the intelligence of England. (loud cheers and laughter).

After Admiral Popov spoke, William Pearce replied at length, providing a little more information about how the contract came into being:

> I was invited to go to Biarritz and there I had the opportunity of meeting his Imperial Highness the Grand Duke Konstantin who, together with Admiral Popov, gave me some further ideas in regard to the construction of the ship. I am glad to say that I have been very successful in carrying them out and today we have succeeded in putting this ship into the water.

> During the time the vessel has been in course of construction, a very great and gradually increasing interest has been shown in her amongst the engineers and naval architects of this country. The *Livadia* is looked upon by them as a great experiment. I look upon it, and I'm quite sure the designer, Admiral Popov, looks upon it no longer as an experiment.[35]

While *Livadia* was fitting-out in September, newspapers were gripped by the story that 'Russian Nihilists' were intent on destroying the ship by planting two nitro-glycerine bombs somewhere onboard, probably in the coal bunkers. Police were posted at the yard and the underwater hull checked by divers. However, no bombs were found and no arrests made.

On 6 October, *Livadia*, with Admiral Popoff in command, left Fairfield basin to run trials. One disadvantage of her great beam was immediately obvious when the river channel had to be closed to all traffic to allow her passage. Trials consisted of a six hour run at various speeds including a run between the Cloch Light and Cumbrae Light at a speed of 15 knots. The following day, a series of six, full-power runs over the measured mile took place, showing a mean speed of 15.864 knots and mean indicated horsepower of 12,383. This was 2000 hp more than the contract guarantee. On 15 October 1880, at 11am, *Livadia* was handed over to the Russian Navy and left the Tail of the Bank for Sevastopol with Admiral Popoff again in command. On board were Admiral Sir Houston Stewart, Chief Constructor, Sir E J Reed, formerly Chief Constructor and William Pearce.

So impressed were the Russian authorities with the speedy construction and machinery performance of *Livadia* that Elders were given a bonus of £40k in addition to the final stage payment. They were also given the contract to re-engine the ironclad turret ship *Petr Veliky* (Peter the Great). On 14 July 1881, the Russian battleship arrived in the basin at Fairfield, remaining there until

Livadia. This line image from *The London Illustrated News* gives some idea of the launch of this highly unusual ship. While many shipbuilders would have been reluctant to take on such a contract, the novelty and celebrity appealed to William Pearce.

The Guion liner *Alaska*, second of the 'Fairfield Flyers' which improved on *Arizona's* speed by reducing the Atlantic crossing to under one week, albeit at enormous coal consumption.

February 1882, while the old machinery was stripped out to be replaced with new vertical compound engines.

While everything had gone so well for *Livadia* – the speed of construction, performance of her machinery and successful trials etc – her subsequent career would prove to be less than glorious. Sailing from the Clyde to Brest in fair to moderate seas, she did indeed, as trials suggested, prove to be a very stable ship, justifying the claims of her designer. However, while crossing the Bay of Biscay, she ran into a severe gale which resulted in damage to parts of the underwater hull. *Livadia* put into Ferrol for repairs which would have taken a few days under normal circumstances but, with no dry dock large enough to accommodate her, the vessel did not reach Sevastopol until June 1881, by which time Alexander II had been assassinated.

Richard Barnwell

In 1880 William Pearce made a significant addition to his senior shipyard staff with the appointment of Richard Barnwell to look after the commercial side of the business. Barnwell, born in Canterbury in 1849, had been trained at the Union Bank of London. His wife, Annie, maiden name Sowter, was a sister of Dinah Elizabeth Pearce, wife of William Pearce. Barnwell arrived at Fairfield at a time when the Company was reaching new heights through the construction of fast ocean liners subsequently known as the Fairfield Flyers.

The Fairfield Flyers

While *Livadia* captured the imagination for its novelty, by the time the vessel arrived in Sevastopol, John Elder & Co, were about to launch *Alaska* for the Guion line, which, it was hoped, would be faster than *Arizona*. The record for the fastest crossing of the Atlantic, set by *Arizona,* was improved on by *Alaska,* even if she would prove to be less than economical to run. A third liner, *Oregon*, would be the fastest of the three, reducing the crossing time from *Arizona*'s 7 days 6 hours and 20 minutes to 6 days, 10 hours, 10 minutes. On 19 September 1883, *Oregon* had made a trial trip from Ailsa Craig to Cumbrae at an average speed of 20 knots. A newspaper report published on 22 September stated:

> We believe this speed has never yet been approached by any oceangoing steamer and that it fully realises the expectations of the designers. The revolutions per minute when that rate was attained were 62, the steam pressure was 110 lbs per square inch and the vacuum 28. It is almost impossible to give a sufficient idea of the appearance of the vessel's enormous engines in full operation. The great width of the ship has permitted the various parts of the engine being placed in relation to each other with more ease and with less tortuous connections than is sometimes necessary. The great height of the structure, the ponderous cranks and pistons and the multitude of moving parts inspire the onlooker with a species of awe but, notwithstanding the grandeur, it is difficult to imagine that their movements condense into so small a space a power equal to that of 12,000 horses.[36]

According to figures published at the time, the shipbuilding output for the Clyde in 1883 was almost 420,000 tons, 28,000 more than the year before. By the end of 1883 however, the industry was stagnating, in part attributed to dull trade and the high cost of new vessels. This resulted in over 4000 men on Clydeside being paid off by December. In January 1884, when the steamer *Ruapehu*, for the New Zealand Steamship Co, was completed, 800 men were dismissed at Fairfield. At the same time, wage reductions of 10% were imposed.

Alexander Carnegie Kirk and the Triple Expansion Engine

In 1870, while Messrs Ure, Jamieson and Pearce were coming to terms with their responsibilities running the yard, a new appointment was made to the management of the engine works in Centre Street. This was 40 year old Alexander Carnegie Kirk, a marine engineer who had served a five year apprenticeship at Robert Napier's Vulcan Foundry in Glasgow, as had so many talented engineers before him. Kirk had worked in a variety of places and at various roles including with London based marine engineers Maudslay, Sons and Field as Chief Draughtsman. Kirk's inventive flair had resulted in a number of appliances concerned with refrigeration for ships, as well as a number of papers delivered before the Institution of Engineers in Scotland. In his new role at Centre Street and, subsequently, at the new engine works at Fairfield, Kirk began to look at the compound engine, making numerous detailed improvements. One of John Elder's patents had referred to compounding in several stages, not just two but three (triple expansion), and four (quadruple expansion). Kirk began to develop these ideas which led to the first successful triple expansion engine. The first iteration of his engine with three cylinders and three cranks was fitted in the screw steamer *Propontis* in 1874. The development work for this pioneering steam engine was undertaken by Kirk at the new engine and boiler works on the Fairfield site.[37] However, as boilers of the day were unable to produce steam at the high pressure Kirk's engine required, *Propontis* was not a success. In 1877, Kirk left Fairfield to become a partner at Robert Napier & Co's Govan shipyard.

In 1881 Napier's yard contracted to build the steamer *Aberdeen* for the Aberdeen Line. By this time, boiler technology had progressed, particularly with the use of steel, enabling higher pressures to be achieved. *Aberdeen* was fitted with Kirk's triple expansion engine, similar to the *Propontis* engine but with boilers of the new kind. This time, the engine was a great success, recording on trials in 1882, a consumption of coal at the rate of 1.28 lbs per horsepower.[38] This was yet another dramatic increase in fuel economy and lead to the widespread use of the triple expansion engine for decades to come. This breakthrough also maintained the Clyde's reputation as the centre of steam marine engineering technology.

Alexander Carnegie Kirk 1830–1893. Kirk perfected the fuel efficient triple expansion steam engine in 1874 which then became the principal steam marine engine until the advent of the steam turbine but, even then, continued to propel ships for many decades into the twentieth century.

William Pearce 'A Big Boom Person All Round'

Pearce was undoubtedly a man of great energy and ability, able to pull off deals and make things happen for his business. It seems that, whatever else his workers may have thought of him, there was genuine regard for his abilities. At the beginning of 1884, Pearce became seriously ill and had to retire from business life for a period of five months. When he returned to the Fairfield Works on 28 May, he was greeted by 3000 people, effectively the entire workforce, greatly reduced because of the depression in shipbuilding then underway. On a platform raised for the occasion, his shipyard and engine managers J Shepherd and Bryce Douglas, were seated with a few other officials and a workers' representative. After a few introductions, the representative, Mr Eaglesham said, with reference to Pearce:

'I am sure I speak the sentiments of all before me when I say that we rejoice to have him in our midst again and have an opportunity of looking him in the face.' (cheers) Continuing in this vein

William Pearce. Talented, energetic and a shrewd businessman, Pearce continued the trajectory begun by John Elder in taking the Fairfield yard to the top through the design and building of the fastest ships afloat.

The three masted topsail steam schooner *Lady Torfrida*, completed in 1889 for Sir William G Pearce and not to be confused with three earlier yachts with the name *Torfrida* also built at Fairfield for and by the designs of Willian Pearce.

for some time, he then read out a declaration written by the workmen:

> Dear sir, we the undersigned on behalf of the fellow workmen employed in Fairfield Works desire to express our respect to you, and our deep gratitude and thanksgiving to Almighty God for His boundless mercy towards us in the restoring of your health and bringing you safely amongst us once again. The friendly relations which have always existed between us have led to this expression of our best feelings towards you at this time, and while we rejoice in the hope that you may have a continuation of good health, happiness and prosperity, we earnestly hope that you may be long spared to come in and out from among us, that you may add fresh lustre and renown to the great firm of John Elder and Co of which you are the head.

To this Pearce replied:

> I cannot express to you the great satisfaction it is to me to be welcomed back in the way you have welcomed me today. The address you have been good enough to present I shall always appreciate, and I shall keep it by me as long as I live. I wish I could have come back to you in better times. I wish the trade of this country had not been so depressed as it is at the present time. . . you will excuse me from making a long speech to you, because I'm not able to do so today. I may however invite you on some future occasion to meet me in more pleasant manner than we are met today . . . when I shall have more to say to you upon this and more pleasant subjects. But, for your encouragement, let me tell you that I think you have evidence I have not forgotten you during the time I have been away. I can give you my assurance that I shall do everything in my power to keep this place fully employed, and I can tell you that I have hopes that I shall be enabled to do so.[39]

It was a remarkable exchange between worker and employer in an industry not renowned for its good relations. The poor state of trade and the fear of further layoffs were no doubt part of the motivation behind the workers' declaration, with the hope that Pearce might yet pull a rabbit out of the hat.

Pearce, a member of the Conservative Party, had political aspirations, first revealed unsuccessfully in 1880. In February 1884, because of ill health, he had to decline a visit to his old place of work, Chatham, where it had been hoped he might stand as a Conservative candidate at the next General Election.

The description of William Pearce that appeared in the *Sporting Gazette* of 9 August 1884 also offers an interesting portrait:

> less than 15 years ago, he was a draughtsman in one of the Government yards. Now he is a millionaire, and a big boom person all round. Of the many steam yachts seen on the Solent this week, his was decidedly the smartest.

However, the journal *Justice* noted on 14 June 1884, with reference to *Umbria* and *Etruria*:

> it is said that William Pearce, sole partner in John Elder & Co, will clear £100,000 profit on the two steel steamers he is building for Cunard. This is partly by reason of the fall in the cost of materials, but chiefly by the reduction in wages. We desire to call the attention of working men who are not yet socialists to this cool admission.

In 1884, with orders for new vessels hard to find, employee numbers dwindled. To provide employment for his men, in November of that year, William Pearce offered to lay down a 5000-ton ship, then a large vessel, at his own expense to keep between 1000 and 1500 men employed. He met with workers' representatives with whom he proposed a rate of pay equal to 50 shillings per week which was lower than current rates but which would enable him, he thought, to build a ship, the cost of which would be attractive to shipowners. Pearce wanted a guarantee that, if he laid down such a ship at his own expense, men would stick to this reduced rate of pay for 12 months. Although the men's representatives recommended Pearce's offer, the men were having none of it, demanding 60 shillings per week.[40]

Prins Hendrik, the second ship of that name built at Fairfield for Dutch ferry operator Stoomvaart Maatschappij Zeeland. This paddle steamer was built for the Flushing–Queenborough/ Folkestone service.

Etruria, seen here, with her sister ship *Umbria*, were built for Cunard after some slick negotiation by William Pearce with his Cunard counterpart, John Burns, Chairman of the shipping line.

Umbria and *Etruria*

Cunard, having made the decision to build two new express steamers for the North Atlantic service, decided that John Elder & Co would be their builder of choice, despite Cunard's most recent steamers, *Servia* and *Aurania,* having been built by J&G Thomson at Clydebank. Negotiating the contract was left entirely between the two principals, John Burns, Chairman of Cunard, and William Pearce of John Elder & Co. Over the course of two weeks, the terms of a contract were struck by both men who were located within 30 miles of one another – John Burns, residing in Wemyss Castle at Wemyss Bay and Pearce at Fairfield. It was from Wemyss Castle that the details were thrashed out, with Burns sending frequent letters and telegrams to Cunard directors in Liverpool. What follows gives an indication of how negotiations proceeded for what would be the largest ships afloat.

From the outset, part of the cost of the new vessels was to be offset by Burns asking Pearce to accept the old Cunard vessels, *Parthia* and *Batavia,* in part exchange. As part of his bargaining strategy, Pearce indicated that he might well enter into contract with the Guion Line, Cunard's competitor, for a new steamer should it not be possible to enter into a contract with Cunard.

Surviving documents, letters and telegrams in the Cunard Archive, detail the final round of negotiations. These, and the manner in which they were conducted, are described here in some detail, as they show something of the personalities involved and, in particular, that John Burns might well have been somewhat outmanoeuvred by Pearce.

5 July 1883. In a letter to his Directors in Liverpool, John Burns wrote:

> I have been a long time with Mr Pearce and I will come to this shortly and . . . he will not take less than £310,000 for each ship and will not give more than book value for the *Parthia* and *Batavia*, the ultimate date of delivery for both ships being 31 Dec 1884. My strong advice is to accept those terms, if not, you must make up your mind to Mr Pearce going into a large transaction with Mr Guion tomorrow – If you agree to Mr Pearce's terms, he will agree not to build for Guion before 1 Jan 1885. Under the whole circumstances of the case, I have no hesitation in

> recommending to close with Mr Pearce – I send a copy of this to the London directors desiring them to telegraph me early tomorrow morning.

6 July from Burns to Directors at Liverpool

> I can assure the Directors that I have had a very heavy day's work in negotiating with Mr Pearce and after a full knowledge of all the circumstances of the case, and the exceptional difficulties surrounding the position, I am prepared to assert that it was absolutely impossible to make a better bargain than I have done – I fought every inch, till at last Mr Pearce distinctly stated that he would not go a step further as he had great misgivings in regard to going into the transaction at all, owing to the question of taking old ships, which he said was most distasteful to him.
>
> I do not think that anyone can appreciate the obstacles which presented themselves unless being on the spot and I feel that I have done my very upmost in the best interests of the Company.

Burns detailed the basics of the negotiation:

> 2 ships with dimensions of 500x57x40 with a price of £310,000 each with stage payments as follows:
> One eighth: ship in frame
> One eighth: ship plated
> One quarter: when launched
> One quarter: when engines on board
> One quarter: when vessel completed and delivered.

On 7 July, a preliminary contract was countersigned by both parties.

12 July. From the Glasgow office of Cunard (30 Jamaica Street), a letter was sent to Directors at Liverpool describing an important issue that had arisen. This concerned the spacing of frames proposed by Pearce. While the latter specified 30 inches, Cunard's Superintendent Engineer, Mr Logan, wanted 27 or 28 at most. When this was communicated to Pearce along with several other additions, Pearce responded the following day with:

13 July

> Dear Mr Burns,
> I find that the good faith upon which I signed the preliminary contract for the two steamers has not been responded to. In the first place it was a good deal strained when asked to accept the terms of payment and the terms on which the old ships were to be taken over.
>
> I may state here [that] the cost as got out for these ships was for each £318,247. This [is] now increased by:
>
> | | |
> |---|---|
> | Interest on instalments | £2000 |
> | Extra cabins on lower deck | £3500 |
> | Electric lighting | £2000 |
> | Extra spare gear | £1250 |
> | | total £327,000 |
>
> I asked only £310,000 without any idea of having to take over old ships above their value and I cannot afford to reduce that price by £8750.
>
> There is another matter – I understand that the spacing of the frames at 30 inches is now objected to.
>
> I must not make a mistake in my shipbuilding even for the honour of building for the Cunard Co, and I positively refuse to build a ship of this size for anyone with the spacing less than 30 inches.

After signing this letter, Pearce added at the bottom:

> Afterwards, in order to keep everything on an amicable basis, I agreed to fit cabins on the lower deck for first class passengers instead of for steerage and cargo as was understood by my Managers and estimated for by them. Now we are asked to do what was never contemplated by us such as fitting electric lights throughout the ship and extra engine spare gear.

14 July
A meeting of Cunard's Executive Committee was held at which it was reaffirmed that frame spacing should not exceed 27 inches but 'in case of need might be 28 inches'. It was agreed to leave the matter in the hands of the Chairman but noted that if he was compelled to accept 30 inches that the shell plating must be increased in thickness by 1/16 inch on four strakes for a length of 300 feet amidships.

17 July
John Burns and William Pearce met again, in an attempt to finalise a contract. Burns wrote to his Directors in Liverpool, noting that:

> I have had fully three hours with Mr Pearce, the upshot of which I shall narrate as briefly as possible.

He absolutely declined to build the ships if we insisted upon spacing the frames less than 30 inches (the same as *Oregon*) and stated that his credit was as much at stake as that of the Cunard Company, and that he has fully considered the question of strength and argued that he obtained it in a greater degree by the plan which he proposed – and I am bound to say that both Captain Watson and I consider there is great force in his argument.

1. In lieu of making the space less than 30 inches Mr Pearce has agreed to the following alterations on the specification and upon the plan:-

Strakes 'N O' to be increased 1/16″ thickness. Strake 'P' main sheer strake to be increased 1/8″ thickness and doubling strake dispensed with and 1″ butt straps substituted.

2. The terms of payment for each ship to be as follows, viz:- In four equal instalments in cash or by bill or bills at 6 months for the total or portion or portions of the amount of instalment due, adding interest at bank rate but not exceeding 5 percent. (Instalments as in final version below)

3. Mr Pearce agrees to pay the cost of electric lighting viz:- £2000 for each ship, but as I consider it is better for the Company we should ourselves make the contract for electric lighting, Mr Pearce has agreed to reduce the price of each ship from £310,000 to £308,000 which comes to the same thing and I think puts the matter upon a satisfactory footing.

Mr Pearce said it was utterly useless to try to keep the matter longer secret as the newspaper people had already got wind of it.

17 July 1883. The offer from Fairfield:

The ships to be 500x57x40 feet and the price of each is £308,000. Payment to be four equal instalments in cash or by bill:

1 – when vessel is in frame

2 – when beams, keelson, stringers, tie plates and steel decks are in place.

3 – when the vessel is riveted, decks laid and caulked and the vessel launched.

4 – when the vessel is completed and delivered.

The first ship to be delivered on 1 February 1885 and the second on 1 March 1885.

Fairfield undertook not to contract or build for Mr Guion before 30 June 1885 and also to take the steam ships *Batavia* and *Parthia* not later than 31 December 1884 at their book values on 31 Dec 1882 viz, £31,500 and £51,000 respectively, less depreciation at the rate of 5% per annum until the dates of their delivery to us. (Cutlery, bed ware, consumables etc not part of deal.) The price of each vessel when taken delivery of, to be imputed to the instalment or instalments of the new ship or ships next due.

17 July. Telegram from Burns to Cunard, Liverpool.

All settled with Pearce and specification signed – see letter – show this telegram to the Liverpool directors.

19 July. Letter from Cunard secretary to Messrs John Elder & Co, accepting their offer of 17 July.

Umbria was completed well ahead of the contract date, making a trial voyage down to Ailsa Craig and back on 4 October 1884, with many guests on board including Lord Shaftsbury and Lord Balfour. Her maiden voyage to New York began on 1 November 1884. *Etruria* was completed in March 1885 and began her maiden voyage on 25 April that year.

In the General Election of 1885, William Pearce's political ambitions were fulfilled when he was elected MP for the newly created constituency of Govan. This prompted him to make a change to the financial structure of his shipbuilding business as well as to give it a new name.

ERA TWO 1885–1965

The Fairfield Shipbuilding & Engineering Co Ltd

In 1885 Pearce converted the business into a limited company under the title The Fairfield Shipbuilding & Engineering Co Ltd. His election to Parliament partly explains the move to make his privately owned business, John Elder & Co, a limited concern, in order to avoid a clash of interests given his access to Government business and future warship contracts. The Fairfield name had been used to describe the shipyard from the outset, by virtue of the yard's location at Fairfield Estate, but was now officially incorporated into the name of the company and would remain so for the next 83 years.

At the same time, Pearce also registered a second name, The Fairfield Shipbuilding, Engineering and Ordnance Co Ltd, indicating his intention to extend shipbuilding and marine engineering to include gun mountings and ordnance for warships. Pearce was undoubtedly taking into account the growing trend for warship contracts to be awarded to private shipyards where shipbuilding capacity, allied to gun mounting manufacture, would make a potential contractor more attractive to the Admiralty. The precedent for this had been set by Sir W G Armstrong Mitchell & Co Ltd, on Tyneside in 1883. They would later be followed in 1888 by Vickers Sons & Co Ltd. at Barrow in Furness. The boast made by both companies was that they could manufacture all aspects of a warship within their company.

Should Pearce have progressed with his interest in ordnance manufacture, ground existed at Fairfield for this development. Out of the 70 acres at Fairfield owned either by the Company or William Pearce himself, the shipyard and engine works occupied about 50, with 20 acres unused

Adder at speed on the Firth of Clyde in the summer of 1890. She was built for the Clyde to Belfast service operated by G&J Burns.

to the west of the fitting-out basin.[1] Events would prevent this development taking place.

In 1887, Richard Saxton White was appointed Manager of the Shipbuilding Department following the departure of J W Shepherd. White was well qualified, having served an apprenticeship with Day Summers & Co, in Southampton, builders of small vessels, steam engines and cranes. His first appointment after completing his apprenticeship was in the drawing office at Wm Denny's Dumbarton yard, followed by a period at Robert Napier's and Alexander Stephens' Clyde yards. In 1875, he became Shipyard Manager of the Barrow Shipbuilding Co where he was responsible for the construction of the Inman line's *City of Rome*. In 1883, he was made Manager at Armstrong Mitchell's yard at Walker on Tyneside where he was involved in the design of the first bulk oil carrying ship, *Gluckauf*.

The rise of the steamship over sailing ships continued year on year but, even by 1887, there were still large numbers of the latter being built. In 1887 *The Engineer* reported that 326 vessels were built on the Clyde that year, aggregating 185,360 tons. Of these, 203 were steamships and 123 were sailing ships, the latter amounting to 37,575 tons.[2]

Death of Sir William Pearce

The untimely death of Sir William Pearce prompted a number of actions, principal of which was the reconstitution of the existing Company from a private to a public one. When Sir William Pearce formed the limited liability company known as the Fairfield Shipbuilding & Engineering Company Limited in 1885, he, his wife, and Richard Barnwell, the step-brother of Lady Pearce, were the only significant shareholders. Barnwell, who had joined the business in 1880, became a partner and was made Managing Director in 1885, as Pearce began increasingly to focus attention on his political career, spending much time in London. Knighted in 1887, it was while at his London home early in December 1888, that he became seriously ill. On 18 December, he died suddenly of heart failure at the age of 53.[3] Percy Hillhouse (Fairfield's naval architect from 1902 to 1921), in his lecture celebrating the centenary of John Elder, given in 1924, paid the following tribute to Sir William Pearce:

> [He was] a man of great commercial ability and of untiring energy, and under his influence, the prosperity and prestige of the Company steadily increased. He controlled the fortunes of the yard for 18 years, during which period the outstanding work of the Company was the building of Atlantic liners – the 'greyhounds of the Atlantic'. Famous vessels such as *Arizona*, *Alaska*, *Oregon* and *Etruria* successfully held the Atlantic record, bringing the duration of the voyage down from seven and a half days to six days.[4]

In July 1889, seven months after the death of Sir William, the 'old' company was dissolved and converted into a public limited liability company on the basis that this would allow far greater investment in what was a highly successful and profitable business. The 'new' company was incorporated on 10 July 1889. At the first Board meeting held on the 12 July at Fairfield, the assets, premises, contracts etc, of the old company were transferred to the new. After his father's death, Sir William George Pearce was made Chairman and Richard Barnwell continued as Managing Director. In addition to Pearce and Barnwell, at this time the Board consisted of Sir James Roberton (Lawyer and Professor at Glasgow University), Sir William McOnie (Brassfounder and former Lord Provost of Glasgow) and Richard Saxton White (Shipyard Manager). With the exception of Sir William McOnie, all had been directors of the old company. At the same time, preparations were in hand to float the new company, pay the old company its estimated value of £600,000 for the works, tools and land, and raise an additional £250,000 to fund the future working of the business.

The creation of the new company brought with it some

A fine hand drawn cross section of the Hamburg America liner *Normannia* completed in 1890. The image clearly shows the large proportion of the lower hull required for propulsion – boilers, coal supply and two four-decks-high triple expansion engines, with twin shafts and propellers for a service speed of 19 knots.

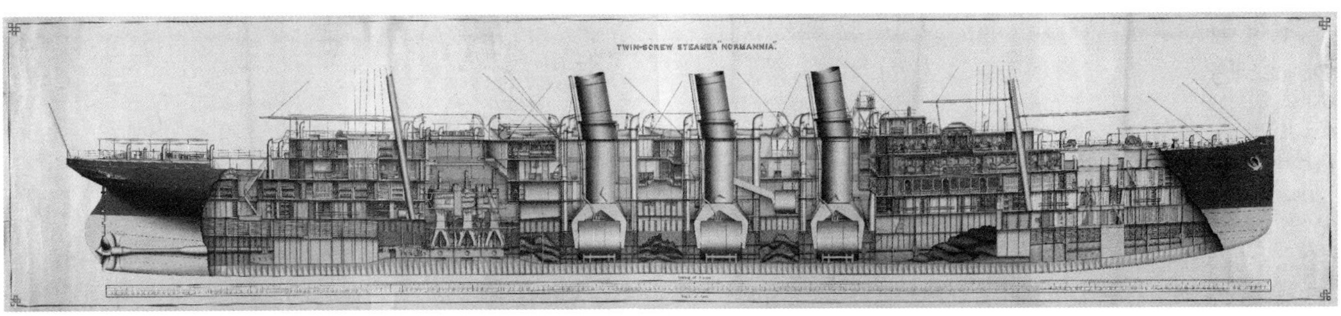

Built for another great German shipping line, Norddeutscher Lloyd, *Weimar* was built for the transatlantic route. *Weimar* was one of the last German liners to be built at Fairfield as, increasingly, the development of the German shipbuilding industry ensured that in years to come German ships would be built in Germany.

exposure to the company's past financial performance as few documents from the earlier period have survived. However, the new company auditors, Price Waterhouse, stated in the prospectus published in 1889,[5] that, between the years 1878–1888, annual profits had not fallen below £46,540, equivalent to £4.577m[6] at 2023 rates. The years 1885–1888 had recorded annual profits averaging £67,463. Indeed, in the first year of trading, the new company recorded a gross profit of £67,670, equivalent to over £7.3m at 2023 rates. Fairfield was highly profitable.

The Fairfield Works, which the prospectus said covered 44.5 acres, was the largest on the Clyde. The prospectus summarised a remarkable output from the yard into five categories, the first of which was the British Government, for whom numerous vessels up to armoured cruisers had been built but not yet battleships; the Atlantic trade, for whom the fastest ships in the world had been built for the Cunard and Guion lines and also Norddeutscher Lloyd; the Australasian trade with ships for the Orient and New Zealand Steamship lines; the Cape and China trade for Donald Currie & Co; and finally, under the heading Channel Steamers, a list of fast paddle and screw steamers for a variety of owners. Altogether, Fairfield was the most successful and prestigious shipyard of its day.

However, the market for the construction of ships for foreign shipping lines, which Fairfield had taken advantage of, was beginning to change, as countries developed their own industrial capacity, something that would happen repeatedly. After 1890, for example, German shipyards were subsidised by their Government as part of the maritime growth of that country.

Grand Offices Befitting

It was almost certainly Sir William Pearce's ambition to demonstrate his company's prestige by constructing a new headquarters building commensurate with his great achievement. It is likely that, shortly before his death in December 1888, he briefed the Glasgow architect John Honeyman, who was about to take John Keppie into partnership, about a new office building to house senior management, ship and engine drawing offices and the counting house, the last where purchasing and wages etc were administered. It was not until 26 July 1889 at one of the first meetings of the newly constituted business that a set of plans, five in number, were presented to the new Fairfield Board by Richard Barnwell. The cost of the building was estimated at between £10,500 and £11,000. Sir William G Pearce instructed that the building should be built as soon as possible.

By March 1892, the new office building was nearing completion. The Board was informed that the final cost had exceeded Mr Honeyman's estimate by £3374 and that the latter's account for professional services was £820. This,

The magnificent Fairfield building designed by Honeyman & Keppie and completed in 1891 was almost certainly proposed by Sir William Pearce as the shipyard's administrative centre. This architectural statement was intended to show just how successful the Fairfield yard had become. He, however, would not live to see construction work begin in 1889. A-listed in 1970, the building is no longer part of the shipyard but home to a shipbuilding museum and business centre. This image shows the building in 2021 virtually unchanged since it opened in 1891.

the Board felt, was excessive and that £600 was more appropriate, leaving it to Barnwell to bring the issue up with Honeyman. It was estimated that a further £7000 would be spent on the building by the Company although this was later reduced to £3488, a figure that included all gas fittings, grates, joiner work and general furnishings.[7] The imposing two storey building with a red sandstone frontage is in the French style. Two stone carved figures by sculptor James Pittendrigh Macgillivray stand at the main entrance depicting the two main areas of the Company – the shipwright figure for the shipbuilding department and the engineer for the engine department.

At the same meeting, the Board agreed to the purchase of new sheerlegs by Fullarton Hodgart & Barclay at the fitting-out basin for £7500, plate rolls by Hugh Smith for £2410 and gas producers to heat the frame and plate furnaces, these to be built by the Company themselves to the designs of W P Ingham of Middlesborough.

In June 1889, Board member James Roberton was knighted but died two months later, reducing the Board to five. As the shipbuilding department was represented by Richard Saxton White, the decision was taken to offer Andrew Laing, Manager of the Engine works, the seat formerly occupied by Roberton. This made sense in giving a voice to the engine side of the business which, after all, had been the main technical driver in quite literally propelling Fairfield to pole position, producing some of the most fuel efficient and powerful steam machinery of the day. Much of this was down to the intelligence and skill of Laing, considered to be one of the finest marine engineers of his day, and upholding the reputation of Fairfield established decades earlier by John Elder.

In April 1891 the passenger cargo liner *Weimar* left the Fairfield yard on completion to begin a regular service between Bremen and Baltimore. *Weimar* was the 15th ship built for Norddeutscher Lloyd (North German Lloyd) and the last. Henceforth ships for this line would be built in German shipyards, an indication of the growing importance of shipbuilding there.

In August 1891 the contest between the two leading Clyde shipbuilders about who built the most prestigious ships was renewed when Fairfield signed a contract with Cunard for two ships for what would be, once again, the largest and fastest ships in the world. This was a result of Cunard's desire to trump the Inman Line's *City of New York* and *City of Paris*, which had been built by Clydeside rivals J&G Thomson in 1888. Having the fastest ship on the North Atlantic was good for business. With the Inman vessels taking the honours on the Atlantic, Cunard had no option but to go one better and regain their leading position.

At least one newspaper picked up on this story one year later when it reported that the hope at Fairfield was that they would wrest the Blue Riband from the present

The triple expansion engines for *Campania* and *Lucania* were among the largest ever made. This photograph shows the lower half of both engines for *Campania* under construction. The completed engines would be approximately twice the height of what is seen here.

holder, the Inman Line's *City of Paris* adding 'In the yard itself, the laudable anxiety to turnout the fastest Atlantic liner is probably exceeded by the keen desire to lower the colours of their neighbours over the way – Thomson, the Clydebank firm who are the builders of the Inman crack liner.' [8]

The ships would become the *Campania* and *Lucania,* which on entry to service took the Blue Riband from the Inman ships. While *Campania* and *Lucania* were successful ships for Cunard, they were a source of great embarrassment and financial loss at Fairfield with serious repercussions in the Board room.

The month after the contract for the Cunarders was finalised, Fairfield made a significant appointment to the Board in the person of Dr Francis Elgar. Like William Pearce before him, Elgar had trained as a naval architect, firstly at Portsmouth Dockyard and then at the Royal School of Naval Architecture at South Kensington, graduating from there in 1867. In 1883, he became the first professor of naval architecture in the UK, at Glasgow University where Isabella Elder, widow of John Elder, had funded the John Elder Chair of Naval Architecture.

With the contract for the Cunarders secured, upgrades to plant and facilities had to be made, most notably an extension to the fitting-out basin costing £12,000 inclusive of new sheerlegs and a new sawmill. The extensive joiner and cabinet making work to be carried out required an expansion of the joiner shop and, as the sawmill was in the way of this, it had to be relocated at a cost of £6000. A shafting lathe, boring machines, a new furnace and new stables were added to this, resulting in a total of £24,420 being authorised.

Andrew Laing, engine works manager and a director of the Company. A highly regarded engineer, Laing joined Fairfield as a draughtsman in 1877 and such was his ability that by 1889 he was manager of the engine works and a director of the company. Allegedly not the easiest of persons to work with, his time at Fairfield ended under a cloud in 1895.

Difficulties for the Fairfield Board

Despite being widely regarded as the famous Fairfield Shipyard and the most important shipbuilding company in the UK if not the world and having taken the contract for two of the largest vessels so far built, the Fairfield Company was facing issues which would come to a head over the next few years.

From the minutes recorded during the early years of the 1890s, it is clear that there were issues surrounding the efficiency of the Company and a general inference of dissatisfaction with the Shipyard, under the management of Richard Saxton White, as well as in Andrew Laing's attitude. Both men were also directors.

Andrew Laing, a person known for his brilliance as an engineer and also for his directness, was asked at the Board meeting held on 23 November 1892 to explain a letter he had sent to the Managing Director, Richard Barnwell, which could be 'read as a reflection on Mr Barnwell's management'. Laing expressed regret that his letter had been misunderstood and asked if it could be withdrawn, which the Board agreed to.

There was further embarrassment for Laing at the same meeting when a letter was read from the Chairman of Cunard, Sir John Burns, complaining about Laing's treatment of the Cunard company's representative. Immediately after this, yet a third letter was read, this time from Barnwell to Laing, in which he advised Laing 'to avoid friction with representatives of companies with whom business was being done'.

Perhaps aware that not everything was as good as it could be in the running of the Company, the Chairman, Sir William G Pearce, turned to Dr Francis Elgar, a Board member since September 1891 and increasingly playing an important role, requesting that Elgar be consulted on all technical matters and that he be considered as the Company's consultant naval architect.

While these events were playing out, an interesting distraction, probably unknown to the Fairfield Board, took place when, in July 1892, Willard I Lansing, employed at a lumber yard in Providence, Rhode Island, was travelling in Europe investigating labour on behalf of his local newspaper, the *Providence News*. While in Glasgow he visited the Fairfield Works and wrote this short description for the newspaper.

> The Fairfield Shipbuilding and Engineering Works are the largest probably in the world. From three to four thousand men are employed. Most of them are engaged through a foreman, and labour on piece

> work. A week consists of 54 working hours, or an average of nine hours a day. Of course their wages depend on their dexterity. You are required to be recommended before gaining admittance and are pretty carefully watched while there. In this way, I was not enabled to investigate as I would have wished. At present, they are building two mammoth Cunard steamers, each of which is 625 feet long, 65 feet wide and 47 feet deep to main deck. In each of these mammoth vessels are to be placed twelve boilers, with a diameter of 16 or 17 feet, two engines and two smoke stacks 19 feet in diameter. Two hours were consumed in passing through the yard and buildings. Everything from start to finish is put on by this company.

After then describing Blackies print works in Glasgow, Willard added 'The labouring classes of Great Britain, but especially here in Glasgow, are improving rapidly, but as one of the Magistrates told me "the greatest enemy they had to fight and feared was strong drink".'[9]

Elgar's Admiralty Connections

Francis Elgar's appointment to the Fairfield Board brought with it connections to the Admiralty, its people and the recognition that, although primarily a builder of merchant ships, Fairfield was equally interested in contracts for warships. During Elgar's period at Fairfield, firstly as naval architect and Director and then as Chairman, Fairfield became one of the Admiralty's premier private shipbuilders.

By the end of the nineteenth century, with the Royal Navy being the largest navy by far, warship building programmes were a significant source of work for shipbuilders in the UK. At that time, the leading UK warship building yard outside the Royal Dockyards was Armstrong Whitworth & Co at Elswick on the River Tyne. Armstrong's naval architect and a Director of the company was Philip Watts, who had formerly been the Chief Constructor for the Admiralty. In this capacity, he would have been well known to Francis Elgar, previously Director of the Royal Dockyards.

In 1886 a young apprentice destined for better things, Eustace Tennyson D'Eyncourt, began work at Elswick and, taking advice from Philip Watts, he studied to become a naval architect, eventually completing his apprenticeship in 1891. D'Eyncourt continued at Armstrong's for a further seven years, working primarily on warship designs for which the Elswick Company had developed a considerable export market as well as becoming a main contractor for the Admiralty. In August 1899 a position as naval architect at Fairfield was created for D'Eyncourt, almost certainly through Elgar's connection with Philip Watts. At Fairfield, D'Eyncourt was able to expand his experience of merchant shipbuilding and, in addition to working on destroyers and cruisers then under construction at Fairfield for the Admiralty, he was able to work on the troopship *Hardinge* for the Royal Indian Marine and the Orient line's *Omrah* and *Orontes*. At the same time, D'Eyncourt took an interest in the latest methods in shipbuilding, and while at Fairfield he had a written a description of what was termed the 'Premium System'. This system, which had been introduced to some departments at Fairfield at the end of 1901, was similar to piecework and promised an increase in wages if a job was completed earlier than in the time allotted. D'Eyncourt returned to Elswick in 1902 after over three years at Fairfield. In 1912, he became the Director of Naval Construction, responsible for the design and construction of the Royal Navy's warships.

However, the issue causing the greatest concern throughout late 1892 and 1893 was the miscalculation of the deadweight capacity of the two Cunarders, *Campania* and *Lucania*, resulting in the ships having reduced cargo carrying capacity. Although both vessels were a great success, having captured the coveted Blue Riband, Cunard held onto the final stage payment until the issue surrounding the deadweight capacity issue was resolved. Added to the fact that it was evident by this time that a major loss would be recorded on both ships, causing the Company to be significantly overdrawn, this naturally focused attention on the person ultimately responsible, Shipyard Manager and Director, Richard Saxton White.

At the end of 1892, White was taken to task concerning the amount of steel actually used in building *Campania* in comparison to the estimates that his department had provided. When asked by Barnwell about excess wages charged against the two Cunarders, White's reply, as minuted, was that 'nothing definite could be said regarding this until the vessel was further on, he being of the impression that probably part of the cost of *Lucania* had been charged to *Campania*'. Whatever the circumstances, such a vague reply could not have augured well for White.[10] Worse was to follow five months later as the scale of the loss became clear.

At the Board meeting held on 14 April 1893, the Chairman, Sir William G Pearce, seconded by Mr Barnwell, asked White to report in detail on two items.

> 1. The various items of labour expenditure and the reason for the increased costs of *Campania* over estimates.
> 2. The Board wished to see the data upon which Mr White formed the opinion that costs on *Campania* were working out favourably as he stated at the board meeting of 22 December 1892.

Lucania, seen here, was slightly faster than her sister *Campania*, taking the Blue Riband as the fastest passenger ship afloat on her second transatlantic crossing and retaining it until 1898.

In what must have seemed to White as a never ending list of implied transgressions, he was, additionally, asked to explain the error in the estimate of deadweight capacity of *Campania* as laid before the Board on 22 December 1892. To this, the unfortunate Shipyard Manager replied that he would go into the matter fully and report to the Board at the meeting of 3 May 1893.

At the 3 May meeting, White delivered his report which the Board thought was too complicated to take in all at once and it was decided to reflect on this until the next Board meeting in a fortnight's time. This Board meeting, held at the London office on 13 May, began with the Chairman saying that the only item he could find in Mr White's defence was that he had made a 10% discount in wages, which the Chairman thought was 'not proper estimating but speculation'. This, the Chairman continued, would in any case, only account for a small amount of the loss. The meeting ended with a resolution, carried unanimously, 'that in consequence of heavy losses incurred in the shipyard, a committee be appointed to enquire into the cost of production and permanent charges with a view to reducing the same.' This Committee comprised the Chairman (Sir William G Pearce), Richard Barnwell and Francis Elgar.

At the Board meeting held on Wednesday 14 June, the agenda covered general business – Cunard would pay for alterations to *Lucania* to minimise vibration; Dr Elgar raised the prospect of building cruisers for the Brazilian Government; issues with the balance of payments on the steamers *Adder* and *Hound* for G&J Burns, and so on; the deficiency in deadweight capacity of *Campania* and *Lucania* was discussed with a figure somewhere between 705 to 650 tons being agreed upon. If these items were routine business, what followed was out of the ordinary.

The report by the Committee into losses in the shipyard was read, after which Mr White was asked if he had any further explanations to make. He did not, but said he thought the report was a fair summary of events. At this point, the Chairman asked Mr White to leave the room. The minutes record the following: 'The Board, after considered discussion, decided it was necessary in the interest of the company that Mr White's connection with it should cease.' The manner in which this should be brought about was then considered and it was decided, in order to show as much consideration as possible for Mr White in bringing about the desired result, that he should be requested to send in his resignation. The following resolution was then carried:

> [In] looking at the report of the Committee of Enquiry and taking into consideration the losses that the company has sustained during Mr White's term of management in the shipyard, he be requested to send in his resignation as Manager of that department.

On Mr White's return, this resolution was communicated to him verbally by the Chairman and the secretary was directed to send him a copy of same. He thereupon asked the Board, out of consideration for himself, to be allowed a short time before sending in his resignation and his desire to know how many days the Board would be willing to allow him.[11]

White's resignation[12] was accepted on 23 June 1893 and a replacement speedily found in Edmund Sharer, who had served his apprenticeship on the Clyde and was Manager of Edward's Shipbuilding Co at Howdon-on-Tyne.

However, White's sacking did nothing to resolve the issue with Cunard. In October 1893, a telegram was received from Cunard, suggesting a meeting by the Directors of both Boards in Liverpool to overcome the deadweight issue. At the same time, a letter from Cunard claiming £165,000 to cover the loss of cargo carrying capacity was considered by the Fairfield Board to be 'monstrous', particularly as the deadweight deficiency had largely been caused by the extra weight of machinery, engines and boilers, in response to Cunard's letters to Fairfield asking for additional speed.

The launch party for *Tantallon Castle*, built for Sir Donald Currie's Castle Line and the South Africa Royal Mail service. Launched by Lady Currie (at front centre), Sir William G Pearce is standing to the left (her right). *Tantallon Castle* was the seventh ship built for the Castle Line. The building behind the launch party is part of Govan Silk Works, the only such works in Scotland.

This situation indicates an insufficiently explicit contract specification, resulting in a 'grey area' with Fairfield confident it was doing its best by fitting more powerful engines but not making it clear to its client that larger machinery (boilers and engines) would take up more space, thus reducing cargo capacity. Both companies were close to taking legal action when the issue was finally settled at a penalty of £30,000 for Cunard's claim on *Campania* and *Lucania* and £60,000 loss made on the contract,[13] a thumping loss for the shipbuilder. The loss on the contract to build *Campania* amounted to £38,514 and *Lucania*, £22,719.[14]

To assist the Company out of this major difficulty, the ordinary shareholders, namely Sir William G Pearce and his mother, Lady Pearce, paid the £90,000 which they had in turn, borrowed from their bankers, the Bank of Scotland.[15]

In September 1893, Sir William Arrol, founder and Chairman of the very successful, structural steel company named after him and renowned for building the Forth Railway Bridge, was appointed to the Fairfield Board in place of Richard Saxton White.

In the summer of 1893, the great Chicago World's Fair opened. This major event which ran from May to October, attracted exhibiters from all over the globe and included a major British section. In what amounted to a roll call of British industry, the British Pavilion included Clyde shipbuilding, represented by Fairfield, J&G Thomson and Wm Denny who sent models to be included in the British transportation section. Fairfield's stand included a model of the *Campania*, *Dunottar Castle* and the yacht *Torfrida*.

Meanwhile, the real *Campania* entered service in May 1893, proving to be another record breaker, taking the Blue Riband from Inman's *City of Paris* when in June she made the westbound crossing in 5 days, 15 hours and 37 minutes at an average speed of 21.12 knots.

At the Fairfield Ordinary General Board meeting of 23 October 1893, the results for the year, unsurprisingly, recorded a greatly reduced profit of £19,754 to which was added £10,000 from reserves. This compares to the profit of £66,772 made the previous year. The annual report noted that the following contracts were completed: *Campania* and *Lucania*, for Cunard, *Pole Star* for the Northern Lighthouse Board, paddle steamer *Royal Sovereign* for service on the Thames, *Hound* for G&J Burns, and the sailing ship *Chiltonford*. Work in hand included a large steamer for Castle Mail Packets, *Tantallon Castle*, machinery for HMS *Hazard* and the sailing ship *Chelmsford*, the twin screw yacht *Giralda*, the large fast paddle steamer for channel service, *Royal Sovereign*, and the paddle steamer, *La Marguerite*, for service on the east coast. The minutes noted that 'no profit was taken owing to the unsatisfactory results of the management of the shipyard'.

Fairfield and New Palace Steamers Ltd

The paddle steamers mentioned in the 1894 annual report, *Royal Sovereign* and *La Marguerite*, were the last two of a trio that Fairfield owned and had built for the Victoria Steamboat Association, who operated them on the River Thames.

In a rather unlikely series of events, by the turn of the nineteenth century, Fairfield, well known for building paddle steamers, owned three of them which operated on the Thames. This complicated and rather obscure chapter in the Fairfield story began in 1890. In that year, Fairfield's Managing Director, Richard Barnwell, on his own account, acquired the paddle steamer *Cobra* from J&G Burns just one year after it had been completed at the Fairfield yard. Barnwell renamed the vessel *St Tudno* and placed her on the Liverpool to North Wales route. In 1891, however, a new company was formed when Barnwell's steamboat interests were merged with his main competitor on that route, the Liverpool, Llandudno and Welsh Coast Co Ltd. The new company was called the Liverpool & North Wales Steamship Co Ltd in which Barnwell had a significant shareholding. On the formation of this company, and for reasons that are unclear, *St Tudno* was sold to the Hamburg Amerika Line, taking back its original name *Cobra*. At much the same time, a new paddle steamer also to be named *St Tudno* was ordered by the Liverpool & North Wales Steamship Co Ltd from Fairfield.

While these events were underway, the London-based Victoria Steamboat Association Ltd that operated a number of pleasure steamers on the Thames, ordered a large paddle steamer from Fairfield in November 1891 to be called *Koh-i-Noor*. Entering service in the summer of 1892, *Koh-i-Noor* was such a success that a second vessel, *Royal Sovereign*, was ordered from Fairfield that year. This vessel was built for the London and East Coast Express Steamship Service Ltd but operated by the Victoria Steamboat Association (VSA) and completed in May 1893. In October 1893, Richard Barnwell secured the contract for a third and larger vessel, *La Marguerite*, for Palace Steamers Ltd, on the basis that this ship together with the other two, would be operated and paid for by VSA. *La Marguerite* was described as the finest and largest vessel of her type afloat, and intended for Channel service between the Thames and Boulogne. It appears that Fairfield had agreed to pay for the construction of these vessels and accept payment over a period of time once the ships were in service. In the summer of 1894 however, Fairfield took legal action against VSA as they had failed to make payments throughout 1893 for *Koh-i-Noor*. By August, VSA was in liquidation and Fairfield took possession of all three paddle steamers. In the wake of the VSA failure, Fairfield set up a company, New Palace

The large paddle steamers *Royal Sovereign* and *Koh-I-Noor* moored by London Bridge about 1893. Owned by Fairfield, the steamers were operated by New Palace Steamers Ltd. Both steamers plied the Thames until the First World War.

Steamers Ltd, incorporated in April 1895, to operate the vessels.

In September 1900 Fairfield had the vessels valued; *La Marguerite* at £54,846, *Royal Sovereign* at £30,000 and *Koh-i-Noor* at £30,000, either for insurance purposes or as a preliminary to selling them. In February 1901 New Palace Steamers purchased *Koh-i-Noor* outright and chartered *Royal Sovereign* and *La Marguerite*. In October 1901, after reviewing New Palace Steamers' figures for 1901, the Fairfield Board agreed that *La Marguerite*, which was expensive to operate, should be sold at a minimum price of £45,000. It was also agreed that William Rhodes be appointed as manager of New Palace Steamers. In February 1902, as *La Marguerite* had not been sold, New Palace Steamers chartered all three boats from Fairfield under Rhodes management.[16] In January 1904 *La Marguerite* was finally sold to the Liverpool & North Wales Steamship Co Ltd in which Fairfield maintained a financial interest. Thereafter, *Koh-i-Noor* and *Royal Sovereign* were chartered annually by New Palace Steamers until the start of the First World War. Some indication of why Fairfield might have continued with this business comes from the operating figures for 1904. In that year, Palace Steamers Ltd made a profit of £9930 of which Fairfield received £8940 equivalent to £896,000 at 2023 prices.

Both ships were laid up during that conflict until March 1918 when they were auctioned. *Koh-i-Noor* was sold to breakers for £6200 and *Royal Sovereign* for £8300 to a series of owners, continuing in service until 1930.

New Palace Steamers Ltd appears to have been in suspension since October 1922, when presumably it ceased operations. The Company nevertheless continued to exist 'in suspension' holding an annual general meeting with directors drawn from the Fairfield Board. Finally, in December 1953, a resolution was passed to wind-up the Company.

In November 1893 Shipyard Manager Edmund Sharer reported that, on examining Company records, he could see that there had been a lack of supervision among riveters. Consequently, he was asked to devise a system that would eliminate the possibility of dishonesty in making piecework returns.

1894: A Year of Mixed Fortunes

The year 1894 saw Fairfield's reputation soaring with the success of *Campania* and *Lucania*. The reality behind the scenes had been quite different with the serious dispute between Fairfield and Cunard and the major dent this would make in the shipyard's finances. To add to financial woes, the years 1893–1894 were depressed, with the shipyard on three-quarter time, although the engine works remained fully employed. The following calendar which is taken from the Minutes for the year 1894, gives an idea of some of the events that Fairfield had to contend with throughout that year.

January

The Board reflected on the tender it had submitted for the cruiser *Powerful* in comparison to the successful tenders.

February

The tender for a *Majestic* class battleship was submitted at £350k for the hull, £3.75k for masts and spars, £84,65k for machinery and £7.73k for auxiliary machinery, totalling £446,129. At the same time, a telegram was received from the Admiralty, confirming that the tender for the cruiser *Powerful* had been unsuccessful, although the tender for three torpedo boat destroyers, *Handy*, *Hart* and *Hunter* had been accepted.

March

Third Sea Lord and Controller of the Navy, Vice Admiral Sir John Fisher, wrote asking if they would accept £417k for the battleship, instead of the £446k submitted. The Company counter offered with £430k which Fisher declined. However, the Company was invited to tender for two second class cruisers, subsequently *Venus* and *Diana*.

April

The tender for the two cruisers at £206k each was accepted by the Admiralty. They also approved the design of hull for the torpedo boat destroyers with work 'to begin at once'.

The steering tillers on both *Campania* and *Lucania* were defective which the Fairfield Board considered to be an issue for the manufacturers, Brown Brothers of Edinburgh.

The minutes noted that there was friction between two Board members, Andrew Laing and Edmund Sharer. It was agreed that they should both meet every morning at 9.30 in an attempt to improve relations.

May

North German Lloyd expressed interest in a large cargo steamer. The issues with Cunard regarding *Campania* and *Lucania* were resolved amicably and all legal proceedings dropped.

June

Letter with cheque received from Cunard in settlement of all issues regarding the construction of *Campania* and *Lucania*. This would be the final stage

HMS *Hart* was the second of three '27-knotters' built for the Admiralty between 1894 and 1896. Designed by Fairfield to an Admiralty specification, these torpedo boat destroyers were followed a year later by three 30-knotters.

payment plus extras involved in strengthening the after part of both ships to minimise vibration.

North German Lloyd declined the offer of a cargo ship for the South American trade.

Lady Pearce and Sir William G Pearce have stepped in to make good the £90,000 loss on *Campania* and *Lucania*.

July

The Admiralty invited tenders for machinery for the battleships *Caesar* and *Illustrious*.

Andrew Laing and the Chairman, Sir William Arrol, have had a disagreement over the paddle steamer *La Marguerite*.

Dr Elgar wrote regarding a cable ship for Japan per Messrs Bolling & Lowe.

August

The Victoria Steamboat Association went into receivership, owing Fairfield significant sums. The Board proposed to operate the *Koh-i-noor* on the Thames until the end of the year.

Dr Elgar advised that the hull of steam yacht *Giralda* should be as smooth as possible to get the speed required. Andrew Laing agreed to reduce the profile of various underwater fittings.

The Board agreed to join the National Federation of Shipbuilders & Marine Engineers as requested by the Clyde Shipbuilders Association.

October

A letter received from the Imperial Russian Embassy inviting a tender for a cruiser.

Trouble with the boilers for *Tantallon Castle* and *La Marguerite*. Andrew Laing to investigate.

Managing Director Barnwell said that Director and Manager of the Engine Works, Andrew Laing, conducted the management of his department in a secret way and that he should keep the company fully informed, especially where expense is involved.

November

In highly unusual circumstances, the Board meeting on 20 November, read out a 'requisition' by Lady Pearce and Sir William G Pearce, requesting the Directors in terms of section 66 of [the] Articles of Association, to convene an Extraordinary General Meeting of the Company for the purpose of

The steam yacht *Giralda* was built for Colonel Henry McCalmont, army officer, racehorse owner and yachtsman. *Giralda* later became the Spanish Royal Yacht.

> considering and if, so resolved, passing the following extraordinary resolution viz:- That Mr Andrew Laing be removed from the office of Director.

Laing duly resigned as a Director, remaining as Engine Works Manager until the natural termination of his contract in June 1895, which he chose not to renew. Richard Bramwell wrote that 'Matters have come to such a point through Mr Laing's obstinacy that my Co-Directors have unanimously decided, with myself, that another man must be appointed as the Manager of the Engine Department.'[17] Andrew Laing had just designed and manufactured the most powerful steam engines so far built and overseen their installation in the fastest ships in the world. Difficult as he

The grand staircase of *Regele Carol 1*. The magnificent woodwork in this interior designed by the architect William Leiper was manufactured in the Fairfield joiner and cabinetmaking shops.

The cruiser HMS *Diadem* by the 120-ton sheerlegs designed by Andrew Laing. During this period of warship construction in private yards, it was normal practice to have armament fitted at a Royal Dockyard and not by the shipbuilder. *Carisbrook Castle* can be seen at the west side of the fitting-out basin.

might have been, his departure was a great loss to Fairfield and a win for the Wallsend Slipway & Engineering Co Ltd on Tyneside where he became General Manager in 1896. Laing was replaced as Engine Works Manager by Alex Gracie who had been Under Manager at J & G Thomson's Engine Works at Clydebank shipyard.

The minutes for the four and a half year period January 1895 to July 1899 are unfortunately missing. During that period, Fairfield built a number of well-appointed vessels including *Dunvegan Castle*, *Tintagel Castle* and *Avondale Castle* for Castle Line and the magnificently appointed *Regele Carol 1* for the Romanian State Railways, with interiors by Glasgow architects William Leiper and William James Anderson. There was also a steady increase in Admiralty work, with contracts for three destroyers and two cruisers.

On 7 March 1898, Richard Barnwell died at his home, Elcho House in Balfron, at the age of 49. Like Sir William Pearce ten years earlier, Barnwell's estate at £82,538[18] (equal to £8.863m. at 2023 rates),[19] shows how well rewarded senior management figures were in such a successful and profitable business.

The Barnwells, Fairfield and Aviation

Of the six children born to the Barnwells, two, Frank and Harold, would be closely associated with early pioneering aviation activities in Scotland. Frank served an apprenticeship at Fairfield between the years 1898–1904 which he combined with studies at the Technical College in

Fairfield's fitting-out basin in 1898. The vessel at bottom right is *Regele Carol 1*, built for the Romanian Government. The large vessel at left is the armoured cruiser *Argonaut* which is flanked by two second class cruisers, *Hermes* and *Highflyer*. At *Regele Carol's* stern is the two masted steam yacht *Atmah*, built for Baron Edmund de Rothschild. The bridge carrying the public footpath across the mouth of the basin can be seen as can the newly constructed tenement buildings in the north side of the Clyde.

This late 1890s contemporary illustration of the Fairfield Works shows ten building berths and extensions to either side of the original engine works at top centre. The illustration also shows the inconvenience of the public right of way for which the company had to provide a ferry, previously a swing bridge, to enable members of the public to cross the mouth of the fitting-out basin.

Glasgow and Glasgow University gaining a BSc in naval architecture. Both Frank and Harold were pioneering aviators who experimented with gliders and later powered craft. This they did initially at Balfron in the grounds of the family home and later near Bridge of Allan where they had established the Grampian Engineering and Motor Company, at Causewayhead in 1906. The Barnwells were not Fairfield's only connection with pioneering aviators. In 1887, Percy Pilcher left the Royal Navy to work in Fairfield's engine works, completing an apprenticeship there on 8 July 1889. Pilcher experimented with a series of gliders before moving on to powered flight. He died in September 1899 flying the glider *Hawk*.

The Lockout of 1897

The engineers' lockout and strike of July 1897–January 1898, which lasted for thirty weeks had its origin in London, was, ostensibly about the drive for an 8 hour working day. When the Associated Society of Engineers called its members out on strike, employers through the newly formed Engineering Employers' Federation, imposed a lockout across the UK. The dispute was not resolved until January 1898 when the engineers returned to work without achieving their objective. With a very full order book including several Admiralty contracts, Fairfield management were at first reluctant to impose the lockout as it would have an immediate effect on 900 of their employees. Their reluctance to follow the Employers' Federation resulted in their exclusion. However, in late December, the Company decided to impose the lockout although by this time many of the engineers employed by the Company had signed private agreements to work on. Nevertheless, in January 1898, 600 engineers were paid off – although they were soon back at work after the fire.

Fire!

At 9am on the morning of 4 July 1897, fire broke out in the Polishing Shop which was located at the westernmost end of the long run of shops at the head of the building slips. The fire spread quickly eastwards for a distance of 500 feet before it was put out two hours later by the local fire brigade and the yard's own fire tender. The damage was extensive, destroying, in addition to the Polishing Shop, the Joiner Shop, Admiralty Store, Pattern Shop, Riggers Loft, Mould Loft and Paint Store. The damage was estimated at £50,000 and caused most of the 4000 men employed to be laid off. Most were able to return after a few days. At that time, the cruisers *Diadem*, *Hermes* and *Highflyer*, the steam yacht *Giralda*, two destroyers, the paddle steamer *Empress Queen* and the *Carisbrook Castle* were in various stages of construction.

Looking Back

As the end of the century approached, Fairfield could look back on a great trading record with a number of leading shipping lines including:

Guion Line
Castle Mail Packets (Donald Currie & Co)
Norddeutscher Lloyd
Hamburg American Line
Orient Line
Cunard

Fairfield also built magnificent, fast, paddle steamers for a number of companies including:

Isle of Man Steam Packet Co
London, Chatham and Dover Railway.
London, Brighton and South Coast Railway Co
Liverpool and North Wales Steamship Co
Victoria Steamboat Association

Moreover, Fairfield was now firmly established on the Admiralty list as builders of cruisers and destroyers as well as for propelling machinery.

Gross profit for the year to June 1899 was £104,754 of which £73,530 was for distribution.

Ships under construction were: four first class cruisers; two 30 knot torpedo boat destroyers; the mail steamers *Kinfauns Castle* and *Kildonan Castle*, and *Hardinge*, a large troopship for the Indian Government.

The paddle steamer *Empress Queen* was built for the Isle of Man Steam Packet Company in 1897. She was the largest and fastest vessel on the Liverpool to Douglas service.

Kildonan Castle seen here in 1899 on completion. Built for Donald Currie's Castle Line for service on the Southampton to Cape Town route. In March 1900 the Castle Line merged with the Union Line to form the Union Castle Steamship Co Ltd.

The troopship *Hardinge* built for the Royal Indian Marine.

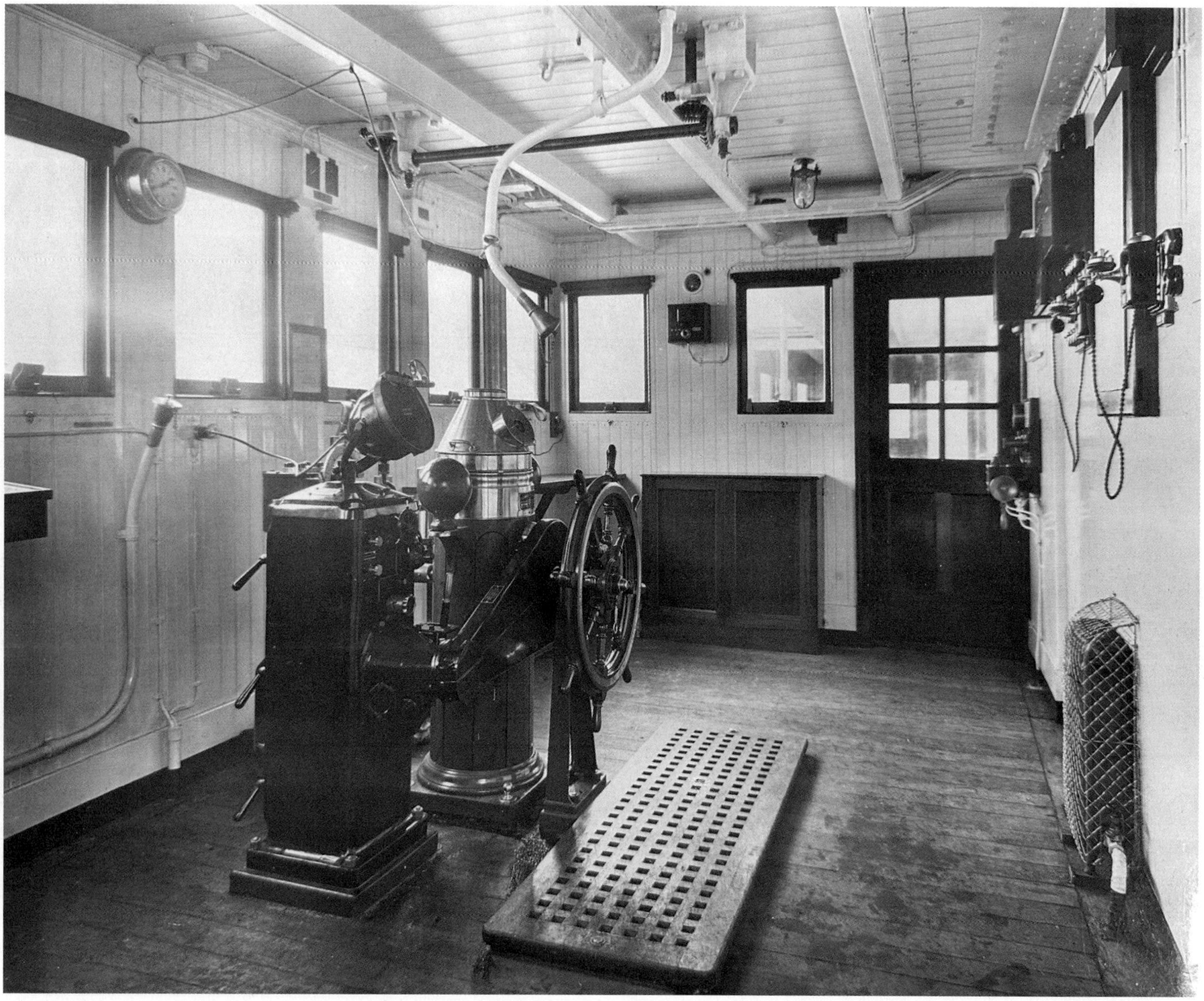

The bridge of the troopship *Hardinge* showing, at centre, the magnetic compass binnacle, and the ship's wheel with repeater and telemotor steering. The gearing on the deckhead could be part of an emergency steering system. The voice pipes will most likely connect to the engine room, captain's quarters and the 'monkey island' (the platform above the bridge). The string coming from the bulkhead and tied to the voice pipe is likely to be for ringing bells for the watchkeepers. Out of view will be a revolution counter and rudder indicator.

In May 1900 Shipyard Manager and Director, Edmund Sharer resigned to take up an appointment as Shipyard Manager at William Beardmore & Co's newly acquired shipyard at Govan, formerly that of Robert Napier & Sons. Sharer's new role with Beardmore was to plan out a new shipyard on ground which had been acquired at Dalmuir to the west of the John Brown yard at Clydebank. Here, a vast new yard, named appropriately the Naval Construction Works, was laid out. This would be the largest yard on the Clyde, and one of the last major green field shipbuilding developments in the growth of the industry on the river. While perhaps not bound by strict confidentiality agreements at that time, Sharer was nevertheless required to return documents in his possession before leaving Fairfield, all of which he was entitled to have, but which would have been of interest to his new employer. Consequently, he handed over drawing office information and general arrangement plans of yard workshops; launching particulars and stability experiments of vessels ranging from Ship No 247 *Alaska* to 408 *Kildonan Castle*; six plans showing arrangement of various machines throughout the yard; 16 books containing costs of vessels; midship section of proposed Isle of Man steamer; book with particulars of trial trips, extras claimed and general

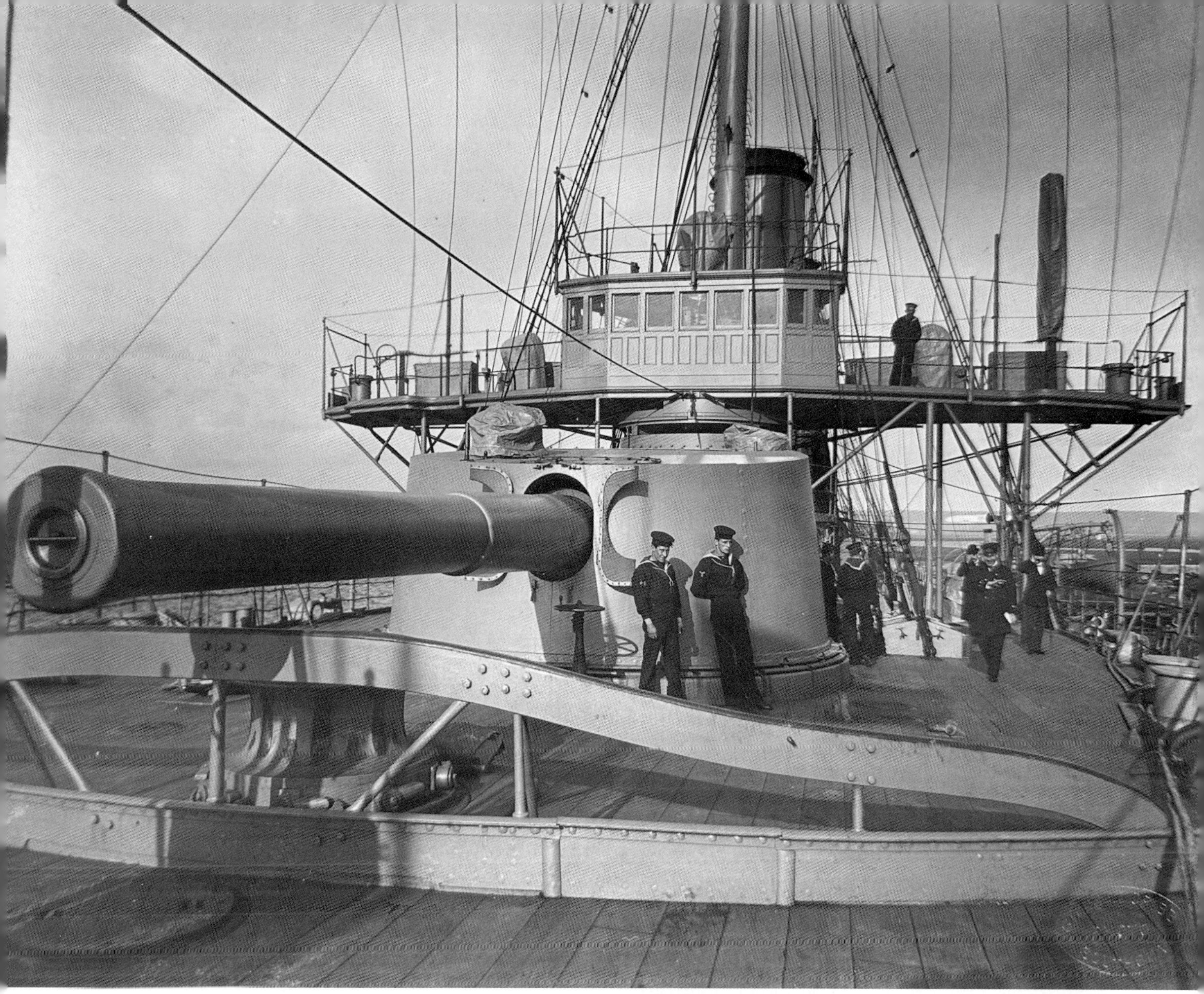

The forward 9.2-inch gun of the armoured cruiser HMS *Good Hope,* which was launched by Ethel Elgar, wife of Francis Elgar, on 21 February 1901. Note the depression rail below the gun barrel, fitted to prevent the gun firing into the hull. *Good Hope* was sunk at the Battle of Coronel in November 1914 with the loss of all 926 men.

information relating to the Ship Department; hull specifications of the cruisers HMS *Good Hope*, *Cressy* and *Aboukir*.

Sharer had been paid a salary of £5000 which included commission, before starting his new job at Govan, where contracts for the cruisers *Berwick* and *Carnarvon* had begun. Beardmore's Govan works were transferred to Dalmuir in 1905.

A posed photo of an officer on the bridge wing of HMS *Good Hope* sighting on a distant object.

NYK Line's *Kumano Maru* on the Firth of Clyde in 1901, the first ship to be fully air-conditioned. The air-conditioning plant was designed and supplied by Thermotank Ltd, inventors of marine air-conditioning and a company also based in Govan.

In 1901 the passenger cargo ship, *Kumano Maru*, built for the Japanese Nippon Yusen Kaisha (NYK) Line, was completed at Fairfield. Although not widely reported at the time, the ship was the first to be fully air-conditioned, although that term had still to be coined. Marine air-conditioning was invented in 1899 at Clydebank shipyard by the electrical engineer Alexander Stewart who later, with his two brothers, founded the firm Thermotank to exploit the invention.[20]

Shipyard Expenditure and Extensions 1901–1905

In July 1901, expenditure of £26,800 was authorised to be spent on new machinery, a new mechanics' shop for the Shipyard Department as well as new machines for the Engine Department.[21] In the following year a further £25,000 was authorised for improvements to the works including the purchase of ground to the east of the yard with a river frontage of 450 feet. This ground was purchased for £16,500. At the end of 1902 it was agreed that a new building slip, pointing upriver, should be constructed on ground to the immediate west of the fitting-out basin. Excavation began but by April 1903, it was decided to abandon the project on cost and the excavated area was used as a timber pond. By October 1903, a total of £53,150 had been spent on the works. In 1904 the fitting-out basin was in need of considerable reconstruction, particularly at the head, and that coupled with the need to provide electricity to the works required additional finance to be agreed by a stock issue.

In April 1902 Fairfield's tender to build the *King Edward VII* class battleship HMS *Commonwealth* (named after the Commonwealth of Australia) was accepted. The placing of this order was significant, representing admission into the small group of UK shipbuilders entrusted to build battleships, then the largest and ultimate form of warship.

The *Monmouth* class armoured cruiser HMS *Donegal* was completed in November 1903.

The *King Edward VII* class battleship HMS *Commonwealth* at the Tail of the Bank on completion in the early months of 1905. Battleships of this type were made obsolete by the arrival of HMS *Dreadnought* in 1906 and reclassified as pre-Dreadnoughts.

Fairfield and the New Cunarders

In 1901 the big news in the British maritime world was that Cunard had decided to build new express liners for the coveted North Atlantic run. As the nation's favoured steamship line, Cunard was expected to reassert British ascendency over the Atlantic in the face of the German challenge which had been presented by North German Lloyd in the form of their record-breaking *Kaiser Wilhelm der Grosse*. This 24,000-ton, 650ft liner took the honours on the North Atlantic in 1897 and not without a degree of indignation in the British press, who expected Cunard to respond. It would be a few years before Cunard rose to the challenge when, on 27 June 1901, the Cunard Board instructed their technical staff to prepare specifications for a 700 foot long, 24 knot steamer for the New York trade.[22]

Cunard's deliberations on the specifications, particularly the machinery for its new ships, extended over a lengthy period, giving rise to conjecture about what these new ships might be like and who might build them. Speculation in the technical press was that Fairfield, given its reputation and track record, would build at least one of the vessels, with the possibility that Vickers Son & Maxim at Barrow or John Brown at Clydebank might build the other. Some pundits speculated that Fairfield alone might build both as had been the case with the sisters *Campania* and *Lucania* ten years earlier. It was not until 21 February 1902 that an outline specification for two large steamers, one of 23 knots and the other of 24 knots, for the Atlantic trade was issued to Fairfield, John Brown and Vickers Son & Maxim.

If events were, so far, in line with expectations, the following statements from both the Cunard and Fairfield Board meetings describe what happened next, during what would be a long run-up to the tendering process.

Fairfield Board Meeting, 7 March 1902

It was agreed that Dr Elgar would see Cunard with regard to the 23 and 24 knot vessels.

Cunard Executive Committee Meeting, 15 October 1902

The Board received a letter from Swan Hunter & Wigham Richardson stating their intention to tender for one of the ships.

Cunard Board Meeting, 23 October 1902

The Board agreed that new outline specifications be sent to Fairfield, John Brown, Vickers Son & Maxim and Swan Hunter & Wigham Richardson.

Fairfield Board Meeting, 18 November 1902

The enquiry from Cunard was considered by the Fairfield Board and it was left to Dr Elgar to convey the following to the Cunard Board:

That the Company shall tender on the Cunard Company's design, refusing to state a maximum price, but agreeing to build, on the basis of time, material and 20% for charges plus profit.

That payments shall be made monthly and that delivery shall finally take place after the Progressive Trials have been satisfactorily completed.

That the company will not agree to an Atlantic trial and shall not be liable for anything that may take place during the running of the ship thereafter, excepting to replace bad material or to make good faulty workmanship, and especially shall not be liable to do any strengthening that may prove necessary during the subsequent running of the vessel.

Cunard Board Meeting, 20 November 1902

Letters received from the above shipbuilders [the four referred to in the 23 October minute] acknowledging the specifications. In a letter, Fairfield stated that they hope to provide plans and an estimate during the first week in December.

Fairfield Board Meeting, 4 December 1902

The Board further discussed the question of the tender to the Cunard Co and adhered to the decision which was made at the last meeting as regards to the basis on which the Company was prepared to build, with the exception that it was decided to mention a price which in the company's opinion would cover the cost of the vessel.

Cunard Board Meeting, 18 December 1902

Proposals and plans for building the ships received from the four shipbuilders and now under consideration by the Board

Cunard Board Meeting, 16 January 1903

Directors from John Brown, Vickers Son & Maxim and Swan Hunter & Wigham Richardson attended a meeting with Cunard to discuss their proposals to build the ships. (Fairfield representatives were not in attendance.)

Cunard Board Meeting, 8 May 1903

Shipbuilders were interviewed by Cunard directors the result of which was that John Brown and Swan Hunter & Wigham Richardson were unofficially told that orders would be placed with them.[23]

From this, it is clear that Fairfield had effectively excluded themselves from the running some time after the meeting of 18 December 1902, in contrast to relative newcomers Swan Hunter, who were determined to join the short list of prestigious Atlantic liner builders. Cunard made a commercial decision based on price while Fairfield was clearly not interested in the contract sufficiently to reduce theirs.

On 30 April 1904 John Brown and Swan Hunter & Wigham Richardson were officially awarded the contract to build the turbine driven ships *Lusitania* and *Mauretania*. These ships, easily the largest in the world on completion, went on to win the Blue Riband and restore British prestige on the North Atlantic. For both builders, the final build cost was well in excess of the 1903 estimates. Fairfield never built another Cunarder although, in April 1903, they tendered unsuccessfully for the intermediate Cunard liner *Caronia*.

Fairfield's withdrawal from the bid to build the giant Cunarders brought into focus the rivalry between Fairfield and the John Brown yard at Clydebank, formerly J&G Thomson & Co. Until this time, both Clyde yards had been equally successful in building the largest express liners on the North Atlantic, even if Fairfield had built the most recent. Henceforth, in the ongoing development of shipbuilding on the River Clyde, John Brown & Co, would go on to build a series of ever larger liners including, for Cunard, *Aquitania*, *Queen Mary*, *Queen Elizabeth* and *QE2*. The mantle of 'leading' Clyde yard passed from Fairfield to John Brown just as it had passed from Robert Napier to John Elder decades earlier. Fairfield's approach to these contracts, perhaps influenced by difficulties experienced with Cunard over the contract for *Campania* and *Lucania* in 1893, may have influenced the Fairfield Board to take a rather uncompromising and risk averse line in the face of the conditions applied by Cunard.

Given the dramatic increase in the size of liners as exemplified by *Lusitania* and *Mauretania*, what might have been considered to be a large vessel by the standards of the 1890s would be considered, by the first decades of the twentieth century, an intermediate sized vessel and the market on which Fairfield would concentrate in future.

In July 1904 Dr Elgar informed the Board of his decision to resign as a Director and Naval Architect of the

The paddle steamer *Duchess of Fife* for the Caledonian Steam Packet Company on completion in June 1903. She operated on Firth of Clyde routes and was requisitioned by the Admiralty in both World Wars. She took part in Operation Dynamo, the evacuation of troops from Dunkirk.

An important person in the development and success of Fairfield during the 1890s and early 1900s as Director and Chairman, Dr Elgar was the first to occupy the naval architecture chair at Glasgow University which was funded by Isabella Elder.

Company. This news was received with 'very great regret' by his fellow directors who asked if he might retain the position of consulting naval architect to which he agreed.[24] Such was the high regard in which Dr Elgar was held that at the October Board meeting the Chairman, Sir William G Pearce, again asked Dr Elgar if he might not reconsider his decision to resign. His reply in the negative was again 'deeply regretted'. Board minutes noted that Dr Elgar would be paid his salary to the end of the year plus 4% of the profits for one half of the total profit for 1904–5. Additionally, Dr Elgar was to be paid 4% for all of the profits made on HM ships *Commonwealth*, *Narcissus*, *Argonaut*, *Forward* and *Foresight*.[25]

In January 1905 Alexander Gracie was appointed Managing Director. Gross profits for the year to June 1905 were £146,779 which included £41,360 brought forward from 1904. Once the various commitments had been paid out, it left £60,195 for distribution including £1950 director's fees, £4000 to the income tax account and £25,000 to the works improvement account.

The Merryweather & Sons fire engine acquired in 1904, the same year in which an electric locomotive was purchased for £1200 to bring steel plate and section among other items from Govan Goods Yard using the rails and catenary of the Glasgow tram system.

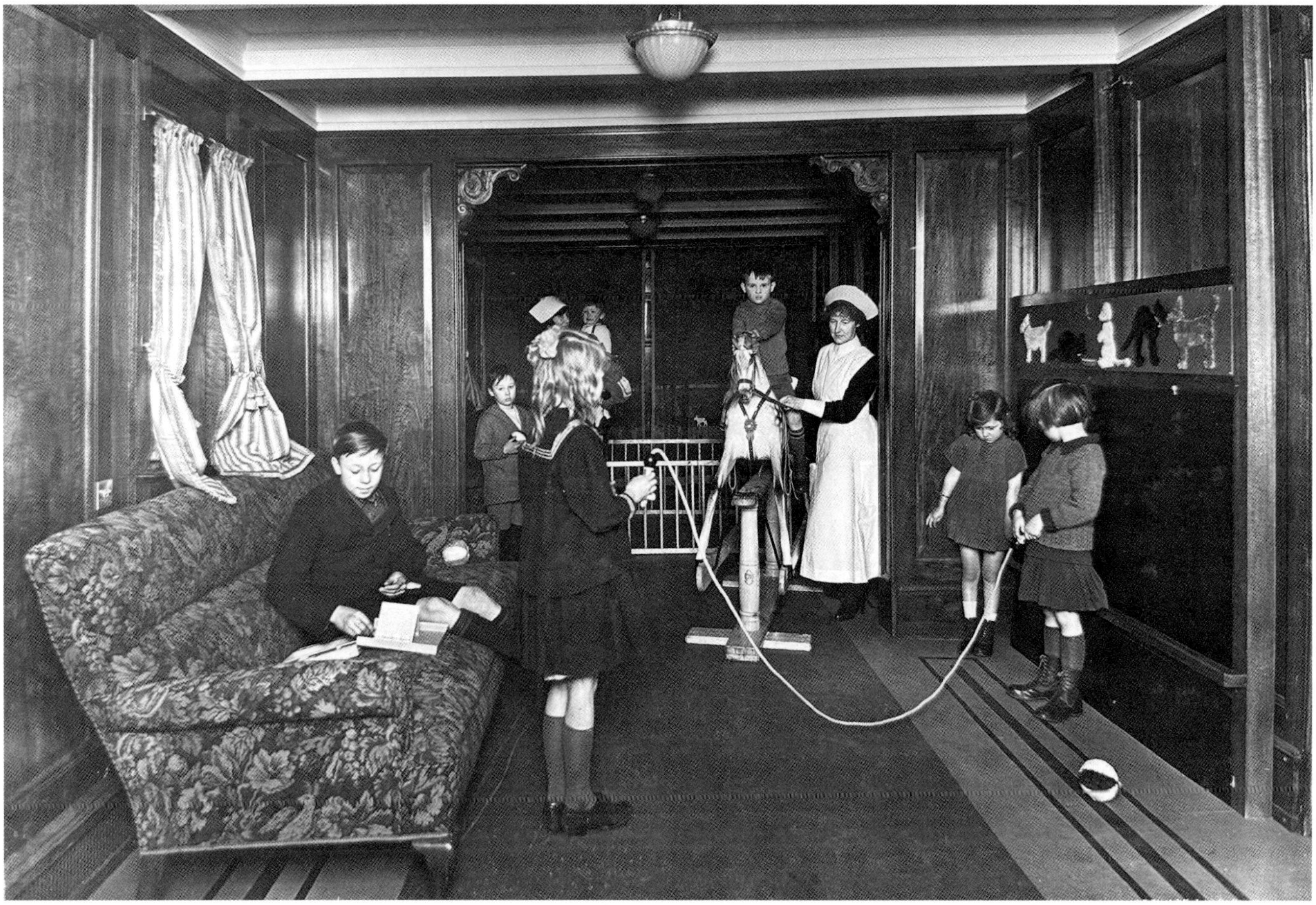

The children's room on the Hamburg America Line's *Fürst Bismarck*.

Fürst Bismarck running trials on the Firth of Clyde in 1905.

Coventry Ordnance Works

Fairfield's association with the Coventry Ordnance Works arose because of the duopoly in the manufacture of heavy naval ordnance enjoyed by Armstrong Whitworth and Vickers Sons & Maxim. Both of these companies could offer the Admiralty armour plate and gun mounting manufacture as well as shipbuilding. Vertically integrated works such as these were seen as an advantage. With no ordnance capacity and, in Fairfield's case, no armour capacity either, shipbuilders John Brown, Fairfield and Cammell Laird, were at a distinct disadvantage. The general background to this and the commercial prospect it offered, was the dramatic expansion in the Royal Navy following the Naval Defence Act of 1889 and, more particularly, the response to the growth of the German Navy from 1897 onwards following the passing of their Naval Laws.

The process that led to Fairfield acquiring heavy gun mounting manufacturing capacity began in 1903 when Sheffield steel maker Charles Cammell & Co acquired the business of Mulliner Wigley Ltd, ordnance manufacturers with large premises in Coventry. Later that year, Cammell acquired Laird Brothers, shipbuilders in Birkenhead, to form Cammell Laird Ltd. The next stage was the creation of a company, the Coventry Ordnance Works, centred on the Mulliner Works. In 1904 John Brown & Co took half of the shares in the Coventry Ordnance Works and in the following year, Fairfield and Cammell Laird each acquired a one fourth interest for £187,500. Alexander Gracie, Engine Works Manager and Director since March 1901, was appointed to the board of the Coventry Ordnance Works as Fairfield's representative.[26]

While the Ordnance Works were being established, on 31 October 1905, an exchange of shares between Fairfield and Cammell Laird was agreed, with Alexander Gracie and William Marshall Rhodes elected to the Cammell Laird Board and John McGregor Laird, (Chairman of Cammell

Laird) and Restal Ratsey Bevis (formerly Managing Director of Cammell Laird Ltd) elected to the Fairfield Board.

In January 1906 the first of what would be many applications for cash to fund the development of the Coventry Ordnance Works was requested from the triumvirate. Of the £50,000 requested, Fairfield's share was £12,500. In April 1907 requests for a further £120,000 were made of the Syndicate of which Fairfield's share was £30,000. The following month a further temporary loan was requested to meet expenditure. Fairfield's contribution was £40,000 at 5% interest.

To add to the productive capacity of the new Company, in October 1906, 20 acres of ground was acquired at Scotstoun by the Clyde where large shops were laid out. Here, complex gun mountings, complete with turrets and guns, much of it manufactured at Coventry, could be assembled in the gun pits and tested. To enable heavy items to be shipped, a dock was cut at right angles to the river, served by a 100-ton gantry crane where disassembled mountings could be transferred into coasting vessels adapted for the task and then transferred to Govan, Clydebank, Greenock or Birkenhead.

Despite the initial capital sunk into the Coventry Syndicate by the three shareholders, it was far from an instant success, suffering from manufacturing and management issues in addition to resistance from Armstrong Whitworth and Vickers. From 1907 onwards until the outbreak of the First World War when the Works finally became profitable, there was a steady inflow of cash from the three shareholders to buy plant and working capital for the business. This became such a drain on Fairfield with no return in sight that, by 1909, the Company approached their bankers, the Bank of Scotland, to increase their overdraft by £30,000, secured by personal guarantees of the directors and a further £50,000 secured by Lady Pearce, although one year later these guarantees were no longer required.[27] In 1911, the Board had resolved to sell its shares in the Coventry Ordnance Works to Cammell Laird. However, this did not take place until much later.

The scout cruiser HMS *Forward* in silhouette on trials looking rather austere with her armament still to be fitted.

Canadian Pacific's *Empress of Britain* seen here in 1906 was completed just two months ahead of her ill-fated sister *Empress of Ireland*.

In the end, Fairfield had little return from its investment in the Coventry Ordnance Works, as these Works were not in production when Fairfield's first dreadnought, the battlecruiser HMS *Indomitable* was built and its main armament mountings, together with those for Fairfield's second dreadnought *New Zealand,* were manufactured by Armstrong at Elswick. The battleship *Valiant's* mountings were manufactured at Barrow and only *Renown* had its mountings manufactured by the Syndicate.

The results for 1906 showed a gross profit of £172,707 of which £97,899 was available for distribution.

Death of Sir William G Pearce and Return of Dr Francis Elgar

On 2 November 1907, Sir William G Pearce died suddenly at his London home at the age of 46. Just 27 years of age when he took over the Chair at Fairfield following his father's death, his board of directors consisted of much older and more experienced men. A barrister, he did not display the outgoing characteristics of his father but nevertheless managed the fortunes of the Company for 18 years. He was Conservative MP for Plymouth for the years 1892–1895, and the bulk of his estate valued at over £15,000 was left to Trinity College, Cambridge where he had studied.

On 14 November 1907 at a Board meeting at the London office in Leadenhall Street to which Dr Francis Elgar was invited, the Chairman, Sir Digby Morant with fellow director Colonel Mosely, proposed, that Dr Elgar be re-elected to the Board as Chairman. However, in October 1907, Dr Elgar had accepted the position of Chairman at Cammell Laird, where significant managerial issues had to be resolved.[28] At the same time, Elgar felt a duty of responsibility to Fairfield where he had spent 13 years and consequently accepted the Chairmanship. By this time,

Built for service on the Great Lakes, the small passenger cargo ship *Keewatin* was completed for Canadian Pacific in September 1907. *Keewatin* has been preserved as a museum ship since 1967, latterly at Kingston, Ontario.

both Fairfield and Cammell Laird were linked through their financial interest in the Coventry Ordnance Works, as well as an exchange of shares and Directors on their respective Boards. While the Directors on the Fairfield Board were undoubtedly pleased to see a man of such eminence return to Govan, the task ahead of him was considerable. He was heavily involved in the reconstruction of Cammell Laird and was overseeing issues arising from shipowners who had defaulted on major contracts at Fairfield. On 17 January 1909, while in Monte Carlo recuperating from an accident, Dr Elgar died suddenly aged 63.

Indomitable

On 19 March 1908 while HMS *Indomitable* was conducting a basin trial running her engines, two ships that happened to be passing Fairfield's basin at that time collided with one another claiming that the disturbance in the river caused by the basin trial was responsible. Fairfield accepted liability for the damage.[29]

When HMS *Indomitable* left Fairfield on completion and sailed to Portsmouth to be commissioned into the Royal Navy, she was the very first of a new class of warship, the battlecruiser. Soon after arrival at Portsmouth, *Indomitable* was made ready to take the Prince of Wales to Quebec in celebration of the city's foundation. On the return journey, the battlecruiser was worked up to make a

Launched on 16 March 1907, HMS *Indomitable* was one of three of the *Invincible* class, the first ships of the battlecruiser type. Note the drag chains bolted to the ships' side.

A view of the fitting-out basin about the autumn of 1907. At the west wharf (right of shot), the liner *Cairo* is nearing completion with sister ship *Heliopolis* ahead of her. At the east wharf the battlecruiser HMS *Indomitable* is by the 130-ton sheer legs. To *Indomtable's* side is the Canadian Governments icebreaking ferry *Stanley*, originally built at Fairfield in 1888 but returned to have new boilers fitted. Ahead of *Indomitable* are the sister ships *Keewatin* and *Assiniboia*, both for Canadian Pacific.

Indomitable running at high speed during trials on the Firth of Clyde in 1908. The fore funnel was raised soon after completion to avoid smoke and soot from causing havoc on the decks below as seen here.

high-speed endurance test of the ship's machinery. From Belle Isle (Labrador) to Fastnet (off southern coast of Ireland), *Indomitable* covered the three-day passage at an average speed of 25.3 knots, which, newspapers were quick to point out, was faster than the record set by the Cunard liner *Lusitania*. It was a great success not just for the *Indomitable* but also for Fairfield's reputation and resulted in Managing Director Alex Gracie receiving the MVO (Member of the Royal Victorian Order) from King Edward VII.

While the Fairfield Works had been associated with innovation in marine propulsion in the past, licences to manufacture the latest development, Charles Parson's

The clean lines of the turbine steamer *Viper* built for G&J Burns' Ardrossan to Belfast service seen here in the Clyde about to be manoeuvred into Fairfield's fitting-out basin after running trials.

Built for the Egyptian Mail Steam Ship Company in 1907, *Heliopolis* and her sister *Cairo* were recovered by Fairfield when the shipping line defaulted on payments and sold to Canadian buyers and renamed *Royal George* and *Royal Edward*. *Royal George*, formerly *Heliopolis* is seen here in 1910.

steam turbine, had to be acquired. This was also the case with the Brown-Curtis turbine which was based on the designs of Charles Curtis of General Electric in the US and developed by John Brown & Co at Clydebank. The bulk of steam engines manufactured at Fairfield over the coming decade would be turbines of the Parsons or Brown-Curtis type.

Heliopolis and *Cairo*, September 1909

In addition to the financial difficulties imposed on the Company in 1909 through its shareholding in the struggling Coventry Ordnance Works, the Egyptian Mail Steam Ship Co Ltd of London defaulted on payments for the large passenger cargo ships *Heliopolis* and *Cairo*, which, against strong competition, had not been a success on the Marseille to Alexandria route. Fairfield recovered the vessels and brought them back to the Clyde, while Alex Gracie, Fairfield's Chairman and Managing Director, left for Toronto and sold them to MacKenzie Mann & Co Ltd for £180,000 each.[30] The vessels were renamed *Royal George* and *Royal Edward* and began service on the North Atlantic in 1910.

Although builders of the largest merchant vessels and warships, Fairfield also specialised in paddle steamers and small ferries such as this one. *Bosphorus No 47*, was one of many ferries built by Fairfield over a 50-year period for service on the Bosphorus at Constantinople/Istanbul.

Turkish Business

In 1903 Fairfield built four passenger ferries, *Bosphorous 47* to *50* inclusive, for the Bosphorous Steam Navigation Co (Chirket-i-Hairié), the first of ten to be constructed by Fairfield for this company between the years 1903 and 1910. In September 1909, however, in a separate business venture, Fairfield requested a concession from the Turkish Government through London based shipbrokers Weddel Turner & Co, to become involved in the reorganisation of the Ottoman Steam Navigation Company. For Fairfield, the deal meant the construction of up to 35,000 tons worth of coastal vessels to operate between Constantinople and the Sea of Marmara. However, despite Fairfield and Weddel Turner being granted the concession, paying a deposit of £5000 each, and with their intention of acquiring a significant shareholding in the Turkish company, the deal fell through, largely because of the prior interest of German steamship company North German Lloyd. In August 1910 Fairfield, through solicitor Sir Edwin Pears, requested the Turkish Government return its deposit plus expenses.[31]

The passenger ferry *Mecklenburg* built for Stoomvaart Maatschappij Zeeland, launched on 25 October 1909.

The Fairfield Crane

In 1910 the Company placed orders with structural steel engineers William Arrol & Co Ltd, at Bridgeton in Glasgow for four, electric berth cranes of the derrick type and a giant cantilever crane for fitting-out. These tall structures represented the latest developments in material handling at the building berths in shipyards and soon came to dominate the skyline, not just at Fairfield but along the banks of the Clyde, as shipyards from Govan to Greenock were equipped with them. The giant cantilever crane in particular was a massive structure, erected beside the fitting out basin to lift very heavy loads such as machinery and boilers into hulls. Recent contracts for battlecruisers with heavy armament persuaded the Board that such a crane was necessary and in September 1909 it was agreed that a 200-ton crane would be ordered at an estimated cost of £47,225. On completion in 1911 the giant cantilever crane was tested to 250 tons. This enabled the Fairfield company to claim that it was the crane with the largest lifting capacity in the world.

Fairfield's main gate in 1909, either at lunch time or the end of the working day. The main office building is to the left of the gate and the new drawing office building is to the right. A four storey tenement building, ubiquitous in Govan and Glasgow, is at extreme right.

The new 'Fairfield Crane' tested to 250 tons on completion in 1911. The ever increasing size of ships meant larger and heavier machinery, although the principal reason driving the need for a larger crane was the main armament for battleships and battlecruisers. Designed and built by Sir William Arrol at his Dalmarnock Works, on completion this crane was briefly the largest capacity crane in the world.

Warships

Between the completion of *Indomitable* in 1908 and the start of hostilities in August 1914, the Admiralty placed contracts with Fairfield for 11 destroyers, 1 battlecruiser, 1 battleship and a survey vessel. Two further orders for destroyers were placed by Australia, the first of which was HMAS *Parramatta*, the first vessel to be built for the fledgling Royal Australian Navy. Launched by Margot Asquith, the wife of the British Prime Minister at Fairfield on 9 February 1910, *Parramatta* was completed in August of that year and sailed for Australia, arriving there in December.

In March 1911 the Board authorised Alexander Gracie to proceed with negotiations regarding a Canadian shipbuilding scheme. This was followed the following month by the approval of a draft agreement, with the instruction to send it to the British and Canadian Shipbuilding and Dock Co Ltd for approval. There was no further mention of this in the Fairfield minute books, although the minutes for John Brown & Co Ltd, Clydebank, in September of the previous year, mention a Canadian proposal for a shipyard as a joint venture with the Fairfield Company.[32]

The Coventry Ordnance Works would prove to be the source of great regret for the Fairfield Board during the

The launching ceremony for the Australian destroyer HMAS *Parramatta* underway on 9 February 1910. *Parramatta* was the first warship constructed for the newly created Royal Australian Navy. The ship was launched by Margot Asquith, wife of the British Prime Minister H H Asquith.

The *Warrior* class armoured cruiser HMS *Cochrane,* running trials early in 1905.

Parramatta running trials in the late summer of 1910. She was removed from service in 1928 and destined to be dismantled. However, and for various reasons, part of this ship still exists on the Hawkesbury River, while her bow stands at the Royal Australian Navy Heritage Centre, Garden Island, Sydney.

The 'Town' class cruiser HMS *Glasgow* making her way down the Clyde past Renfrew in September 1910. (*Ballast Trust*)

The Platers' Shed ran along the head of the building berths separated by the works railway lines. Here, platers stand by the large plate rolls used to impart the correct curvature to shell plates. A large number of men was necessary to manhandle plates as in this instance there is no adjoining crane for that purpose.

period up until the start of the First World War. Almost every month since its incorporation, the Coventry Ordnance Works asked for significant cash payments from the three syndicate members in proportion to their shareholding. By the Autumn of 1912 the Fairfield Board asked the secretary, Robert Strachan, to write to the Coventry Ordnance Works, saying that the Fairfield Board viewed with great alarm the repeated applications for cash advances. It was estimated that, by the end of 1912, the Coventry Ordnance Works would owe Fairfield £190,000. So bad had Fairfield's position become that, in October 1911, Cammell Laird agreed to pay Fairfield's portion of calls for cash. Serious losses revealed in February 1912 on the contracts for *Maunganui* for the Union Co of New Zealand and the *Normannia* and *Hantonia*, channel steamers for the London & South Western Railway Co, compounded the problem. Additionally, the first national strike by miners, which began in February, caused disruption nationwide, while late delivery of armour for HMS *New Zealand* delayed progress on the battlecruiser. Shipyard Manager, Alexander Sampson, in trying to account for the losses made on these vessels, called his two under-managers to give account, which resulted in one resigning.

Fairfield's Shipyard Drawing Office photographed about 1909. There was a separate Engine Drawing Office, an Admiralty Drawing Office as well as separate offices for the all-female tracing office. The original drawing offices proved to be insufficient in size resulting in new drawing offices being built in 1903.

Govan

In 1912 an ever expanding city of Glasgow moved to annex the adjacent districts of Govan and Partick. Govan's population had grown from 9000 in 1860 to 89,605 in 1912, largely because of the growth of the shipbuilding industry within its borders. That year, Glasgow's population stood at 784,496. The city's predatory moves were initially resisted by Govan Town Council, who asked the Fairfield Company, one of the largest industrial concerns in Govan, if they would support their stance. Fairfield chose to remain neutral and, in the event, Glasgow did swallow Govan and Partick.

The Crest of the Wave

1913 was a record year of output for the River Clyde. 370 vessels were launched, with a tonnage of 756,976, including the liners *Aquitania*, *Alsatian* and *Calgarian*. Port Glasgow shipbuilders Russell & Co headed the list, with 16 vessels of 83,876 tons. Fairfield was ranked tenth, unusually low, with six vessels totalling 34,080 tons, but second after John Brown & Co in machinery output, which, for both reciprocating and turbine machinery, totalled 202,200 ihp. *The Glasgow Herald* noted that an annual output of 300,000 tons had been reached in 1881, 400,000 tons in 1893, 500,000 tons in 1901 and 600,000 tons in 1907, stating that 'never before has there been anything like 760,000 tons launched within twelve months'. Under the headline 'The Crest of a Wave', the newspaper said 'there are indications that the apex of prosperity has been reached'.[33]

Empress of Russia, the handsome Canadian Pacific liner moving slowly past Clydebank in March 1913. Note the bucket dredger and hopper are working on the river channel.

Fairfield's second battlecruiser HMS *New Zealand*, funded by the Government of New Zealand. She is seen here in the river Clyde in 1912 photographed by Fairfield's great Clydeside rival, John Brown & Company, who, unlike Fairfield, had house photographers.

The following year, 1914, would change all of that, and, for Fairfield, it began rather inauspiciously, when, in February, Hugh Allan, Chairman of the Allan Line, wrote a letter of complaint about the late delivery, stability and deadweight of the passenger cargo liner *Calgarian*. In April, when it transpired that a heavy loss had been made on building this ship, the Board's response was to ask for a report from Fairfield's Naval Architect, Percy Hillhouse, and the Shipyard Manager, Alexander Sampson, who was also a Director. At the meeting of the Board in May, Hillhouse gave his report. He said that he had reported to Mr Sampson that the midship section required by the Allan Line involved additional weight over that which he had allowed in his calculation, but that there was a margin of weight that he considered would meet the requirement.

The result of the Board's investigation showed that Mr Hillhouse had been exceedingly negligent in not making a special report about this departure from the terms of the specification and, likewise, negligence on the part of Mr Sampson, for not taking steps to protect the company's interests by allowing the vessel to be built to Allan Line specification instead of the specified British Corporation requirements. As Hillhouse was on the point of leaving to join the *Calgarian* to report on her first voyage from Liverpool to Canada, he was off the hook. Sampson then gave his account of why such a heavy loss had been made, after which the Board adjourned to the following morning. At this meeting, which Sampson was asked not to attend, previous losses in the shipbuilding department were referred to, in addition to the issues surrounding *Calgarian*. The Board took the view that, in protecting the Company's interests, they should ask Sampson to resign. When Sampson was asked back into the boardroom, the Chairman, Alexander Gracie, delivered the verdict. On hearing this, Sampson gave the curt reply 'the Board can take my resignation now'. He followed this by walking out of the meeting. Subsequently, Sampson wrote to the Chairman, expressing regret at his discourteous departure and saying that, if reinstated, he would resign from the Board and devote time to reorganising the shipyard department. The Board was not inclined to agree to this and his dismissal stood. He was given an additional £1k on his salary in May. Alex Gracie felt obliged to contact William Hitchens, Chairman of Cammell Laird, to advise him of these events because of the significant interest Cammell Laird held in Fairfield.

At the same meeting, the Board also decided to consult their lawyers regarding the Allan Line claim about *Calgarian* and so began a legal process that would drag on for two years. It was not until November 1916 that Sir Philip Watts, formerly the Director of Naval Construction at the Admiralty, who had been appointed arbiter, found in Fairfield's favour.

First World War

In the months preceding the outbreak of war, the rapid pace of naval construction at the yard continued, although this was not on the same scale as the avalanche of orders that would follow. In June 1914, with the battleship *Valiant*, the cruiser *Undaunted* and several destroyers under construction, the Admiralty accepted Fairfield's tender for a battleship of the *Royal Sovereign* class to be named *Renown*. That month, a Shipyard Manager to replace Alexander Sampson was found in Hugh McMillan, who had occupied a similar post with Belfast shipbuilders Workman Clark.

The Allan liner *Calgarian* making her way down the River Clyde in March 1914. She is heading towards 'Dalmuir Bend' where, at William Beardmore's Naval Construction Works, *Calgarian*'s sister ship, *Alsatian*, had been completed in December 1913. Note the very high tide and extensive flooding on Newshot Isle at left, where the old river channel had been.

Taken on 24 April 1914, this photograph shows some of the drillers at Fairfield with their pneumatic drilling machines at front.

On 8 July King George V and Queen Mary arrived at Fairfield where the Royal party was met by a crowd of 20,000, mostly workers and their families. The Royal party then toured the works, walking under the hull of the battleship *Valiant* which was then nearing launching stage. They left Fairfield by paddle steamer which took them down river to William Beardmore's Naval Construction Works at Dalmuir.

The declaration of war made on 4 August was mentioned at the first meeting of the Fairfield Board on 18

Alexander Gracie. M.V.O. J.P
Chairman and Managing Director

George Strachan. J.P
Director and Secretary

James Hamilton
Shipyard Manager and Directc

Capt. Sir Douglas Brownrigg. Bart., R.N. Director

Admiral Sir George Digby Morant K.C.B

Restal Ratsey Bevis
Director

The Fairfield Board during the First World War. While the top three were full-time employees of the company, the bottom three attended mainly for Board meetings. The latter were there because of their business expertise and connections, especially to the upper echelons of the Admiralty. While the Shipyard Manager was a member of the Board, the Engineering Manager was not at this time, although both the Engineering Director, Alexander Cleghorn, and the Naval Architect, Percy Hillhouse, were asked to join Board meetings on occasions.

August, when it was noted, that the Admiralty had asked the Company to accelerate construction of HMS *Undaunted,* together with the machinery for the cruiser *Arethusa* which was under construction at Chatham and the Bolinder diesel machinery for the RFA tanker *Ferol* under construction at Devonport.

The completion of *Calgarian* for the Allan Line and *Llanstephan Castle* for Union Castle earlier in 1914 effectively brought to an end the construction of merchant vessels and, in particular, passenger cargo ships, as the resources of the Company were concentrated on warships. There was one exception however: the Union Steam Ship Company of New Zealand's *Aotearoa,* a 14,750-ton passenger cargo liner for the Australian-Canada route. This vessel had been laid down in late 1914 and launched on 30 June 1915. From an engineering standpoint, *Aotearoa* represented the latest ideas in propulsion, with geared turbines running on superheated steam generated in oil-fired water-tube boilers. She was taken over by the Admiralty for service as an armed merchant cruiser and renamed HMS *Avenger.* On 14 June 1917 *Avenger* was torpedoed and sunk west of the Shetlands with the loss of one life.

On 30 October, the return of Admiral Jackie Fisher to the Admiralty as First Sea Lord brought with it immediate action when, on 3 November 1914, a meeting with various shipbuilders took place at the Admiralty. Fisher initiated a massive programme of naval shipbuilding totalling 612 vessels of various classes including 64 submarines. Fisher considered that submarines were of the greatest need and gave out orders for twenty to be built immediately. Most British shipbuilders, including Fairfield, had stood by as the early development of the submarine in the UK took hold at Chatham dockyard and at Vickers' yard in Barrow. As a result of Fisher's meeting however, Fairfield was given contracts for two, *E37* and *E38* as well as two 'M' class destroyers, *Mindful* and *Mischief,* both of which were fitted with Brown-Curtis geared turbines built by Fairfield under licence from John Brown & Co.

To cope with the growing list of Admiralty contracts, in November 1914, the Fairfield Board, at a cost of £8500,

The Armed Merchant Cruiser HMS *Avenger* passing John Brown's shipyard at Clydebank from the Fairfield yard in March 1916. Laid down as *Aotearoa* for the Union Steamship Co of New Zealand, the ship was fitted with eight 6-inch and two 3pdr guns. She was torpedoed and sunk on 14 June 1917 by *U69.*

With four submarine contracts placed by 1915, it was enough to create a separate submarine engine department where these young engineers would work. Fairfield would eventually be contracted to build thirteen submarines of which nine were completed.

approved the building of two berths at the west end of the yard suitable for the construction of submarines. This was followed by a further £16,000, authorised in December, to upgrade these berths to include a covered building shed with overhead craneage.

HMS *Renown*

Fisher's influence on the building programme had further implications for Fairfield when, in December 1914, and at great speed, he pushed through authorisation for the construction of two battlecruisers, a ship type that he felt was essential to the war effort, particularly after the decisive defeat inflicted on a German squadron by battlecruisers at the Battle of the Falkland Islands that very month. The contract placed with Fairfield for an 'R' class battleship in June was changed to a battlecruiser, HMS *Renown*, while a sister vessel, HMS *Repulse*, was switched to John Brown's at Clydebank as Palmers' yard at Jarrow, where the contract was initially placed, did not have a building berth capable of taking the hull.

To set an agenda for rapid construction, Fisher met with the managing directors of Fairfield and John Brown, Alexander Gracie and Thomas Bell, to impress upon them in no uncertain terms, the urgency of building these ships in the shortest time possible. Fisher's demand was that both

A midships view of the *Queen Elizabeth* class battleship HMS *Valiant* as she passes Clydebank in late 1915 or early 1916. The crew have gathered at the ship's side to see what is happening in John Brown's shipyard and in particular the battlecruiser *Repulse*.

should be built within 15 months, an unrealistically short time for what would be the largest ships in the Navy, especially as their design had barely begun. At a retiral speech made many years later, Thomas Bell recalled that Fisher told him and Gracie that 'I am going to have these ships delivered on time and if you fail me, your houses will be made a dunghill and you and your wives liquidated. I expect to hear tomorrow that you have started preparations for these ships.'[34] And thus the two Clyde yards were once again in a race to build ships, this time against the clock and not one another.

The keels for both *Renown* and *Repulse* were laid on 25 January, Fisher's birthday, although this, according to a note made by Thomas Bell in August 1916, was just for show:

> It is only fair to say that the official date for laying the keel, 25 January 1915, was a paper date only, to enable some officials to wire Lord Fisher that it had been laid on his birthday. The designs were not sufficiently advanced, nor was [there] enough material in the yard to proceed with the work for another six weeks and even then, it was with the greatest difficulty that working drawings could be prepared and approved to keep pace with the ship's construction.[35]

Design work at the Admiralty ship design offices began on 30 December 1914 and by 17 January 1915, both yards had been supplied with sufficient information to start on the midship section of the ship, out to the turn of the bilge.[36]

It was not until early April 1915, that general building drawings were made available to the shipbuilders, although sufficient information had been available to make a start with the hull. By September 1915, *Renown* was running two months behind *Repulse,* with completion set to be closer to September 1916 rather than June. One of the problems

Another event captured by the photographer at John Brown's shipyard. HMS *Renown* sailing past Clydebank on her way either to run trials or to be delivered in September 1916. As the Clyde is off to the left, the river in view in the background is the Cart with Paisley beyond. The puffs of smoke down *Renown's* side are from the two tugs at her bow but out of shot.

causing delay at both yards was the lack of men. At Fairfield, the shortage of skilled men delayed completion of *Renown*'s machinery, while the supply of the three main armament mountings by the Coventry Ordnance Works was running late for the want of 100 fitters. At the same time, steel and copper steam pipes were not being delivered in the quantity required for the ship to raise steam by June 1916.

The issue of late delivery prompted a discussion about Admiralty contracts at the December 1915 Fairfield Board meeting.

> Departmental reports were read and . . . the Board unanimously resolved to place on record that they view with great concern the recent repeated failures to maintain dates stated for progress and completion of contracts of which the position now reported in respect of HMS *Renown* and the delay in delivering HMS *Mischief* are glaring examples and instruct that the works managers be specially informed of this minute.[37]

Whatever instructions were issued as a result of this decision, this did not stop the Secretary to the Admiralty writing to the Company twice over this issue. Both letters were read at the Board meeting held on 3 March 1916 at

Taken from the giant cantilever crane, this photograph shows the building berths of the main yard on a misty day in 1918. Nearest the camera on No 1 berth is the battlecruiser HMS *Rodney*, sister ship to HMS *Hood*. Suspended in March 1917, work never resumed and what you see of her, in the region of 5000–6000 tons, was broken up on the slip in 1919. The tower cranes at left of shot with a maximum load of 12 tons were being introduced, replacing the lattice steel derricks which were capable of a lift of just 5 tons. (*Imperial War Museum*)

Fairfield. The first expressed 'grave concern' about the poor progress being made on vessels under construction. While the second letter, presumably a response to the Company's reply to the first, said that 'it still remains a matter of vital importance that the building and equipping of new ships of war should steadily be maintained at the highest possible rate of production'. The letter ended by requesting that the Managing Director, Alexander Gracie, meet with the First Lord of the Admiralty, Arthur Balfour, (former Prime Minister) on 6 March at 4pm. There is no further reference to the outcome of this issue which most likely concerned progress of *Renown*.

Whatever difficulties were experienced in the yard, *Renown* left the Clyde on 20 September five weeks after *Repulse*. The build time for these ships, if not quite Fisher's 15 months, was a remarkable 20 months and a construction record for ships of their size.

After spending some time on board *Renown*, assistant constructor H S Pengelly began a letter to his superior, the Director of Naval Construction, with 'I have not written for some time, because, as you suggest, the ship is in the happy position of having no defects worth mentioning' and he ended with, 'I have thoroughly enjoyed my 12 months at sea in the best ship in the Navy.'

At the Board meeting held on 16 October 1916, correspondence was read concerning a proposal by Canadian Pacific Railway about the possibility of establishing, not for the first time, a shipyard in Canada. This proposal appears to have been relayed through the Coventry Ordnance Works. The Fairfield Board agreed to leave this proposal in the hands of Alexander Gracie. When the proposal submitted by the Coventry Ordnance Works was found to be unacceptable to Canadian Pacific, the Fairfield Board agreed to go it alone with the proposal and to invest up to £100,000 in the project. Nothing appears to have come of this scheme however.[38]

HMS *Rodney*

The most significant contract received at Fairfield during the First World War was for the *Hood* class battlecruiser *Rodney*. Ordered in May 1916 and laid down on 9 October, *Rodney* was one of four battlecruisers of the same class, the others being *Anson*, awarded to Armstrongs on the Tyne, *Hood* to John Brown's at Clydebank and *Howe* to Cammell Laird at Birkenhead. Initially, all four were to proceed at low priority, although that changed in February 1917 when Admiralty instructions were for *Hood* only to proceed 'with all despatch'. The following month, *Rodney* and her two sisters were suspended. Despite the scale of the British shipbuilding industry, the plain fact was that, at this stage in the present emergency, other classes of warship, especially destroyers and submarines, were more important than capital ships. Of the key resources determining the pace and scale of warship construction, labour was at its limit while gun mounting manufacturing was under severe pressure given the sheer volume of mountings required.

While progress on *Hood* at Clydebank continued at an increased rate, at Fairfield, alterations at the head of No1 berth went ahead to accommodate *Rodney*'s 860 foot long hull in anticipation that work on the ship would be resumed. By May 1918 the Admiralty were still contemplating restarting work on two of the three suspended battlecruisers and it was not until February 1919, that the decision was made to cancel the ships.[39] Steelwork amounting to approximately 6000 tons on Fairfield's largest berth was subsequently broken up and the shipbuilders compensated for the loss of revenue that might otherwise have been made from that berth.

Munitions of War

The realisation that the war would be one of industrial output as much as conflict on the battlefield led to the passing of the Munitions of War Act in July 1915 and the creation of the Ministry of Munitions under David Lloyd George. The aim was to co-ordinate, regulate and maximise the manufacture and supply of war materials. Under the terms of the Act, factories engaged in war contracts such as the Clyde shipyards were declared Controlled Establishments which, in the case of shipyards, gave the Admiralty control of what was built and where. The Act also introduced dilution, permitting the introduction of semi-skilled workers into the factories, banned strikes and

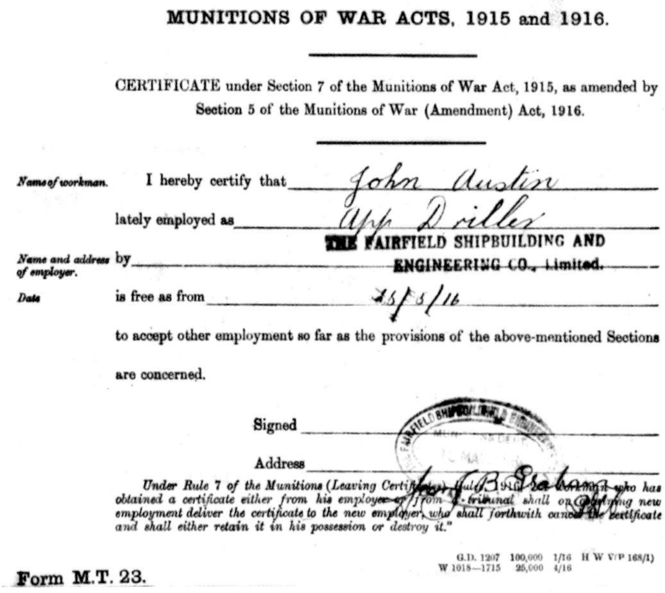

MUNITIONS OF WAR ACTS, 1915 and 1916.

CERTIFICATE under Section 7 of the Munitions of War Act, 1915, as amended by Section 5 of the Munitions of War (Amendment) Act, 1916.

Name of workman. I hereby certify that John Austin

lately employed as App Driller

Name and address of employer. by THE FAIRFIELD SHIPBUILDING AND ENGINEERING CO., Limited.

Date is free as from 25/5/16

to accept other employment so far as the provisions of the above-mentioned Sections are concerned.

Signed

Address

Under Rule 7 of the Munitions (Leaving Certificates) Rules, 1916, "a workman who has obtained a certificate either from his employer or from a tribunal shall on obtaining new employment deliver the certificate to the new employer, who shall forthwith cancel the certificate and shall either retain it in his possession or destroy it."

G.D. 1207 100,000 1/16 H W V/P 168/1)
W 1018—1715 25,000 4/16

Form M.T. 23.

The Leaving Certificate issued to John Austin, an apprentice driller. Without a Leaving Certificate, a shipyard worker would not be employed anywhere else for a period of six weeks.

introduced a system of leaving certificates which discouraged workers from moving from one job to another without permission. Cutting across rights hard won over many years, much of this was anathema to the trade unions and lead to significant disruption.

On 12 July Fairfield became a Controlled Establishment and the conditions of the Act applied in full. In defiance, some of the first strikes were at Fairfield. On 27 July coppersmiths went on strike over the attitude of Fairfield management in their interpretation of the Act. At the subsequent Munitions Tribunal, the coppersmiths were fined 2/6 (12.5p) which was paid by the Union who regarded the men's action as quite legal.[40]

Other strikes in the Glasgow area were also motivated by a deep sense of injustice. With the population of the City surging beyond one million and housing in short supply, rising prices, evictions and particularly rent increases set the scene for discontent where the wives and mothers of fighting and working men challenged the landlords, leading to a series of rent strikes. In May 1915 resistance to eviction and rent strikes in Govan was led by housewife Mary Barbour. The threat of industrial action by shipyard and munitions workers in support of the rent strikers became another compelling reason for the Government to take action and on 23 December 1915 the Increase of Rent and Mortgage Interest (War Restrictions) Act was passed, which forbade landlords from raising rents or increasing the mortgage interest rate for the duration of the war and six months after.[41]

Dilution and Employment at Fairfield

By August 1916 the Ministry of Munitions could report that dilution was operating in 150 of the largest of the 300 controlled engineering and shipyard establishments on the Clyde, with 14,000 women employed. The majority of these women were employed in the new munitions factories, with only 1800 of the 14,000 directly engaged in shipbuilding, general engineering, and metal trades.[42] In the Clyde yards, records show that 1001 women were employed by September 1916, the two largest employers being Beardmore and Fairfield with 370 and 134 respectively, accounting for 50% of the total. By November 1918, there

A posed photograph showing two female workers at the head of the fitting-out basin. The standard ship *War Patriot* is in the background while the destroyer is most likely to be either *Walrus* or *Wolfhound*. Dominating the scene is the Arrol giant cantilever crane with the name Fairfield fixed to the top of the cantilever. The photograph was taken between February and April 1918. (*Imperial War Museum*)

One of the bays in the engine shops showing engine parts, shafting and a host of other parts for various ships under construction Taken at the same time as the other photographs in the Fairfield yard by war photographer Horace Nichols in 1918, it shows the galleries on two levels where light machining work was carried out. The heaviest machines were on the ground floor including the large lathe at bottom right with its spinning face plate. Behind that is a vertical boring machine and elsewhere more lathes slotting machines and surface table. Parts include shafting for the destroyer *Sikh* and the standard Ship No 543 which was later cancelled. (*Imperial War Museum*)

were 31,000 women employed across the UK-wide shipbuilding industry, equal to 6.6% of the workforce. At Fairfield, overall employment figures as quoted in the *History of the Ministry of Munitions,* state that this number fluctuated between 7500 and 13,500.

Alcohol Control

As the factories and shipyards had been declared Controlled Establishments so too, in a sense, were the pubs across industrial areas in the UK such as Glasgow and the West of Scotland. In July 1915 the Ministry of Munitions established an oddly named organisation termed the 'Central Control Board (liquor traffic)'. Put bluntly, this Board had as its aim the reduction in the number of hours men spent in public houses and thus, it was hoped, result in more productive time at work. Opening hours for public houses were restricted to 4.5 hours on weekdays and 4 hours on Sundays

Another initiative taken by the Central Control Board (liquor traffic) to increase output was the creation of canteens in factories and shipyards where good, nutritious food could be served at lunch time. While there were nutritional benefits for workers, the real reason was to stop men going to the pub at the lunch break, after which it was known that productivity dropped.

The canteen at Fairfield was built during the last half of 1916 at a cost of £16,000 (paid for out of excess profits tax) and was initially staffed by the Women's Legion. However, it seems the canteen did not get off to the best of starts as, by August 1917, Fairfield was in debt to the Women's Legion for £167. Moreover, the women subsequently decided to withdraw from the management of the canteen which then had to be managed by a workers' committee.

In August 1915 a situation arose at Fairfield which threatened to bring the whole river to a standstill. At 7.30 on the morning of 26 August, 15 men were found smoking and reading papers on board the battleship HMS *Valiant*. Two of them were sacked for 'not attending to work' these words being written on the clearance certificate which the Act required the men to be given to pass on to future employers. The men objected to this with the result that the shipwrights went on strike in defiance of the Act. Management duly notified the Ministry of Munitions who took the issue to a Tribunal. Seventeen shipwrights were fined but, as three failed to pay the fine, they were imprisoned in Duke Street prison. It was not until November that the men were released, their fines having been paid by the Union. In October 1915, the Clyde Shipyards Joint Trades Vigilance Committee said: 'The harsh and unjust administration of the Munitions Act had given rise to an immense volume of righteous indignation.'[43] A comment from Christopher Addison at the Ministry of Munitions was that 'There is no doubt that the men have genuine grievances and that the management of the Fairfield yard is very bad.'[44]

From the Shipyard Diary

The following account from Fairfield's *Shipyard Diary* gives some idea of activity in the yard for the months of August and September 1916.

2 August. HM Submarine *E48* launched and towed to Messrs Beardmore, Dalmuir, for completion by them as per Admiralty order.

10 August. *Renown* Ship No 503: Conference held re progress and completion of.

11 August. Board meeting at Fairfield.
No 518 HMTB *Orcadia* left Fairfield 8.35pm for Govan Graving Dock.

18 August. Three sons of the Sultan of Johore, a cousin and Abdul Rahman CMG, visited the Works.

19 August. Holidays: Works closed at noon to reopen on Monday 28th at 9.45am except, as requested by the Admiralty:
For men required on No 503 [*Renown*] who will receive intimation from their foremen.
Men not engaged on No 503, but who may desire to remain at work and give their names to their foremen.
For men who remain at work on No 503, the holidays will commence on Saturday 16th September.

29 August. Mr Lyndon McCassey, KC, Mrs McCassey and son visited the Works, and lunched with Chairman and Managers.

5 Sept. First Lord of Admiralty (Mr A J Balfour), Third Sea Lord (Admiral Tudor) Commodore de Bartolomé, CB (Naval Secretary), Mr G H Steele (Private Secretary), Captain Brian Barttelot (Captain Superintendent Clyde District) and members of Government Dilution Committee, Mr McCassey KC, Sir Thomas Munro and 2 other gentlemen, visited the Works today and were taken around by the Chairman, Secretary and Managers.

11 Sept. HMS *Avenger* (No 499) left Fairfield Dock for Govan Dry Dock.

16 Sept. HMS *Renown* at 2.30 left Fairfield Dock and proceeded up river to turn. Berthed at Princes Quay thereafter. Departure on Sunday was postponed due to weather.

18 Sept. Monday 4.30pm [*Renown*] proceeded down river; passed yard at 5pm, arrived Tail of the Bank 7pm.

20 Sept. [*Renown*] Trials completed.

21 Sept. [*Renown*] left for Portsmouth 8.30am

21 Sept. British and neutral journalists visited the Works (party of 19) arrived 10.20am left 11.40am

22 Sept. [*Renown*] arrived Portsmouth 7.40am

23 Sept. (Saturday). HMS *Oriana* launched 11.15am
No 518 HMTB *Orcadia* out on preliminary trials; returned to Fairfield 3.15pm. Satisfactory.

25 Sept. Monday. No 518 *Orcadia* out on full power trial returned to Fairfield 5pm. Satisfactory.

27 Sept. Admiral Funikoshi and 4 staff officers, Imperial Japanese Navy, visited the Works today at 10am.

29 Sept. No 518 HMTB *Orcadia* left 7.23am for Acceptance Trials; trial successful and vessel taken over by Admiralty officers 2pm.

The Tragedy of *K13*

One of the most tragic events to befall Fairfield during the First World War was the sinking of the steam submarine

Looking aft along the inner bottom of standard ship *War Pioneer*. Note the winch at right and the rivet fire at left attended by the rivet boy. The frames are in place and the deck beams are being fitted. (*Imperial War Museum*)

K13 in January 1917. The 'K' class submarines, by far the largest at the time, were designed for a surface speed of 24 knots in line with Admiralty operational requirements that they should accompany the battlefleet on the surface. Existing submarines used diesels on the surface and electric propulsion submerged. As diesels in their present state of development could not achieve the required surface speed, the only way for 'K' class submarines to achieve 24 knots was with the use of steam turbines. Creating steam meant having oil fired boilers and a means of venting the products of combustion through funnels. Thus 'K' class boats running on the surface had two funnels which had to be sealed off and retracted into the submarine's casing when preparing to dive. Additionally, large fans in the boiler room, necessary to draw air down from the outside to enable combustion, had also to be closed off and made watertight.

In September 1915 the Admiralty placed contracts for two 'K' class submarines with Fairfield, *K13* and *K14*. By early January 1917 *K13* was complete and ready to begin a series of basin trials which included submerging. On 18 January, *K13* left Fairfield to conduct speed and diving trials on the Firth of Clyde. On 20 January, more diving trials were conducted in the Gareloch, when the submarine went down to 65 feet, staying there for one hour. On Monday 29 January *K13* left Fairfield early in the morning to run final acceptance trials. That morning, with Commander Herbert in command, she went down to 85 feet, staying there for 2 hours while various checks were made. The boat broke surface shortly before 12 noon, at which point it was decided to stop for lunch. Herbert then decided that he wanted one short and final dive to conclude the trials programme. At 3pm, Herbert ordered half speed ahead both, dive 20 feet. On board were 80 persons from a normal in-service complement of 59. The expanded complement included 55 naval officers and ratings, 11 Admiralty sub-contractors and 14 Fairfield managers and men. Among the Fairfield men was the company's naval architect, Percy Hillhouse, who would subsequently write an account of the ensuing ordeal.[45] Soon after *K13* disappeared under the water, Commander Michell of the submarine *E50*, also running trials that day, noted a disturbance on the surface where *K13* had just gone down. At 3.30, sensing that something was wrong, Michell sent a man ashore to ring for help. And so began a 57 hour ordeal which claimed the lives of 32 men. A subsequent Board of Enquiry held in Fairfield's Boardroom found that the four large 37 inch diameter vents through which fans drew air down into the boiler room and which should have been closed on diving, were still open, allowing a torrent of water to fill the boiler and engine rooms and all compartments aft. As *K13* flooded, Herbert ordered watertight doors shut. This included the door in the boiler room bulkhead and effectively sealed the fate of those aft of this bulkhead, but saved the lives of those forward. The bow of the stricken submarine was eventually raised sufficiently for a hole to be cut through the outer casing and pressure hull for the 48 survivors to get out. The Board of Enquiry found the accident was caused by human error.

Taken from the top of the cantilever crane in 1918, this shows at right the steelworking shops and, closest to the camera, the joiner shop. (*Imperial War Museum*)

In February, the Fairfield Board authorised £3350 to be distributed among the relatives of the six Fairfield employees who were lost in the accident. This sum included £500 given to the Admiralty for distribution among the relatives of HM servicemen who lost their lives.

On 15 March *K13* was raised and returned to Fairfield five days later. She was refitted, given a new 'swan' bow and returned to the Navy for service as *K22*.

On 14 June 1917 Muirhead Bone, the first British War Artist, arrived to start work on a number of drawings recording activity in the yard. He completed eight in number showing various views throughout the works.

The working year ended on Saturday 29 December at noon. The total roll of employees on that day was 8383, with 5584 in the shipyard and 2799 in the engine works. Work resumed on 7 January but it was several days before all employees returned.

Fairfield emerged from the war in good shape, with draft agreements in place with Canadian Pacific, the Anchor Line and the Orient Steam Navigation Co. Not only that, yard plant and facilities were up to date, with approximately £464,000 having been spent over the course of the war years, at least 40% of this figure in the form of grants from the Admiralty.[46]

Profits had soared from naval work too as the results for the war years show.

Gross profits before deductions.

1914	£46,552
1915	£100,644
1916	£107,046
1917	£139,306
1918	£160,046

By August 1919 the Company had paid back over £400,000 in excess profits duty to the Exchequer.

War Output

Company records show that, in total, 45 vessels were built during the war period, including one battleship, one battlecruiser, light cruisers, submarines, destroyers, minesweepers and barges. The combined tonnage was 144,777. The output of steam and diesel machinery for the above, plus for ships building elsewhere, amounted to 1,166,035 hp. See appendix D for details of war output. Additionally, 47 vessels were either repaired or refitted, some of them twice.

In July 1917 the Company was contracted by the Admiralty to construct 50 Sopwith T.1 Cuckoo torpedo

Dazzle painted *Train Ferry Number 3*, making her way down the Clyde in October/November 1917. Built for the War Office, she was a roll-on roll-off train ferry with four sets of rails intended to speed up the transportation of war materials between the UK and French ports.

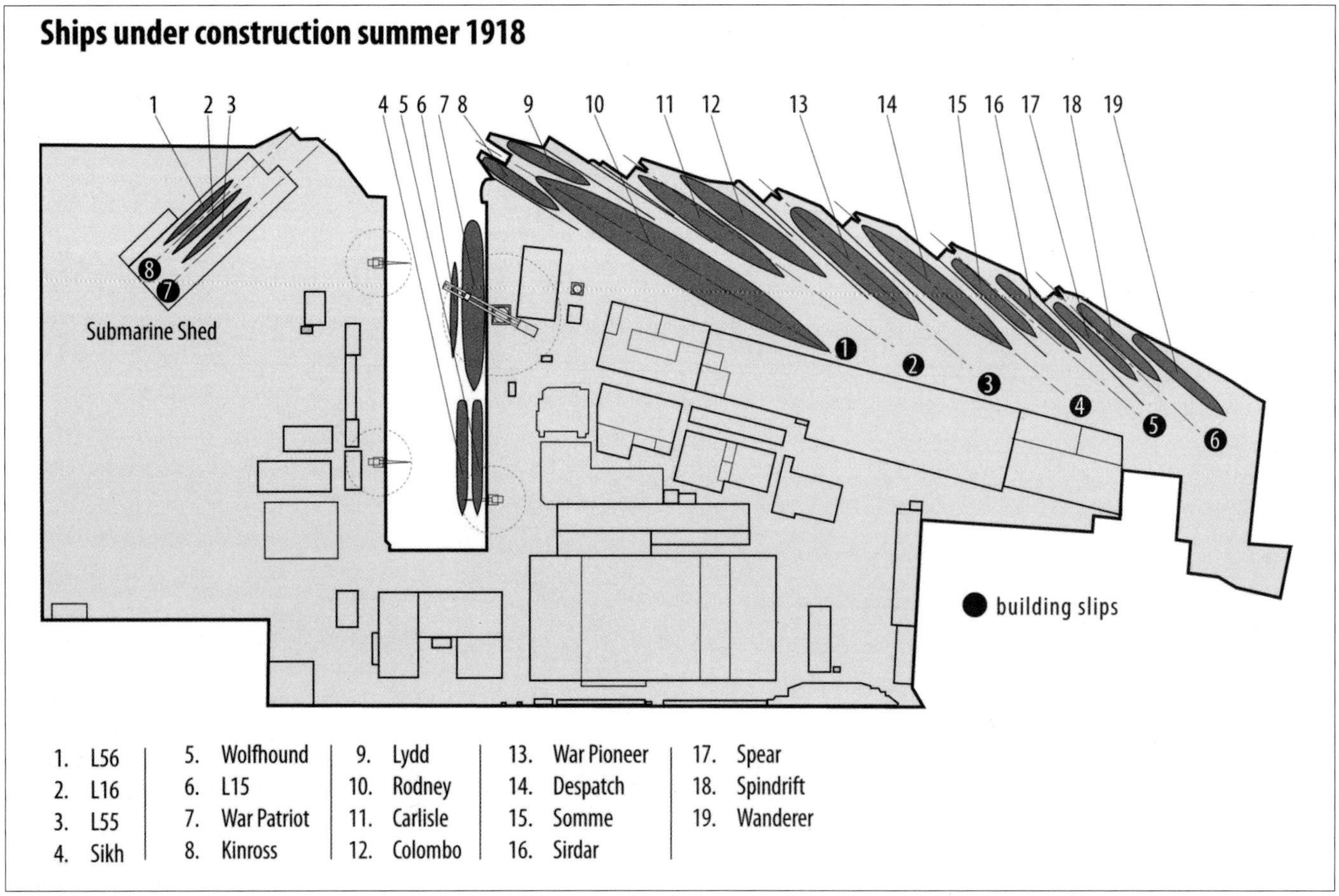

Plan of the Fairfield yard in the summer of 1918 showing 19 vessels under construction and their location.

bombers under sub-licence from the Sopwith Aviation Company. The first of these, N7001, left the yard on 28 July 1918 for Renfrew.

The last two months of 1918 saw not just the end of the war on 11 November but also the completion of the last war contracts. The destroyer *Spear* was launched on Saturday 9 November, while on the 12th the cruiser *Carlisle* departed Fairfield at 7.30am to run trials. That day, the Works closed at 12am and did not resume until 9.45am on the 13th, although very few workers reported for work. On 14 November, 14 newspaper men were received at the Works by Naval Architect Percy Hillhouse. Between then and the end of the year, the submarines *L55* and *E38* left the yard, the minesweeper *Lydd* was launched, the destroyer *Spear* left to be docked at Govan and the cruiser *Colombo* was launched. The General Election, the first in eight years, took place on Saturday 14 December and resulted in a landslide for the coalition government of Lloyd George. This was the first election in which all men over the age of 21 could vote, Women had to be 30 years of age and property owners.

The Fairfield Works closed on Saturday 28 December 1918 and reopened on 6 January 1919 at 7.55am. As was the custom throughout Clydeside, few men reported for work on the official starting date and it was not until several days later that full numbers were achieved.

However, more fundamental issues were set to overtake the workplace because of concerns among the workforce that men demobilised from the forces, coupled to the halt in munitions and war work, would lead to unemployment. Reducing the number of hours worked per week would make more jobs available. This, together with the general trade union objective of reducing the number of hours worked and a sense that new, post-war working conditions should be established, led to the 40 hours movement on Clydeside.

At the beginning of 1919, the working week had been reduced from 54 to 47 hours without a reduction in wages. The hours worked per day were from 7.55 to 12.15 and then from 1.15 to 5.30. Saturdays were worked from 7.55 to mid-day

The 40 hours movement, led by the Clyde Workers Committee that had led strikes early in the war, pressed for an eight hour day and, although not officially backed by the trade unions, called for a strike to begin on Monday 27

January. Things came to a head on Friday 31 January when a mass gathering in George Square led to a forceful police response, with the Government concerned that industrial militancy in Glasgow might precipitate widespread support. Thereafter, there was a slow return to work, but the strike was not called off until 10 February although it continued until 18 February when the 47 hour week was formally ratified.

Support for the 40 Hours Strike at Fairfield

This table shows the total number of employees on the Company roll on three dates of the strike action, beginning, middle and end, and the number of employees on strike on those dates:

	Monday 27 January	Friday 31 January	Tuesday 18 February
Shipyard	5417/1825	5109/4935	5112/771
Engine Works	1943/613	1923/1719	1991/201

The completion of wartime contracts enabled the yard to resume work on the type of vessels it specialised in, namely, passenger and passenger cargo liners. By 1919 the following ships were on order:

Empress of Canada and *Montrose* for Canadian Pacific Railroads.
Tuscania and *Transylvania* for Anchor Line (Henderson Brothers) Ltd.
Athenia and *Letitia* for Anchor-Donaldson Line.
A mail and passenger steamer for the Union Steamship Co of New Zealand Ltd (eventually built as *Aorangi*).
Lady Denison-Pender a cable repair ship for the Eastern Telegraph Company.[47]

New construction was accompanied by reconditioning work in 1919. This included *Armadale Castle*, *Kinfauns Castle* and *Friederichsruh* originally built at Fairfield as *Fürst Bismarck*.

In 1919 yet another manager and director of the Company was asked to leave. Hugh McMillan, who had replaced Alexander Sampson when he too fell foul of the Board, was asked to resign following expenditure of over £24,700 on various items in the yard without authorisation. Having thus lost the confidence of his fellow Directors, he was asked to resign as Shipyard Manager and Director.[48] This McMillan duly did, leaving to form the Blythswood Shipbuilding Company on ground immediately to the east of Yarrow's Scotstoun yard.

On 5 November 1919 all powers of control over the works which had first come into operation in July 1915 under the Munitions of War Act were revoked.

Coventry Ordnance Works 2

At the end of the war, when it was clear that the manufacture of ordnance on the scale reached during the conflict would no longer be required, it was decided by the three controlling companies that, in order to survive, the manufacturing capacity of the Coventry Ordnance Works should be transformed from armaments to other areas such as railway engineering and power generation. To enable this, a combination was agreed with Dick Kerr & Co Ltd and Phoenix Dynamo Manufacturing Co Ltd, manufacturers of generators, dynamos and electric equipment for railways. This combination was renamed the English Electric Co Ltd. The shareholding held by Fairfield was sold off and the

The 'C' class cruiser HMS *Carlisle* at the Tail of the Bank, probably just after the end of hostilities in November 1918. She was built in just over one year.

proceeds put towards the construction of a new shipyard at the west end of the existing yard.

Fairfield West Yard

At the end of the First World War there was a general expectation among shipbuilders that there would be a flood of orders for merchant ships to replace those sunk during the war. The immediate future looked good, with orders for passenger cargo vessels either booked or in the pipeline and even the prospect of a major warship order to replace the cancelled battlecruiser HMS *Rodney*. In anticipation of this, the Board decided to lay down new slips beside the submarine shed at the west end of the yard for small to medium sized vessels. This would leave the main yard free to concentrate on larger vessels. As early as 1914, the Board had purchased 15 acres of ground from the Pearce family for £14,000. In 1915, the submarine sheds, which

Fairfield yard at its maximum extent in 1921 with six berths in the main yard and four in the West Yard.

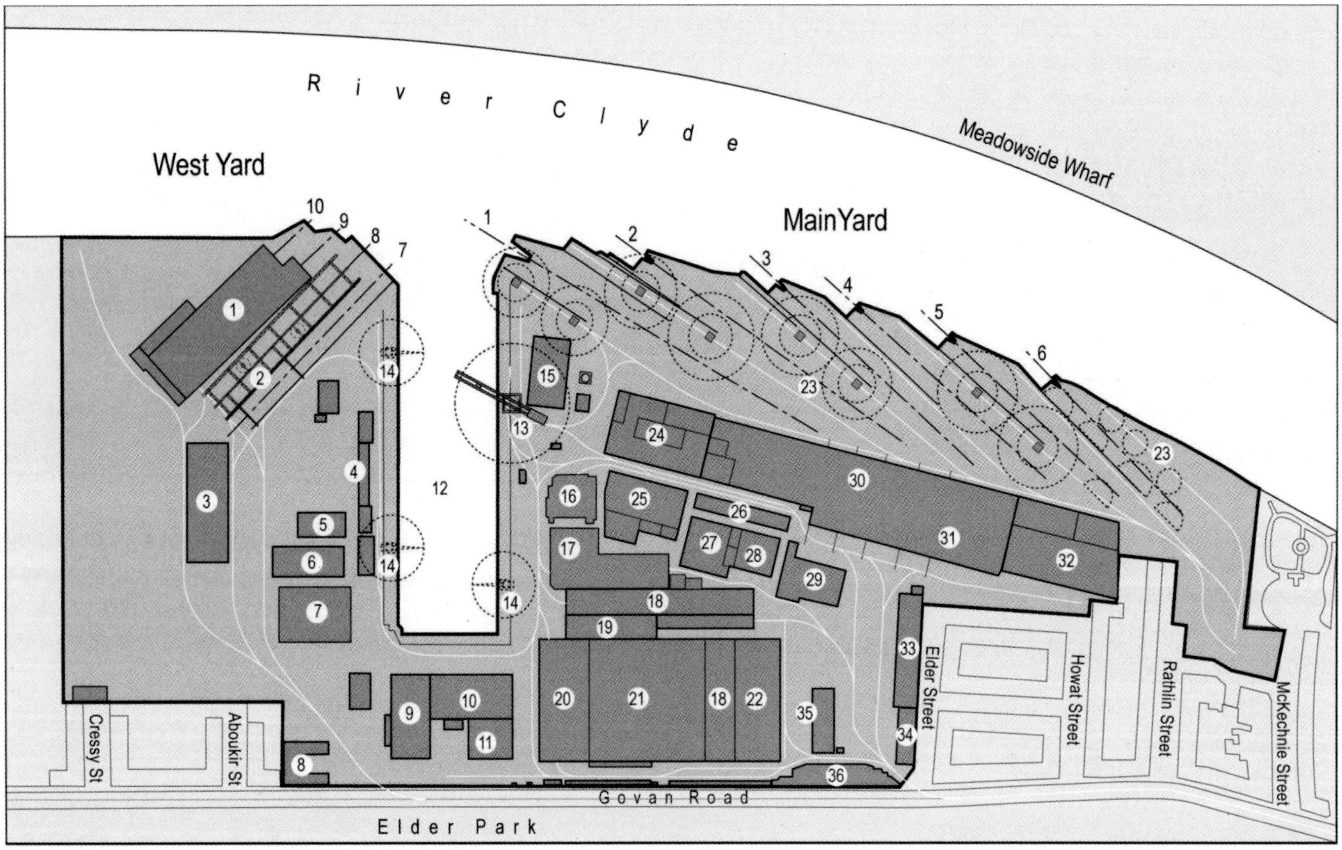

Fairfield Shipbuilding & Engineering Co Ltd 1921

Feet 250 500 1000
Metres 75 150 300

1. Submarine sheds
2. Shipbuilding gantry
3. West platers shed
4. Workshops
5. Timber shed
6. Canteen
7. Sawmill, boatshed and pattern shop
8. Stables and garage
9. Brass foundry
10. Coppersmiths shop
11. Pipe shop
12. Fitting-out basin
13. 200ton cantilever crane
14. 10 and 20ton travelling cranes (3)
15. Engine dock workshop
16. Power station
17. Smithy
18. Boiler shops
19. Funnel shop
20. Fitting shop
21. Engine shop
22. Boiler forge and plate furnaces
23. 12ton fixed tower cranes and 5ton derricks at east end of yard
24. Joiners shop
25. Mechanics shop
26. Stores
27. Sheet iron shop
28. Plumbers shop
29. Angle smiths shop
30. Platers shed
31. Beam shed
32. Frame shed
33. Mould loft
34. Drawing office and model shop
35. Electrical workshop and stores
36. Main office and drawing offices

Building berth dimensions

1. 900x106
2. 720x85
3. 600x72
4. 600x72
5. 540x66
6 540x66
7. 500x65
8. 500x65
9. 350x48
10. 350x48

comprised four berths served by overhead cranes, had been built to the immediate west of the fitting-out basin. The new West Yard as proposed in 1919, envisaged removal of the submarine sheds to be replaced with four new slips of between 430 and 500 feet in length, served by a gantry system of overhead cranes. The West Yard would be funded in part by the sale of shares in the Coventry Ordnance Works. In October 1919, the cost of the West Yard was calculated at £116,408.

In the event, the submarine shed was retained, with its four berths reconstructed as two wider berths. To the side of the shed, two additional berths were created, served by an overhead gantry system built by Edinburgh based structural engineers Redpath Brown & Co Ltd. However, by the end of 1920, shipbuilding had entered into what would be one of the worst recessions so far experienced. Altogether, including vessels built during the war in the submarine shed, only 13 hulls and 18 barges were launched from the West Yard before the slump in orders took effect. The last vessel launched was the oil tanker *Santa Cruz* on 22 June 1921.

All Change: the Sperling Combine

In 1918 London based merchant bankers Sperling & Co Ltd began a series of shipbuilding acquisitions, probably in the belief that the expected post-war boom in shipbuilding would deliver significant profits. Sperling achieved this ambition through the initial acquisition of the Northumberland Shipbuilding Co Ltd at Howdon on Tyneside, a small shipyard dating from 1898. The next step in Sperling's acquisition spree via its Northumberland proxy was the Belfast shipyard Workman Clark.

These events coincided with Fairfield's uncoupling of its association with Cammell Laird and the winding up of the Coventry Ordnance Works which would be reformed as the English Electric Co Ltd. As part of this process, Cammell Laird indicated that it would be disposing of its significant shareholding in Fairfield. The Northumberland Shipbuilding Co was then able to pull off a significant coup by acquiring these shares which, together with others it had purchased, enabled it to become the majority Fairfield shareholder and thus in control of the Company. The Sperling Combine eventually included the Northumberland Shipbuilding Co, Workman Clark & Co, William Doxford & Sons, Blythswood Shipbuilding Co and Fairfield, the last being by far the most prestigious.

This acquisition of Fairfield, which could not have been foreseen, had an immediate impact on the composition of the Fairfield Board which had to be reformed to reflect new ownership. The existing Board consisted of:

Sir Alex Gracie (Chairman)
Sir Douglas Brownrigg
Admiral Sir Digby Morant
Restal Ratsey Bevis (formerly Managing Director of Cammell Laird)
Alex Cleghorn (Engine Works Manager)
James Hamilton (Shipyard Manager)
George Strachan (Secretary).

First to go, in October 1919, were Admiral Sir Digby Morant and Restal Ratsey Bevis. By December 1919, the new board included:

Robert Alfred Workman (from Workman Clark), Chairman,
Edward McKay Edgar (Senior partner in Sperling & Co Ltd)
Sir John Esplen (naval architect)
William Orr Workman (from Workman Clark)
George Strachan
Alexander McAusland Kennedy (Director and Managing Director from July 1920)

Sir Alex Gracie, Alex Cleghorn and James Hamilton remained until July 1920, when they too resigned. With the exception of George Strachan, the Fairfield Board had been completely remade.

The most significant departure from the outgoing Board was Sir Alex Gracie who had been Fairfield's Managing Director since 1905 and Chairman since March 1909. Gracie's replacement as Managing Director was Alexander Kennedy, who had held a similar post at the Northumberland Shipbuilding Co since 1916. Kennedy, born in Dumbarton in 1860, had served an apprenticeship at McMillan's Dumbarton yard and had occupied various management roles including at William Hamilton & Co's Port Glasgow yard. His move to Fairfield was a significant step up in both scale of operation and prestige. In October 1923, following poor results by the Northumberland Shipbuilding Co's group of companies, its shares in Fairfield were sold at a great loss given the very poor state of trade at that time.

This, in turn, resulted in changes in the Board as members of the 'new' Board resigned, including Chairman Robert Workman who was replaced by Alexander Kennedy.[49]

Despite these changes, the day to day running of Fairfield continued as before. The Company's membership of the Combine had come at a significant financial liability, as the outlook for shipbuilding plummeted. Fairfield was obliged to pick up various major cancellation fees, such as for steel contracts entered into with steel makers Baldwins Ltd by Workman Clark. On a lesser scale, Fairfield also had

The launch party for the cruiser HMS *Despatch* photographed on 10 September 1919. The First Lord of the Admiralty, Walter Long was in attendance along with a company of army cadets. Unfortunately, after being christened, *Despatch* refused to budge and had to be shored up. She was successfully launched on 24 September.

to support its subsidiary, the Monmouth Shipbuilding Co at Chepstow which, during the early 1920s, defaulted on contracts for ships and machinery.[50] However, the prolonged depression in the shipbuilding industry that took hold late in 1920 thwarted Sperling's profit-making ambitions for its shipbuilding empire. The Northumberland Shipbuilding Co went into liquidation in 1926 following the failure of the Sperling Combine partnership which had been dissolved in November 1925. A new company, the Shipbuilding Investment Company, registered in December 1926 backed by Sperling and the merchant bank Kleinwort, took over the Combine's four yards. The Northumberland Shipbuilding Co was reformed as the Northumberland Shipbuilding Co (1927) Ltd with Sir Alexander Kennedy as Chairman and Managing Director in addition to his role at Fairfield.

Chepstow

To understand how Fairfield came into possession of a shipyard in Wales, a little back-tracking is necessary. Despite the size of the shipbuilding industry in the UK, it was unable to cope with the demands placed upon it during the Great War. This applied equally to the replacement of merchant ships being lost to the U-boat offensive, and naval building programmes, such as the battlecruisers ordered in May 1916 which were suspended and later cancelled. Lack of labour was the principal cause, together with the high demand placed on certain items such as armour and gun mechanisms. In May 1917 merchant losses to U-boats had reached a critical stage, resulting in the proposal from the First Sea Lord, Sir Eric Geddes, to establish three national shipyards where standard merchant ships could be assembled from parts fabricated elsewhere. One of these yards was at Chepstow, Monmouthshire on the River Wye, where shipbuilding, together with various other iron and steel manufactures, had been undertaken since the 1850s. Much of this was carried on by Edward Finch & Co and included railway trucks, dock gates and bridges, the last giving rise to the name Bridge Works for the site. Between 200 and 500 persons were employed.

In late 1915, a new shipyard adjoining Finch's Bridge

Works was established, with backing from Lord Inchcape (James Mackay) and others. It had eight slips, capable of building ships up to 600 feet in length. This yard, now amalgamated with the Bridge Works, was acquired by the Standard Shipbuilding Company as National Shipyard No 1. At the end of the war, development of the shipyard stopped. In total, 15 standard ships were built at Chepstow until the summer of 1921. In January 1920 the yard was acquired by the Sperling Syndicate who renamed it the Monmouth Shipbuilding Co. On the collapse of the Sperling Combine in December 1924, Fairfield acquired the Monmouth works for £25,000. With shipbuilding in the midst of a deep depression, the Monmouth yard was not required to build ships but railway wagons, for which some of the plate working plant at Monmouth was suitable. Other exclusively shipbuilding items such as berth cranes were sold off. Fairfield entered into a five-year agreement with the Cambrian Wagon Co Ltd of Cardiff to act as agents to secure work for Chepstow. By January 1925, a sample 20-ton steel wagon had been made. Over time, the Chepstow works began the manufacture of other railway related products such as foot bridges and mainline bridges for LNER and LMS. The Chepstow works appear to have

Chepstow shipyard with seven building berths by the River Wye, Monmouthshire during the early 1920s. Fairfield acquired the shipyard in December 1924 following the collapse of the Sperling Combine of which Fairfield and Chepstow were part. Rather than build ships, Fairfield developed the yard into a structural steel business initially to build railway wagons and railway bridges.

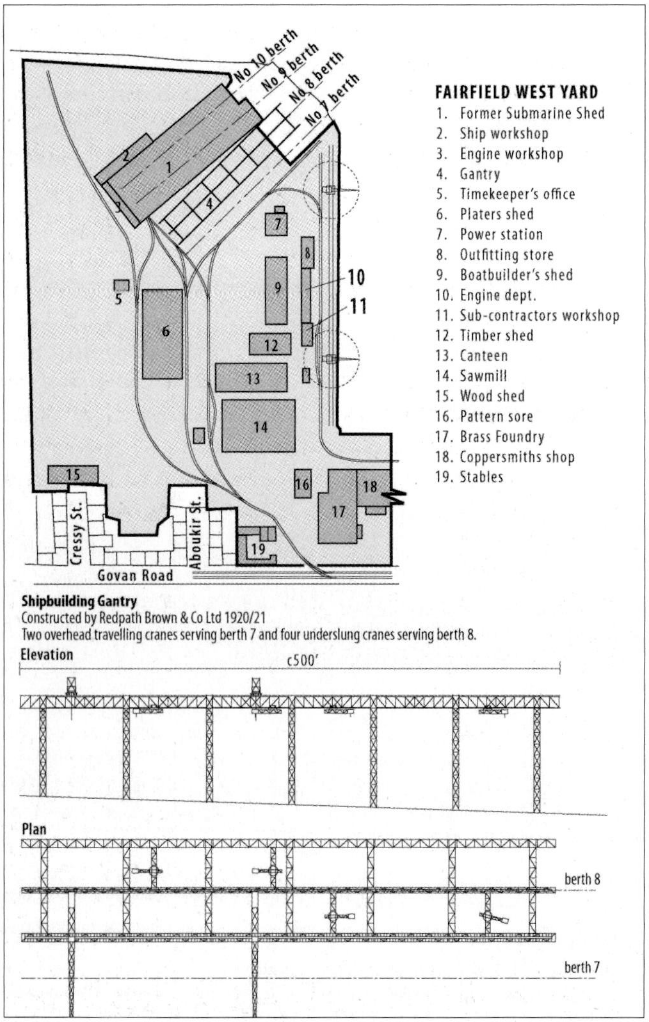

Planned at the end of the First World War when prospects for merchant orders seemed plentiful, the new West Yard included the submarine shed where four original berths were reduced to two and the new shipbuilding gantry by Redpath Brown served both new berths. Only eight ships were built in the West Yard, its usefulness cut short by the onset of the Depression. The West Yard was demolished in 1937–38.

traded under the title Fairfield Shipbuilding & Engineering Co and, although managed locally, the general manager usually attended the monthly board meetings at the Fairfield shipyard in Glasgow, to give account of contracts and financial performance. Major items of expenditure had to be sanctioned by the Fairfield board.

The 1920s

Far from being the decade when shipyards would capitalise on new building in the spirit of a post-war era of enterprise, the 1920s were characterised by stagnation, depressed markets, the suspension and cancellation of contracts and, for Fairfield, the challenges resulting from being part of a shipbuilding combine. The 1920s would prove to be a difficult decade for not only Fairfield but across industry in Britain. Fairfield's shipbuilding programme for the post-war period which had begun well soon began to suffer from the economic slump which began to take effect in 1920 when costs rose and the shipping trade began to shrink.

On 18 August 1920 the liner *Empress of Canada* was launched, becoming the largest passenger vessel so far built by Fairfield. After the vessel was successfully in the water, the launch party were able to tour the *Empress of Britain*, built in the yard in 1908, which was back at Fairfield to be converted from coal to oil firing.

Sir Thomas Fisher, General Manager of Canadian Pacific Steamships, gave some indication of the difficulties facing the shipbuilding industry when he said at the launch that the *Empress of Canada* cost £1,700,00 to build and that, before the War, the figure would have been one third of that. The great increase in running costs had caused first class fares for crossing the Atlantic to rise from £19 to £50.50, second class from £11 to £28 and third £6.50 to £19.

Early stages in the construction of the Anchor liner *Tuscania* on 18 November 1919. Preassembled floor beams lie in the foreground waiting to be erected.

The cable ship *Lady Denison Pender.* built for the Eastern Telegraph Company, at the Tail of the Bank on completion in September 1920. To be based in Aden, the cable ship subsequently worked in the Indian Ocean and down the east coast of Africa.

Canadian Pacific's *Empress of Canada*, built for service in the Pacific. After just six years the ship was returned to Fairfield to be re-engined with more fuel efficient single reduction geared turbines.

By early 1921, 70% of the contracts placed had either been suspended or cancelled. In April 1921, Anchor-Donaldson asked for work on *Letitia* to be suspended. In May 1921, Anchor Line (Henderson) asked for *Tuscania* and *Transylvania* to be suspended. At the eventual launch of the 17,000-ton Anchor Liner *Tuscania* on 4 October 1921, Sir Alexander Kennedy, referred to the world-wide slump causing shipbuilders to submit tenders with no profit and to the enormous cost of vessels such as *Tuscania*, a sum which was out of all proportion to market values.

On 24 October 1921 the Admiralty placed contracts for four 48,000-ton battlecruisers, three of them on the Clyde.[51] This was a major boost not just to the shipyards concerned but to the armaments industries that had been greatly extended during the War and which were now suffering from the cessation of war work. These large, complex and very expensive warships offered years of employment to thousands of people but, after just three weeks, on 16 November, the contracts were suspended. This was a consequence of the Washington Conference which began on 12 November 1921 between the major naval powers. The aim was to prevent yet another naval arms race, principally between the United States and Japan. The outcome was a series of restrictions across future warship construction, most notably the so called 'Battleship Building Holiday', an agreement that no battleships or battlecruisers should be built for a period of ten years, subsequently extended to fourteen. The Washington Naval Treaty was signed on 6 February 1922 and the battlecruiser contracts were cancelled three weeks later on 21 February. If the cancellation of the battlecruiser contracts was understandable and a positive act in the interests of peace, it hastened the collapse of the armaments industries in the UK.[52]

While the Anchor-Donaldson Line's order for two sister ships was placed in the last months of the war, the first, *Athenia*, was launched on 28 January 1922. The second, *Letitia,* was suspended for two and a half years and lay on the stocks until the Line's trading position improved. She was not launched until 14 October 1924.

By September 1922, employment at Fairfield had dropped to a low of just 250, 180 of whom were in the shipyard department and 70 in the engine works.[53] However, in the same month, Alexander Kennedy was able to announce that the contract with the Union Steam

Fairfield's highly skilled cabinetmakers photographed on 21 February 1922.

Built for the Union Steamship Company of New Zealand, the well-appointed *Aorangi*, seen here soon after her launch on 17 June 1924, was the largest diesel propelled passenger vessel at that time. *Aorangi* was built for service in the Pacific between Vancouver and Sydney.

Ship Co of New Zealand for the motorship *Aorangi* had finally been signed. Fitted with Fairfield Sulzer diesels, *Aorangi*, on completion, would be the largest motor passenger ship afloat.

In October 1922, the Orient Line was released from its obligation to maintain a berth reservation which it had agreed with Fairfield in February 1918. Despite the fact that no ships were launched from Fairfield in 1923, the first year in the history of the yard that this had happened, at the Annual General Meeting in October of that year, Sir Alex Kennedy was able to announce a profit of £59,105 to which £169,586 from the previous year was added. He noted that 'during the last two and a half years, the industry has passed through a period of unprecedented industrial depression.' He continued, 'Orders are few and far between and those actually taken have in most cases to be taken at unremunerative prices. In view of these circumstances and taking into consideration the difficulties which Fairfield, in common with other ship building concerns has had to face, he thought they would agree that their position as shown by the figures in the balance sheet, was very satisfactory.'[54]

The aft end of *Aorangi's* engine room showing Fairfield Sulzer single acting two-stroke diesels.

In October 1923, Sir Alex Kennedy became Chairman of Fairfield while two directors associated with the Sperling Combine, Robert A Workman and Edward McKay Edgar, resigned.

At the end of 1923, with work running out again, the Holder Line contracted Fairfield to build the refrigerated

Man and boy posing on the upper deck beams of HMS *Berwick*. The only safety measure visible is that the three planks have been dogged together to avoid slippage. The deck beams are temporarily supported by timber shores until the rest of the hull's structure is in place.

A group of shipwrights photographed on 31 October 1929. The two men at centre front are foremen, most likely Head Foreman and Assistant Foreman.

motor cargo ship, *Upwey Grange*. In 1924 the Admiralty endeavoured to spread what contracts it had been voted funds for by supporting its key private yards. This included, at Fairfield, the order for the 'County' class cruiser *Berwick* and in the following year, the machinery for the cruiser *London*, which was under construction at Portsmouth Dockyard. In December 1927, the refrigerated motorship *Taranaki* for the Shaw Savill & Albion Line was launched, at which Sir Alexander Kennedy said – incorrectly as it turned out – that he thought the worst of the depression was over. Another cruiser, *Norfolk*, would follow in 1927, but Fairfield's most prestigious ship of the decade was the 26,000-ton *Empress of Japan* for Canadian Pacific.

An intriguing painting of the launch of *Empress of Japan* on 17 December 1930. So thick was the fog that the ship disappeared soon after entering the water. She is seen here re-emerging to be taken by a tug into the fitting-out basin. Artist unknown.

Empress of Japan

Launched on 17 December 1929, *Empress of Japan* eclipsed the *Empress of Canada* as the largest passenger ship yet launched at the Fairfield yard. She joined three other Fairfield built Empresses, *Russia*, *Asia* and *Canada* on trans-Pacific routes and was the twelfth vessel built for Canadian Pacific. Launched in a thick fog, the launch party watched as the ship disappeared halfway down the slipway, with the continuing sound of drag chains the only indication that the ship had been successfully launched. Renowned for magnificent interiors and soon to be named 'Queen of the Pacific', *Empress of Japan* ran successful trials in May 1930, making a speed of 23 knots. On 7 June 1930 she departed from Fairfield for Liverpool and her maiden voyage to Quebec. Thereafter, she returned to Southampton before setting out to Hong Kong and the start of service on the Vancouver, Yokohama, Kobe, Shanghai, Hong Kong circuit.

If it appeared that, in the late 1920s, the depression in

The handsome steamship *Empress of Japan* running trials in the early summer of 1930.

Looking aft from HMS *Berwick's* fo'c'sle in the summer of 1927; final touches being applied to the all but complete cruiser. The merchantman to the right is probably Bibby Line's *Cheshire* while at left the West Yard gantry can be seen.

The party assembled to launch the cruiser HMS *Norfolk* on 12 December 1928. The sponsor was the Countess of Leicester, Alice Emily Coke. A notebook held by Fairfield noted that, as with all sponsors, the Countess of Leicester was presented with a gift. In this instance the gift comprised jewellery in a case together with two hand painted medallions – the Leicester coat of arms and the Fairfield Seal at a total cost of £128 and six shillings. The Reverend G Duncan of Govan Manse is leading a group of choir boys.

shipbuilding was lifting, it was in reality set to deepen significantly following the Wall Street crash in October 1929. This was followed by the onset of the Great Depression. On 7 April 1931, at the launch of the paddle steamer *Jeanie Deans*, Sir Alexander Kennedy said that three quarters of building berths in UK shipyards were idle. In recognition of the limited amount work on hand, on 29 April notices were posted throughout the works announcing a reduction in the hours to be worked daily. These would be 9am until 12.15pm and 1.15pm until 4.30 with no Saturday working. This was followed on 30 April with a 10% reduction in the salaries of officials and foremen. At the end of 1931, the works closed for what would be the longest ever New Year holiday, closing on 25 December at 5.30pm (Christmas Day was a working day until the 1960s) until 7.55am on 11 January 1932.[55] Short time working ended in October 1932 but was reintroduced again on 30 January 1933 until 5 March 1934 when full time working was resumed.

HMS *Norfolk* launched into the Clyde on 12 December 1928.

Rationalisation

With no apparent end to the Depression, moves were made by leading shipbuilders to reduce capacity, by eliminating shipyards which were either idle, had few orders or were financially weak. The overall intention was to reshape the industry in line with anticipated demand post-depression, giving remaining shipbuilders the chance of booking contracts at remunerative prices. The result was the formation, in 1929, of National Shipbuilders Security Ltd, under the chairmanship of Lower Clyde shipbuilder, Sir James Lithgow. Once acquired, ailing yards had their equipment and facilities sold off and fixed structures dismantled under the legal provision that shipbuilding should not be undertaken on that site for a period of 40 years. By this means, shipyards, great and small, were closed and, of a reported 684 shipbuilding berths in existence in the UK in 1930, 116 were neutralised or dismantled.[56] Casualties on the Clyde included Beardmore's large Naval Construction Works at Dalmuir, McMillan's at Dumbarton and Fairfield's West Yard, as well as Palmers' at Jarrow and the Northumberland Shipbuilding Co on the Tyne, which was sold off in January 1932.

Conceived during the optimism at the end of the First World War that spurred some shipbuilders into fresh investment, the West Yard proved to be a costly mistake. The yard lay idle for 12 years until purchased by National Shipbuilders Security Ltd in 1933 for £30,000. The submarine sheds and shipbuilding gantry, the latter completed just 13 years earlier, were demolished in 1938.

An aerial view of the fitting-out basin in the spring of 1929. From left, the Blue Star cargo liner *Arandora Star* in for repair, the heavy cruiser *Norfolk*, the Bosphorus ferry *Altinkum* and *Empress of Canada* returned to Fairfield for reconditioning (new machinery). Note the deserted West Yard.

An excellent view of the fitting-out basin in April 1930 showing the range of vessels constructed at Fairfield. From left, the cruiser HMS *Norfolk*, *Princess Joan* for service on the British Colombia coastal service, and the intermediate liner *Empress of Japan*, both vessels for Canadian Pacific, and lastly, the paddle steamer *Southsea* for the Southern Railway. *Princess Joan* is being guided into the basin by the tug *Flying Condor*.

The Bibby liner *Worcestershire* launched into the Clyde on 8 October 1930.

Although Fairfield had fared better than some UK shipbuilding firms, by 1933 the long term effects of the depression began to impact on the finances of the company, forcing it to approach the Bank of England to draw down from an agreed loan figure of £150,000. This was in addition to £10,000 loaned from the Shipbuilders Investment Company.

One of the lowest points in the history of the yard to date was reached in 1933 when the profit for the year to 31 June was £363, 4 shillings and 2 pence. To this the Company was able to bring forward £23,406 from the previous year. Nevertheless, Chairman Alexander Kennedy, in referring to this unfavourable outcome, said that '[we] have every reason to be satisfied with the result considering the prolonged and severe depression the industry [is] going through'. He noted the assistance that the Admiralty had given to the industry by placing orders where it could. He also said that, when merchant work was at a low ebb, the industry relied on naval programmes to maintain the skill levels of craftsmen, an issue which Kennedy felt was of national importance.[57] To underline the last point, he was able to announce that the Admiralty had placed an order for the sloop *Kingfisher* with the yard.

The financial predicament that Fairfield found itself in was set to deteriorate throughout 1934, despite further Admiralty orders for two destroyers, *Garland* and *Gypsy*, and the paddle steamer *Marchioness of Lorne* for the Caledonian Steam Packet Co. When the results for the year to June 1934 were announced, a loss of £22,806 was declared. Perhaps motivated by the need to recall better times, Kennedy referred to this being Fairfield's 45th Annual General Meeting since it became a public liability company. He went on to say, in words intended for shareholders, that:

> It is of interest to note that on the 30th of last month, the firm completed the first century of its existence. The record of Fairfield's work during those years is

Mrs R B S Reford with Managing Director Alexander Kennedy at her side performing the launch of Canadian Pacific's *Princess Elizabeth* on 16 January 1930. Mrs Reford was the daughter of W R McInnes, Vice President of the Canadian Pacific Railway Co Ltd. Mrs Reford was presented with a diamond and platinum broach as well as the mallet used to launch the ship.

> a matter of history. Fairfield has always been in the forefront of all scientific and technical developments in connection with shipbuilding and engineering and we take pride in the valuable contribution which the firm has made to these industries and to the shipping trade of the world. Your directors are deeply aware of the rich heritage of achievement to which they have fallen heir, and no effort on their part has been or will be spared in order to maintain and if possible, enhance its great reputation.

On a brighter note, work at Chepstow seemed to be improving throughout 1934 with contracts placed for pontoons at Gosport landing stage, a bridge for LNER's St Helen's branch line, work at Fenchurch Street Station, a gas holder for South Molton in Devon and bridge renewals for Great Southern Railways in Ireland.

If the events of 1934 had brought the Company to its lowest ebb so far, then worse was to follow in 1935 when the Anchor Line (Henderson Brothers) went into liquidation and was thus unable to honour bills totalling £145,000 owed to Fairfield.[58] As the funds available at the Bank of England could not be used to discharge these bills and as Fairfield was at the limit of its overdraft, the Company was plunged into crisis. It was at this point that industrialist Sir James Lithgow and his brother Henry, in agreement with the Bank of England, acquired a controlling interest in Fairfield by buying a substantial number of ordinary shares and taking over its overdraft, which then stood at £295,000.[59] Or, as the minutes of November 1935 put it: 'In order to maintain continuity, Sir James Lithgow and Henry Lithgow have undertaken to satisfy the claim of the bill holders.' At a stroke, the fortunes of the Company were restored and the Lithgows had added a famous, if

The three crank diagonal triple expansion engines for the paddle steamer *Mercury* in the engine shops at Fairfield. *Mercury* was built for the London, Midland & Scottish Railway Co, and completed in February 1934. Unlike turbine and diesel machinery, steam engines of this type were a marvel to see in action.

Marchioness of Graham soon after launching under the giant cantilever crane.

ailing, name to their shipbuilding empire. The rescue of Fairfield was in good time, as the long depression that had gripped the country was ending and rearmament about to kick in. Nevertheless, in June 1935, at the launch of the Bibby liner *Derbyshire*, Alexander Kennedy said that only 32% of building berths were occupied and 44% of men in the shipbuilding trades were unable to obtain regular employment. Fairfield's shipyard diary records details of the naming ceremony of *Derbyshire* and subsequent lunch for 152 people held in the Ship Drawing Office. The ship was named by Lady Stephenson, wife of Sir Hugh L Stephenson, Governor of Bengal. The purvey was provided by the Grosvenor, Glasgow at a cost of £105 while Lady Stephenson was presented with a 'diamond, single stone and platinum bar broach in a tortoiseshell and gold case' purchased from Edwards, Glasgow, at a cost of £140.[60]

Over the next few years, work picked up across the range of ship types with which Fairfield was well familiar, including paddle steamers, cargo liners, destroyers and cruisers. The engines and boilers for all of these vessels were manufactured at Fairfield, together with the machinery for ships building or being reconditioned elsewhere such as the main and auxiliary machinery and boilers for the battleship *Queen Elizabeth* being rebuilt at Portsmouth Dockyard.

Engaged to Capacity

On 15 February 1937, with the 'Battleship Building Holiday' now at an end, Fairfield submitted its tender to the Admiralty for the construction of a *King George V* class battleship of which five were to be built. The first two, *King George V* and *Prince of Wales* were laid down at Vickers-Armstrong's naval yard on the Tyne and at Cammell Laird's Birkenhead yard. As was so often the case with major warships, Fairfield and John Brown each received the contract for one. Fairfield's ship, the tender for which was accepted on 28 April 1937, would be in the second group[61] which were ordered under the 1937 programme. Originally named *Beatty* after the late Admiral of the Fleet, the battleship's name was subsequently changed to *Howe* after

The bar on board the Anchor liner *Circassia* completed in 1937.

the eighteenth century Admiral of the Fleet. The original construction dates specified launch in 1939 and completion in January 1941 but this would prove to be wildly optimistic, particularly after the Second World War began, with the immediate requirement for material and industrial capacity that would dislocate shipbuilding programmes.

Designed by the Royal Corps of Naval Constructors in London under the control of the Director of Naval Construction (DNC), at this time Stanley Goodall, once the building drawings for the battleships were ready to be passed to the shipbuilders, on 3 May 1937 an official from the DNCs office was sent with the plans by train from Euston to Glasgow Central, where they were received by officials from the John Brown and Fairfield shipyards. Thereafter, the drawing offices of the respective shipyards would develop the Admiralty drawings while the first contracts for long-lead items were drawn up.

In June 1937 the turnaround of Fairfield was complete, with profits declared at £223,678 plus £11,938 brought forward from the previous year, making a total of £235,617. That year was Sir Alexander Kennedy's last year as Managing Director with his retirement taking effect in March. In January of the following year he also resigned as Chairman of the Fairfield Board to be succeeded by Sir James Lithgow.

In February 1939, the contract was placed for another, very large warship, the fleet aircraft carrier *Implacable*. This brought the order book to: 1 battleship (*Howe*), 1 fleet aircraft carrier (*Implacable*), 1 troopship (*Devonshire*), 2 cruisers (*Liverpool* and *Phoebe*), 2 destroyers (*Juno* and *Kelvin*), the caisson for No 7 Dock at Portsmouth, and the machinery for the reconstructed battleship *Queen Elizabeth*. Interestingly, the Fairfield minutes for April 1939 note that the Company was approached by the American aircraft manufacturers, Consolidated Aircraft, to take out a licence to manufacture their Consolidated Model 28 'airboat', an amphibious flying boat which became the famous Catalina. Although Shipyard Manager and Director George Barr was in favour of taking a licence, the minutes record that, after a discussion, it was decided to decline the offer.[62]

HMS *Liverpool* enters the Clyde past the abandoned and soon to be demolished West Yard on 24 March 1937. The armoured belt has already been fitted.

The Shipyard Diary in addition to noting ship movements also, on occasion, noted various dates concerning people associated with the Company, both past and present.

Saturday 22 January 1938:	Sir James Lithgow's first day at Fairfield as Chairman of the Company.
Saturday 14 January 1939:	Sir Alex M Kennedy, late Chairman of Fairfield Co died at Culzean Park, Gosforth, Northumberland.
6 February 1939:	Reported at Works this date - Mr H E Skinner RCNC, constructor to be principal Ship Overseer, HM Ship *Implacable*, destroyers and HM Caisson.
7 February 1939:	Mr W Blakey, Overseer on TBDs left Fairfield this date.
9 February 1939:	Miss J Wright, Head Engine Tracer, died this date (am) 44 years of service with firm.
8 March 1939:	The Earl of Elgin & Kincardine, KT, CMG, LL D appointed a Director of the Company at Board Meeting held this date.
20 March 1939:	Mr Layton of Admiralty engine overseers staff left Fairfield this date.

By the outbreak of the Second World War, the Fairfield works and its subsidiary at Chepstow were 'engaged to capacity', a comment recorded in the minutes for October, when a gross profit to June 1939 of £286,608 was declared.[63]

The 'J' class destroyer HMS *Juno* running trials on the Firth of Clyde in the summer of 1939. *Juno* was sunk by Italian aircraft off Crete on 21 May 1941.

The Second World War

In the immediate run up to the outbreak of war, the bulk of the work contracted at Fairfield reflected the grim international outlook and was comprised almost exclusively of warships. In the summer of 1939, the last merchant vessel, the troopship *Devonshire* for the Bibby line, sailed on completion for Dartmouth on 5 July.

The declaration of war played havoc with the building programme planned at Fairfield when priorities, understandably, changed in reaction to events at sea. This was brought home almost immediately with the sinking of a Fairfield built cargo passenger ship. Just a few hours after hostilities began on 3 September, the Donaldson Atlantic liner *Athenia* was sunk 200 miles off the northwest coast of Ireland with the loss of 117 lives. En route from Glasgow to Montreal carrying out a scheduled service, *Athenia* was struck abreast the engine room on the port side by a torpedo fired from *U30*, commanded by Lieutenant Fritz-Julius Lemp. *Athenia* was the first ship to be sunk during the Second World War.

As was the case in the previous world war, the shipbuilding industry would be pressed to the limit to replace merchant vessels lost at sea while also building the types of warship most urgently required. One other demand placed on shipbuilders was the need to repair ships damaged as a result of enemy action and also to convert merchant vessels to other roles. Despite Fairfield's exclusively warship building programme, the first months of the war saw a final rush of merchantmen going to the yard for conversion to Armed Merchant Cruisers as the following typed list made at the time describes:

> SS *California*, arrived at Fairfield on 28 August, shifted to Yorkhill Basin on 11 September where the work was done. Sailed 21 September 1939.

Further repairs were carried out and completed on 9 November.
SS *Transylvania*, arrived at Fairfield on 11 September and shifted to Yorkhill Basin on 27 September and later back to Fairfield, sailing on 5 October.
SS *Montrose*, arrived at Fairfield on 12 September for conversion into an Armed Merchant Cruiser. Renamed HMS *Forfar*. Work completed 5 November and sailed on 6 November.
SS *Cameronia*, berthed at Yorkhill for fitting-out as a DEMS [Defensively Equipped Merchant Ship] by Fairfield.
MV *Derbyshire*, arrived at Fairfield on 15 October for conversion into an Armed Merchant Cruiser, sailed on 4 December.
MV *Circassia*, arrived at Fairfield on 17 October for conversion into an Armed Merchant Cruiser, sailed 21 December.[64]

In his book *Merchantman Rearmed*, first published in 1949, Captain Sir David Bone recalls the hasty conversion of the Anchor liner *Transylvania* at Fairfield, of which he had been captain since the ship had been completed 14 years earlier.

> Back in the same shipyard, the same basin in which when new she was polished up for her maiden voyage, I attended to the dismantlement and removal of stores, fittings and furniture that were considered unnecessary in war service. In the haste of emergency, that could not be done gently or with the regard due to the fine furniture or fabric, I was not happy in recalling the care with which these were originally installed and the pride we had in them

The 'O' class destroyer *Oribi*. Originally to have been called *Observer*, the name was changed after the South African Government sponsored construction of the ship, Oribi being the name of a South African antelope.

> throughout the years. But speed in wartime is the urgent need and I had to let the pickaxe, jackscrew and crowbar undo all the interior fitments and panelling once so tenderly built up by prideful craftsmen.[65]

In September 1939 additional Admiralty contracts were taken for a *Lion* class battleship (*Thunderer*), the *Dido* class cruiser *Bellona*, and the destroyers *Oribi* and *Offa*. By December 1939, with the destroyers *Juno* and *Kelvin* having been completed, the order book stood at:

> 2 x battleships, *Howe* and *Thunderer*[66]
> 1 x fleet aircraft carrier, *Implacable*
> 2 x cruisers, *Phoebe* and *Bellona*
> 6 x destroyers, *Napier*, *Nestor*, *Musketeer*, *Myrmidon*, *Oribi* and *Offa*

This significant workload at Fairfield was increased in March 1940 when two further destroyers, *Redoubt* and *Relentless*, were booked. In June the Admiralty told Fairfield to suspend work on *Howe* and *Bellona*. The reason given was the general shortage of materials and the need to

The 'Hunt' class escort destroyer originally to have been named *Haldon*, was transferred to the Free French Naval Forces in 1942 and named *La Combattante*.

The 'O' class destroyer HMS *Onslaught*, completed in the summer of 1942, seen here off Port Glasgow. Later transferred to the Pakistan Navy, she survived until 1977 when scrapped.

accelerate other, more urgently required contracts. However, the workload was set to increase again when, in July, eight landing craft were contracted for, together with machinery and shafting for spares, while in August, four Landing Craft Tank [LCTs] and two 'Hunt' class escort destroyers, *Goathland* and *Haldon*, were added to the list. The LCTs, first of a new type, were substantial craft, each capable of carrying three 40-ton tanks. In August, the Admiralty asked for work on *Howe* and *Bellona* to resume while, in December, the destroyer *Napier* was completed and commissioned into the Royal Australian Navy.

The consequences of such a heavy work load soon became apparent and, by the end of 1940, Admiralty concerns grew about the slow progress being made on various contracts. The evidence for this comes from reports in the minutes of John Brown & Co at Clydebank, as the Fairfield minutes remain silent on the subject. According to these reports, Fairfield had taken on work equivalent to John Brown's yard at Clydebank but, with a smaller workforce, Fairfield was making slower progress.

On 27 January 1941 Claude Hanniford, the Admiralty's Director of Naval Contracts, visited both Fairfield and Clydebank to see how Fairfield's workload could be lightened. The outcome was the transfer of the destroyers *Redoubt* and *Relentless* to Clydebank, with the suggestion that the cruiser *Bellona* might also be transferred, although this idea was dropped.

Difficulties for Fairfield continued as a result of the bombing of Clydeside by the Luftwaffe on the nights of 13 and 14 March 1941. At least one bomb landed in the yard at the head of No 2 berth, while a parachute land mine had landed at the bottom of No 1 berth without exploding.

John Bowen, a draughtsman in the yard working in the destroyer section, gave the following eye witness account of the scene in the yard the morning after the air raid:

> An Australian destroyer was in the fitting out basin at the time and when daylight came, we had to keep the yard gates shut, which caused a riot in the main street, with hundreds and hundreds of men being kept from getting to their work. By that time, senior managers had taken over, and I was walking around the yard with an Australian officer who said, 'did you see the landmine that came down?,' I said 'No, where is it?' 'Come here' – the aircraft carrier, *Implacable*, was on No 1 berth and at the bottom of the berth, three or four landing craft were being built. Then we saw a parachute draped over the landing craft, and I said 'Well, where's the landmine?' and he said 'there,

Men coming off the battleship HMS *Howe* at the end of a shift in August 1942.

A view of the quarterdeck of HMS *Howe* beneath 'Y' quadruple 14-inch gun mounting in the final stages of fitting-out. In the foreground at left, is the stepped interior of a 20mm Oerlikon gun tub. Note the rivet fire in the centre of the photo. In the far distance Alexander Stephen's yard can be seen at the left and Barclay Curle's yard at right. (*Imperial War Museum*)

> down there!' and pointed to where it was, under the parachute – obviously it hadn't gone off so I said 'Oh, thanks, I think we should retire.' But a lot of that parachute ended up with the girls in the tracing office, it was lovely silk.

John described the formation of the Fairfield Home Guard which was attached to the 5th battalion of the Highland Light Infantry. Over 300 men were part of the unit under the command of one of the Directors. They were issued with 30 Lee Enfield rifles and live ammunition for those on night guard duty. The basement in the main office building was designated as an air raid shelter, suitably equipped with chemical toilets. The one practice run in which everyone in the building was obliged to descend into the basement had ended in chaos because of the very restricted space available.[67]

As in the First World War, Admiralty financial support was available to shipbuilders to ensure that plant and

With three tugs ready at her stern and two at her bow, the fully camouflaged *Howe* is carefully eased out of the fitting-out basin, stern first, and into the river channel. (*Imperial War Museum*)

facilities were maintained at the highest level of efficiency. In 1942 a report on the equipment of shipyards and engine works was prepared by Cecil Bentham for the Admiralty to advise them on the allocation of funding. Bentham visited each yard in turn, making observations not entirely restricted to equipment as his comments about Fairfield show:

> These works were very extensive and considerable renewals had been made in the engine works, but there was still room for much improvement; a number of lathes could well be renewed. Gear joggling was a speciality, but the machines had too much work for them to cope with.

He noted that in the shipyard there was a variety of work from aircraft carriers to destroyers. In 1941, five cranes had been destroyed by accident and this had created difficulties; temporary cranes were used, but one 10-ton crane and one 20-ton crane and two small run-about cranes were required.[68] However, Bentham also noted that Management was energetic but lacked a modern outlook and were more inclined to complain than act.[69]

Having been turned at the King George V Dock, *Howe* is now facing down river and on her way to the sea.

HMS *Howe* at speed during trials in August 1942.

The *Dido* class cruiser *Bellona*. When construction was halted for a period, the cruiser was completed to a modified design. She is seen here on completion in October 1943.

Despite intense activity on naval contracts, thoughts of post-war merchant contracts surfaced in November 1942, when the possibility of entering into an agreement with Canadian Pacific Railroad arose. For obvious reasons, it depended on the successful outcome of the war which had still to be determined, so the Fairfield Board struck a cautious note, considering it to be unwise to enter into a formal agreement with CPR at this stage, but instead suggested an exchange of letters should take place along the lines of CPR wish to record that it is their intention to employ the Fairfield company for building new ships and recondition old ships after the war to the utmost extent possible and that the Fairfield company will agree to make every endeavour to carry out such work having regard to the company's general position as effectively as possible and that the company will do everything they can to meet the Canadian Pacific Railway Company as regards terms of payment.

If not a formal agreement, these sentiments were nevertheless an optimistic view of the eventual outcome of a war in progress, as well as a reaffirmation of the strong links between CPR and Fairfield.

The commissioning ceremony for the fleet aircraft carrier *Implacable* under way in Fairfield's basin on 28 August 1944. The band of the Royal Marines are playing with just some of the ship's company in attendance on the flight deck. *Implacable*'s full complement was 2300.

Implacable passing Renfrew as it makes its way down the River Clyde from Fairfields. (*Imperial War Museum*)

Implacable moored at the Tail of the Bank following trials as evidenced by the deteriorating paintwork. The Firth of Clyde was exceptionally busy during the war years. In this view, an escort carrier is astern of *Implacable* and a large merchantman off to the left with various other vessels in the distance. (*Imperial War Museum*)

Aircraft carriers

Fairfield's first aircraft carrier was the fleet carrier *Implacable*, laid down on 21 February 1939 and subject to a protracted building period from then onwards, including a period of suspension in favour of more urgent work. She was launched by HRH Queen Elizabeth, with the King in attendance, on 12 December 1942, and commissioned in Fairfield's basin on 22 May 1944.

The second carrier built at Fairfield, the *Colossus* class light fleet *Theseus*, was ordered on 7 August 1942, laid down on 6 January 1943 and not commissioned until February 1946. However, two other aircraft carriers were scheduled to be built at Fairfield when, in August 1943, contracts were placed for the light fleet carrier *Monmouth*, one of eight *Hermes* class and for the fleet carrier *Africa*, one of four *Malta* class. Neither were destined to be laid down.

The pressure of work on Fairfield meant that *Monmouth* could not be laid down until some time in 1944, circumstances that would also apply to *Africa*.[70] In May 1944 Fairfield was told by the Admiralty to suspend work on *Africa* with the exception of a test boiler.[71] As happened with other Admiralty building programmes, it was a step too far. The intense nature of wartime building programmes, competing demands and changing priorities, as well as indecision about the design of the *Malta* class, proved more than the UK shipbuilding industry could cope with. Both of Fairfield's carrier contracts were cancelled outright on 29 October 1945.[72] Had *Africa* gone ahead, it would have been the largest ship so far built at Fairfield.[73]

In general, the labour unrest that had erupted on Clydeside during the First World War was not repeated during the Second. Fairfield had again been declared a Controlled Establishment and while there were disputes, there was significantly less disruption. When hostilities with

The bow of the light fleet carrier *Theseus* in Govan dry dock for a hull inspection.

A broadside view of *Implacable* during trials on the Firth of Clyde. (*Imperial War Museum*)

Germany ended at midnight on 7 May 1945, the Works closed for two days in celebration, the 8th being VE Day.

The radio played an important part in raising morale during the war years, a good example being 'Workers Playtime' which was broadcast from various factories across the UK. On Thursday 30 August 1945, the broadcast was made from Fairfield's canteen, with Dave Willis, Jimmy Plant, Betty Frankiss and Sonny Farrar performing, and George Myddleton at the piano.

The end of the war saw the immediate cessation of naval contracts, leaving shipbuilders with Admiralty work at various stages of completion. Three 'Battle' class destroyers, all of which had been laid down, were either floated away to be broken-up or dismantled on the slip. The cruiser *Blake* was launched on 20 December 1945 and laid up, incomplete, firstly at Dalmuir, and then at the Great Harbour, Greenock.

Work had advanced on the superstructure of HMS *Blake* before work on the cruiser was stopped, as seen here in this photo which appears to show the ship being brought back to Fairfield after launching.

Fairfield's contribution to the war effort was listed by the *Glasgow Herald* in 1948 as:

1 battleship
2 aircraft carriers
3 cruisers
21 destroyers
16 landing craft
2 sloops
2 transport ferries
1 cargo vessel

At Chepstow a return was made to shipbuilding, where 61

Taken towards the end hostilities, in May 1945, this view across the fitting-out basin shows the light fleet carrier *Theseus* and three destroyers, possibly *Wessex*, *Camperdown* and *Finisterre*, in various stages of completion. The stern of an LST (Landing Ship Tank) can be seen at right of shot.

tank landing craft were built during the Second World War.

See Appendix D for details of wartime output.

The immediate future in which far fewer naval ships would be ordered was compensated for by the resumption of merchant work which, after 1945, began well, not just with new orders but with the reconditioning of ships which had had an arduous war either as troopships or armed merchant cruisers.

Fresh Frozen Foods Ltd 1947

Perhaps one of the most unusual episodes in the long history of Fairfield is its association with a company called Fresh Frozen Foods Ltd. The idea behind the company as it was described at the time was that: 'Fish caught will be gutted, cleaned, filleted and quick frozen. Blocks of fillets – brick hard – will be slipped into cellophane bags, sealed by machine, and taken by conveyor belt to the ship's refrigerated hold.'[74] The fillets would be sold directly to agents and not on the open market. It all sounded rather futuristic.

The vessel to be used to catch the fish was the *Algerine* class minesweeper HMS *Felicity*, which, being surplus to requirements, was sold by the Navy in 1946 to be converted at the Fairfield shipyard into a trawler. This would be fitted out with fast freezing equipment and cold storage, sufficient to hold 230 tons of fish. The ship arrived for conversion in Fairfield's basin on 11 August 1946, leaving with a new name, *Fairfree*, on 11 December to sail to Ardrossan, where Fresh Frozen Foods Ltd was headquartered. The idea for this venture possibly began with Sir Charles Dennistoun Burney, a former naval officer and inventor of the Paravane, who devised much of the equipment on *Fairfree*. There were two other principal backers for the idea in addition to Burney, Lord Wooton, Chairman of the Conservative party and Sir James Lithgow, Chairman of Fairfield. Lithgow used Fairfield to convert the minesweeper into a refrigerated trawler and also to buy 250,000 shares of the Company in a deal struck on 22 October 1947. The idea, billed as revolutionising deep sea fishing in Britain, was ahead of its time. The intention was that, if *Fairfree* was a success, she would be followed by a series of other converted minesweepers.

A series of trials were conducted to investigate the practicality of the idea, with department store Lewis's Royal Polytechnic in Glasgow intended as the destination, among others, where the frozen fish would be sold. However, by June of 1948, the enterprise had run out of money and the concept written off as a failure. *Fairfree* was sold and Fairfield disposed of its shareholding at the end of 1948 to Christian Salvesen of Leith at a significant loss.

In the post-war world, it seemed as if nothing much had changed. Fairfield had a healthy order book despite numerous cancellations of warships ordered under war programmes. These were easily compensated for with orders from Anchor Line, Hudson Bay Company, Bibby, Blue Star, P&A Campbell, the reconversion of passenger cargo ships *Cilicia*, *Llanstephan Castle*, *Empress of France*, and *Circassia*, as well as various machinery contracts. Profits at Fairfield remained high, with £306,641 for the

The Fairfield Doxford diesels in the engine works about 1947. Developing 11,400bhp in total, they were for the Anchor Line's passenger cargo ship *Caledonia*.

The paddle steamer *Cardiff Queen* being prepared for launch which took place on 25 February 1947. *Cardiff Queen* was built for P&A Campbell's excursion service on the Bristol Channel.

year ending in June 1947. Both Fairfield and the Chepstow works were fully employed. Similar figures emerged for the following year, with profits: shipyard £291,627, engineering £73,206 and Chepstow £71,426. Profits were proportionally similar to Fairfield's Clydebank competitor, John Brown & Co, who, with a larger yard, reported average profits for the years 1946–1960 of approximately £450,000.[75]

The broader picture showed that, in 1947, British Shipbuilding accounted for over 50% of world shipbuilding contracts. However, by 1950, this had slipped to 38%. A defining feature of shipbuilding in Britain over coming decades was its static output, not declining but not increasing, set against a rapidly expanding market for ships, resulting in an overall loss of market share.

Another factor weighing against shipbuilding was not just the abrupt end to warship contracts in 1945 but the long term decline in the size of the Royal Navy. In late 1947, with an estimated 520 warships laid-up, the shipbreaking industry was the main beneficiary.

In June 1947 the Government published a White Paper, 'Industry and Employment in Scotland'. This showed that in 1946, Scottish shipyards had produced 38% of the UK total with 1,150,000 tons on order and 47,200 persons employed.[76]

Good causes

Like all major companies, Fairfield received regular, almost monthly, requests for financial support from local and UK wide organisations. Board minutes recorded the following donations made by Fairfield in September 1949:

The Glasgow City Mission	£10
The Earl Haig Fund, Glasgow	£100
Glasgow Old People's Welfare Committee	£25
Scottish National Institution for the War Blinded, Glasgow	£50

Sir Harry Lauder in the kilt at the launch of the Canadian Pacific refrigerated cargo vessel *Beavercove* on 16 July 1946. Originally part of a four-ship order placed in 1944, three were reallocated to sister company Lithgows because of Admiralty work at Fairfield. When *Beavercove* entered the water, the first merchant ship launched at Fairfield since the end of hostilities, Sir Harry Lauder sang his signature tunes 'Roamin' in the Gloamin' followed by 'Keep Right on to the End of the Road'. The sponsor was Mrs J C Patteson, wife of J C Patteson, European General manager of CPR. Sir James Lithgow is at far right.

Convalescent Seaside Homes, Dunoon	£50
Scottish Physiotherapy Hospital	£50
Eastpark Home for Infirm Children	£50
Glasgow Kilmun Convalescent Home	£50
Glasgow and District Coast Homes, Saltcoats	£30
Orphan Homes of Scotland, Bridge of Weir	£100
King George's Fund for Sailors	£100
The Navy League, London	£100
Economic League, Glasgow	£100

Modernisation

The war years had emphasised the importance of welded prefabrication, something for which British yards were not yet well adapted, the rationale for plant layout and facilities being concentrated around traditional riveted construction. By the end of the First World War, the Fairfield yard had been in the forefront of technical change by, for example, installing tower cranes of up to 12 tons lift at the building berths and up-to-date machine tools in the steelworking and engine shops. While these arrangements were entirely adequate for riveted constructions where relatively light loads were the norm, welded fabrication enabled units of up to 40 tons to be assembled in fabrication sheds before being taken to the berth for erection by new berth cranes. The berth cranes were only one element in a shipyard organised for welded fabrication, however, as new and much larger sheds with high capacity cranage, were

George Barr, Deputy Managing Director of Fairfield during the Second World War until made Managing Director in 1944. He was knighted in 1951.

required, together with new plant, etc. It would entail a root and branch approach to shipbuilding from drawing office to fitting-out the completed ship.

Because of the financial traumas of the 1920s and early 1930s, the Fairfield yard that entered the Second World War was little different from the one that had emerged at the end of the First. Now, however, the need for modernisation was pressing as the British industry began the changeover to welded fabrication. The first mention of the need to modernise appears in the Minutes for February 1950, when George Barr, Managing Director, said that plans were in preparation.

The late 1940s and early 1950s had also seen an increase in demand for tankers, a type of ship for which the Fairfield yard was not ideally suited. Tankers consisted mostly of steelwork and a minimum of fitting-out, the latter an element in which Fairfield excelled.

In July 1951, a return to warship construction was made when the Admiralty placed an order for two Type 61 aircraft direction frigates, *Chichester* and *Lincoln*. That month however, despite a healthy order book, the Company was forced to borrow £350k from Lithgows Ltd, a member, like Fairfield, of the Lithgow Group, in order to finance the construction of the tankers *Bomi Hills* and *Enduro*. This situation had arisen because stage payments for ships under construction were not expected for several months, leaving a shortfall in working capital.

There were other reasons for the change in the manufacturing climate. In 1956/7, the closure of the Suez Canal through which much of Europe's oil was shipped, had the effect of accelerating the increase in the size of tankers. Having to navigate round Africa encouraged economies of scale, giving rise to tankers of ever increasing capacity.

The Bibby cargo passenger ship *Warwickshire* after launching on 14 August 1947. She and sister ship *Leicestershire*, also constructed by Fairfield, were built for service to Burma (Myanmar).

The BP tanker *British Realm* launched by Mrs Hector McNeil on 7 June 1951. Fairfield's Chairman, Sir James Lithgow, is by her side. Mrs McNeil was presented with a diamond star broach and case costing £267.

Modes of transport were changing too, where the passenger ship and passenger cargo types in which Fairfield had specialised gave way to air travel. Warship contracts, for long an important part of Fairfield's diet, were becoming more competitive, because the post-war Royal Navy was significantly smaller, with fewer contracts being pursued by many shipbuilders.

During the last half of the 1950s, the trading environment for shipbuilders in the UK had changed significantly from the years after the end of the Second World War, when orders for ships were relatively easy to win. The main issue for Fairfield and other shipbuilders in the UK and Europe was the rise of shipbuilding in Japan, where high levels of productivity, coupled to a completely different approach to management-worker relations and attitude to the workplace resulted in ships built at a price that European yards could not match. In 1957, Japan toppled the UK as the world's leading shipbuilder. The initial response by UK shipbuilders was to accept contracts with very little profit and, on occasion, with no profit at all. Many of these contracts were 'fixed price', which left them vulnerable to rising costs, principally wages, but also materials. Increasingly, credit terms had to be offered to ship owners and often over lengthy periods. This required financing and special arrangements being put in place with banks. As a member of the Lithgow group of companies, Kingston Financial Services (Clyde) Ltd, was formed in 1960 with £7m agreed by the Bank of Scotland to allow member companies, principally Fairfield and Lithgows Ltd, to finance construction of vessels where extended terms had been offered to owners.

The keel laying ceremony for the Canadian Pacific liner *Empress of Britain*. Heavy rain spoiled what should have been a great public relations event. Managing Director, James Lenaghan, at right, and his Canadian Pacific counterpart stand on a keel plate as a rivet is driven home by a hydraulic riveting machine. Note the two men underneath the plate as a drenched James Lenaghan looks on with a rather forced smile.

Empress of Britain on No 1 berth shortly before her launch. Note the funnel under construction at bottom right. Yard modernisation is underway with new Butters Brothers Monotower luffing cranes replacing the tower cranes.

HM Queen Elizabeth looks up at the bow of *Empress of Britain* accompanied by Managing Director, James Lenaghan, at left and Chairman, Lord Elgin behind.

The funnel for *Empress of Britain* being lifted from the quayside to the ship by the 200-ton cantilever crane.

The *Empress of Britain* makes her way down the River Clyde in 1956 led by the tugs *Chieftain* and *Brigadier* at her bow. She would be the last ocean liner built by Fairfield. This ship served under a series of names until beached at Alang, India, for scrapping in 2008.

Henceforth, the issue of how contracts would be financed became the defining characteristic of the UK shipbuilding industry. Market conditions were tough and set to become much tougher. Fairfield had the additional burden of a major shipyard modernisation programme underway that would cost £4m on completion.

However, many yards had modernised, some more so than others. Despite the improvement which new plant and processes brought, the fundamental issue of low productivity remained. During the 1950s, modernisation schemes were put in hand in many British shipyards. Fairfield was among those who pursued a root and branch approach where everything from steel entering the yard to the completion of the ship was changed from a yard optimised for riveted construction to all-welded fabrication.

Launched on 15 December 1955, the troopship *Oxfordshire* emerges from a thick fog as she is brought into the fitting-out basin. Another long-lived Fairfield ship, *Oxfordshire* lasted until 1997.

The commissioning ceremony for the Type 61 frigate *Chichester* underway in Fairfield's fitting-out basin. Note the awning spread on an unusually hot day in Glasgow or, perhaps more likely, to keep out the rain.

The 37,000dwt BP tanker *British Energy* at the Tail of the Bank in 1958.

The Type 61 aircraft direction frigate *Chichester* during trials on the Firth of Clyde in the summer of 1958.

Built for Nordic Tankships A/S, jointly owned by BP and the Danish East Asiatic Co, the 35,000dwt *Nordic Hawk* is seen here in deep load during trials at the end of 1958.

The Early 1960s

Against this background, a major contraction of the shipbuilding industry in the UK began. Between 1962 and 1964, William Gray & Co's yard at West Hartlepool, Wm Denny & Company at Dumbarton, Simons-Lobnitz Ltd, Renfrew, Harland & Wolff at Govan and The Blythswood Shipbuilding Co at Scotstoun all closed voluntarily. As shipbuilding historian Hugh Murphy has said, 'all these yards were spatially constrained, and one could, and should, view these closures as entirely rational business decisions given the general increase in size of ships and the intensity of international competition.'[77]

However, in 1960, these difficulties were still to present themselves. At the end of 1959, Fairfield declared profits of £478,625 for the shipyard, £334,174 for the engine works and £134,645 for the structural steel business at Chepstow. The 50,000-ton deadweight tanker for British Petroleum, *British Guardsman*, had been laid down, nine ferries for service on the Bosphorus at Istanbul had been booked at £3.24m, a staff bonus was arranged for payment at Christmas, while a patent application between Fairfield and Mitchell Engineering Ltd was filed. This patent related to a nuclear powered propulsion system. Fairfield was also in discussions with Combustion Engineering Inc, a leader in nuclear steam supply power systems in the US. The Ministry of Transport was keen to encourage the construction of a British nuclear powered merchant ship and, although nothing would come of it, these were Fairfield's preliminary endeavours in this area, for which the Ministry of Transport would ultimately pay the Company £50,000.[78]

Phase two of the shipyard reconstruction had been completed by 1960, while phase three, costed at £847,000 and lasting for three years, was about to start. The design consultant for the new shipyard layout was Glasgow based Babtie Shaw & Morton Ltd. The steel work for the new Fabrication shops running across the head of the building berths was manufactured in house at Fairfield's Chepstow Works. An 80-ton berth crane of a new type, costing £175,000, was ordered from Babcock & Wilcox at Renfrew, while a £112,000 Asquith horizontal milling and boring machine was purchased for the engine works.

HMS *Blake*

HMS *Blake*, a cruiser of the *Tiger* class, was laid down on 17 August 1942 and launched on 20 December 1945 after a leisurely building time, indicating more pressing work elsewhere. She was launched by Lady Blake, wife of Vice-Admiral Sir Geoffrey Blake. The Earl of Elgin, the Lord Provost of Glasgow and Sir Harry Lauder, who sang two songs, were also present on the launching platform. After an equally slow programme of fitting out, work on *Blake* was suspended in September 1947 and the cruiser was taken down river to the Great Harbour, Greenock, to be laid up. In 1953, she was moved to the Gareloch where, in 1954, a contract for her reconstruction to an entirely new design was placed. The superstructure was removed and she was brought back to Fairfield for completion. The boilers and steam turbines were retained. Contractors trials did not take place until October 1960. Fairfield's element in the final cost of this ship was £7.15m.

Two of the Bosphorus ferries, *Turan Emeksiz* and *Ihsan Kalmaz*, side by side on No3 berth in July 1961. Harland & Wolff's Govan shipyard can be seen in the distance.

Opposite: Watched by shipyard workers, the commissioning ceremony for HMS *Blake* taking place on 18 March 1961 at Fairfield's fitting-out basin.

The cruiser HMS *Blake* in the fitting-out basin in 1960 showing the bridge and forward 3-inch gun. The cable layer *Alert* is at right. Originally launched in 1944, and laid up for over ten years, the cruiser retained her original steam machinery but was otherwise rebuilt to a new design.

The departure of *Blake* saw a reduction in those employed in the fitting-out trades, especially electricians. By the end of 1960, Fairfield had begun the process of acquiring shares in David Rowan & Co Ltd, marine engineers based in Glasgow, and part of the Lithgow Group, as the first step in integrating that business with the engineering side of Fairfield. With that in mind, Fairfield approached the US based Westinghouse to acquire a licence to manufacture their steam turbine machinery. Fairfield's overdraft with its bankers, the Bank of Scotland, stood at £500,000.

An engineer sits on the cruiser *Blake*'s outer port shaft bracket while the ship was in dry dock at Govan for a hull inspection.

Opposite above: A view over the fitting-out basin in 1961 showing the cruiser *Blake* together with four of the Bosphorus ferries and the cable ship *Alert*.

Opposite below: *Blake* is manoeuvred out of the fitting-out basin into the River Clyde on her final departure from the shipyard on 18 March 1961.

C99

In 1962, the Admiralty was about to place orders for a second batch of 'County' class guided missile destroyers, while Managing Director, James Lenaghan, was off to Israel to discuss a potential contract with the Israeli shipping company Zim which, he told the Board, would be 30% of cost paid during construction, with the remaining 70% paid over seven years. At the same time, Cunard was inviting tenders for its replacement *Queen Mary*, known then as the *Q3*. Nothing would come of the Israeli enquiry, while Cunard abandoned *Q3* in favour of the *Q4* a few years later. More shares in David Rowan were acquired at £323,400, the refit of the tanker *Elisabeth* for Petrofina was completed, two passenger cargo ships for Bibby Line were ordered, *Glenogle* and *Glenfalloch*, and a tender successfully submitted to the Admiralty for a logistic ship, *Sir Lancelot*.

On 11 February 1963 a new engineering company, Fairfield Rowan Ltd, was established, merging the interests of David Rowan & Co with the engine department of Fairfield. Henceforth, engine manufacture would be concentrated at Fairfield and the Rowan premises at Elliott Street, Glasgow, would eventually be sold off.

In 1963, concern about the plight of the shipbuilding industry prompted local Scottish Television news to record a 15 minute piece at Fairfields. A number of employees were interviewed, most significantly, Derek Kimber, Deputy Managing Director, who gave a two minute interview in which, in one take, succinctly covered the issues facing the industry. It is reproduced here in full.

> No British shipbuilder would pretend the immediate future is bright. Our best customer, which is a British ship owner, is also having a hard time and he is ordering very few new ships. World-wide, the

Opposite: The stern of the 54,000dwt BP tanker *British Guardsman* shortly before her launch on 12 December 1962. In addition to drag chains, the hull has been fitted with water brakes to help slow the ship down during the launching run into the river.

The handsome Glen liners *Glenogle* and *Glenfalloch* fitting out in 1962. *Glenogle* was completed in September of that year followed by *Glenfalloch* in January 1963.

BRITISH GUARDSMAN
LONDON

capacity for building new ships is double what is really needed and unfortunately it is still expanding, particularly in Japan, where their very modest wages plus high efficiency and a dedicated purpose has helped them cut prices to such a fantastically low level that no European shipbuilder, except perhaps in Spain, could match them without losing money. Sweden, alone among European shipbuilders, has taken up this Japanese challenge because she builds ships mainly for export and can't afford to see her best customers drifting off to Japan. Unlike Japan, Sweden pays high wages but she is helped by a thoroughly common sense and down to earth partnership of management and a single trade union. I think too the obligation to do a fair day's work as a social duty is much more widely accepted in Sweden, which is a rather more serious minded country than many others, including our own. There is no doubt that individual productivity is very high in Swedish shipyards, using equipment which is much the same as our own here in Britain. Then again, everyone has heard of the effect of subsidies and cheap, long term loans and tax concessions which some governments draft rather lavishly to their shipbuilding and shipping industries. But in spite of the dice being rather heavily loaded against British shipbuilders, and remember the French, Germans and Italians and others are in similar difficulties, our more efficient and modernised yards have shown that they can get work and I'm sure they will continue to get it. Our most important asset is not the facilities upon which many of us have spent millions of pounds recently, but the employees who use them and now, so many of them are beginning to realise we have not been shouting wolf, I believe we still have a good hope of getting their full cooperation to pull the industry through this tough period but it's getting later than they think and it will be a tight squeeze for some yards.[79]

Kimber's comment was supported by the unpalatable truth that by 1964, the British shipbuilding industry's world share of output had dropped dramatically, from 50.2% in 1947 to 8.3%, despite the demand for ships soaring worldwide.[80]

This 1962 view shows the Fairfield electric tram departing from the shipyard's main gate heading for Govan Goods Yard. Normally used to carry plate and sections, in this instance it seems to be giving a lift to rather well dressed men. The 1891 Fairfield building is at left and the building at right was erected in 1903 to provide additional drawing office space. It was connected to the main building by the corrugated steel bridge. The 1903 drawing office was demolished in 2009. The Fairfield shipyard sign above the gate was restored in 2023 by Govan Workspace, owners of the 1891 building. (*Stuart Sellar*)

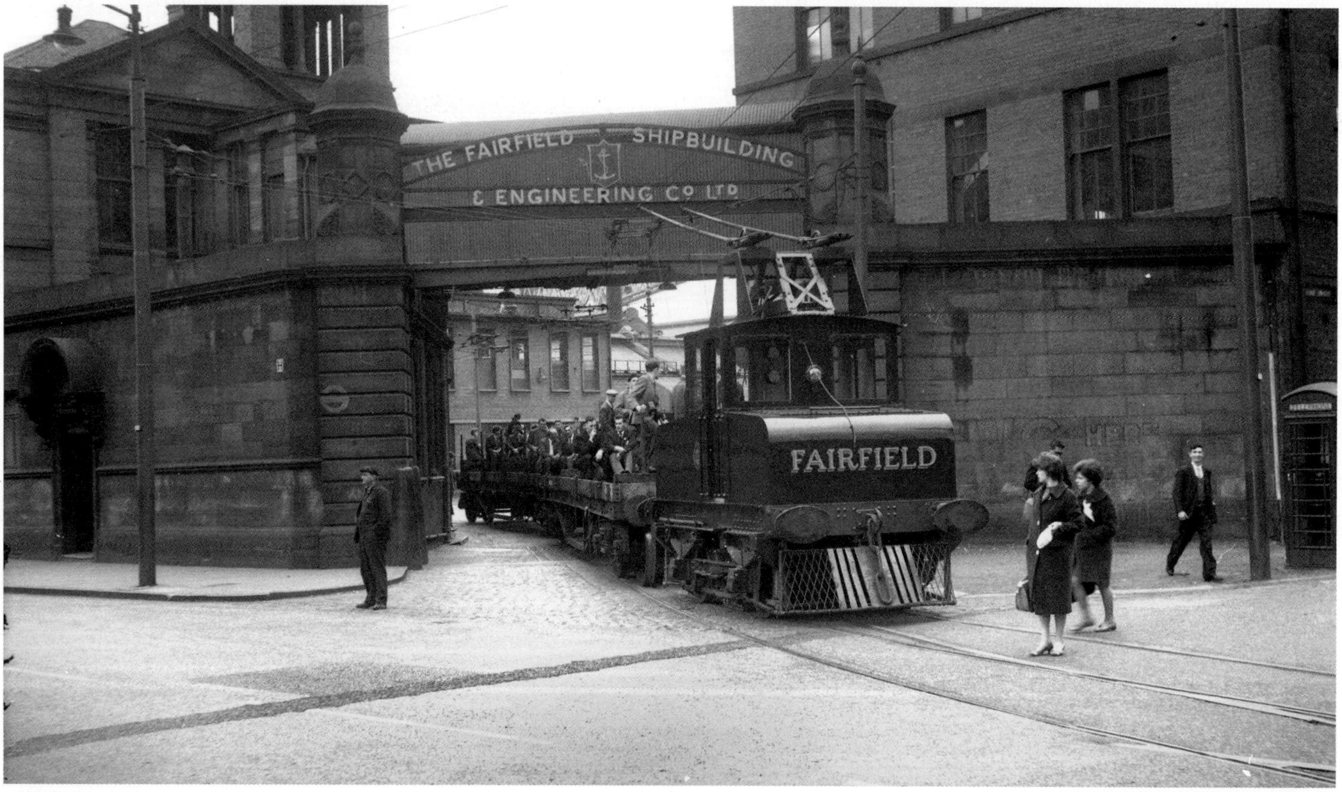

The steam turbine tanker *British Guardsman* on trials in 1963.

A Very Modern Shipyard

On 9 April 1964 the *Glasgow Herald* printed a 12 page supplement devoted exclusively to Fairfield, marking the completion of a £4m modernisation scheme which made the yard one of the most up to date shipbuilding facilities in the UK, if not Europe. It reflected the complete change over to welded, prefabricated unit construction, a process that had taken ten years to complete, during which time ships continued to be built to programme. Fabrication shops replaced the platers sheds, some of which dated back to the late 1860s. The berths, reduced in number to five, were served by travelling luffing cranes capable of lifts of 20, 40 and 80 tons. All of this promised increased productivity at a time when lower costs and shorter build times were essential to face the growing competition coming from Japan and European competitors. The supplement contained articles written by senior management as well as by Joseph Godber, the Minister of Labour, while other articles reflected on a glorious past and looked to a future in which nuclear powered merchant ships might be built.

Behind the public fanfare, the reality was that the yard was in serious financial difficulty, with a host of problems that would coalesce in the following year. As intended, the *Glasgow Herald* supplement was upbeat, but stressed that there was no room for complacency. The challenges ahead were fully recognised. It was nevertheless shocking if not sadly ironic that just one and a half years later, Fairfield would be placed into administration.

Several contracts taken on at this time contributed to an already poor financial position. One of these was the alteration and refurbishment of the former P&O liner *Mongolia*, launched on the Tyne in 1922 and now known as *Acapulco*. This ship had a long career involving several changes of ownership and name changes. In 1961, the ship was acquired by Natumex, a Mexican Government owned line, for conversion to a cruise ship to operate a fortnightly schedule between Acapulco and Los Angeles. Although a low value contract, it demonstrated the high risk involved in certain contracts and probably a degree of desperation by Fairfield management in securing this work to keep fitting-out trades employed. At a contract value of *c*£250,000, the vessel arrived at Fairfield in September 1961, departing for Acapulco at the end of the year, ready for her inaugural cruise on 11 January 1962. With some payments made up front, Fairfield was issued with an 'Irrevocable letter of Credit for sums due on completion backed by the Government of Mexico'. Under these circumstances, the Company felt it was not necessary to insure the contract with the Export Credit Guarantee Department, which would have protected them from non-payment. Despite the promises to pay, Fairfield was left with an unpaid bill of £77,750 which had to be written off as a bad debt. Much worse was to follow. In March 1963, Fairfield signed a contract with Haifa based Nili Somerfin Car Ferries for the construction of a passenger car ferry to be named *Nili.* James Lenaghan, Managing Director, noted that, while the contract for this ship had been taken at a keen price, it should not be a loss maker, words that would come back to haunt him.

Results for the years 1960–64 show Fairfield's deteriorating position.

	Shipyard	Engineering	Chepstow
1960	501,661	347,397	129,494
1961	332,638	93,029	85,360
1962	(259,172)	(355,514)	60,443
1963	224,304	(178,876)	30,434
1964	(26,870)	(453,349)	

Losses in brackets.

A Committee of Enquiry

In February 1965, the plight of the shipbuilding industry, its lack of competitiveness and loss of market share against an expanding world market was recognised by the new Labour Government. In response, Douglas Jay, President of the Board of Trade, appointed a Shipbuilding Inquiry Committee, under the chairmanship of businessman Reay Geddes. Its remit was to explore ways in which the industry could be put on a competitive basis through changes in organisation and production methods. The seven-man committee was not expected to report its findings until February 1966.

While the committee began its task, events at Fairfield began to spiral out of control.

The contract which did most damage was the passenger car ferry *Nili.* By March 1965, the loss on this ship, yet to be completed, had reached £169,000. Lenaghan said the Israeli superintendent staff had been very difficult

Crowds gathering for the launch of the passenger ferry *Nili* on 11 August 1964. The fabrication shops at right displaying the Fairfield name, were manufactured by the Company's structural steel company in Chepstow.

The launch of RFA *Sir Lancelot* on 25 June 1963. She was the lead ship of the 'Round Table' class LSLs (Landing Ship Logistic). *Sir Lancelot* had a long career in various guises lasting until 2008 when she was scrapped. Barclay Curle's Clydeholm shipyard can be seen in the background.

to deal with, while Derek Kimber described the experience of building *Nili* in his forthright manner: 'we have been pushed around by these owners for 18 months and had to put into effect a constant series of alterations both to designs and work already completed. In effect the ship has been designed three times and built twice.'[81] He added that the Holt ship, *Sir Lancelot,* together with *Nili*, showed that the planning and supervision of fitting-out needed a major overhaul if Fairfield was to compete successfully.

A young ship manager, Ian Morrison, described his time on *Nili*:

> That was an experience, she was a passenger luxury cruise ferry. I was put on her for the last five months to finish her off. The [Israeli] superintendent, Juli Yakir, had driven my colleagues and senior colleagues nearly to a nervous breakdown. The owners chopped and changed their minds the whole time, [they] altered the whole of the accommodation. The sea trials were quite a headache. I think I had four hours sleep during three days at sea. We came to the Tail of the Bank and the inspector came down

> and said yes, she will go to Helsinki tonight. We had something like 350 men onboard finishing her on the way to Helsinki. I was supposed to be on board for five days but spent three and a half months on her.[82]

In early May 1965, Fairfield's Chairman, J E Boyd, said that it was clear the Company's activities were too extended for the slender resources at its disposal and it was a matter of some urgency to reduce these activities and relieve the financial pressure on the Company. It was agreed that the Chepstow works should be sold[83] and that loss making Fairfield Rowan should dispose of its works in Elliott Street in Glasgow and concentrate solely on the engine works at Fairfield. Given that many of the diesel engines for ships built in the yard were manufactured elsewhere, giving up engine manufacture was an option.[84]

Meanwhile, completion of the bulk carrier *Cluden* was running two months late because of the breakdown of the Babcock 80-ton berth crane. This required an increase in overtime, adding to the cost. Derek Kimber commented that all contracts in progress were suffering from too little information too late arriving from the drawing office, the shortage of labour in the outfit trades and that, despite shipyard plant being adequate, there simply were not enough men for the volume of work on hand. It was not feasible to construct six ships simultaneously with just 853 men in the steel trades.

Sea trials of HMS *Fife* had been put back six weeks because of machinery modifications.

In May, the Bank of Scotland summoned representatives from the Company to headquarters in Edinburgh to explain why the delivery of *Nili* had been put back, thus disrupting the whole shipbuilding and stage payment programme.

Alarmed at the financial position of the Company, in June 1965, the Bank of Scotland issued an instruction that Fairfield Rowan had to be sold within four months. In August, Reardon Smith ordered a second bulk carrier at a break-even price of £1,900,000. The Sulzer diesel for this ship could be supplied by Hawthorn Leslie or from Sulzer in Winterthur at a substantially lower price than Fairfield Rowan. The loss on *Nili* had accelerated to £499,000 and, when asked if this runaway situation might also apply to the New Zealand ferry *Wahine*, Kimber thought not, as very few alterations had been asked for in this ship unlike *Nili*, which had been 'altered and altered constantly, taking all the steam out of shipyard efforts'. Jim Lenaghan, Managing Director, noted in the minutes that he was 'dissatisfied with planning and the level of production,' a situation over which he presided. The situation seemed hopeless.

The end came on 14 October 1965, when a letter was received from the Bank of Scotland requesting the repayment of all sums due by the Company. Fairfield's Chairman, J E Boyd, replied that the Company was unable to do so. There was no alternative but to appoint a Receiver.[85] Chartered accountant, Alexander Irvine Mackenzie, was duly appointed and so the Fairfield Shipbuilding & Engineering Co Ltd., which could trace its origins back to the time of Sir William Pearce had passed into history.

Collapse

As members of the Committee of Enquiry were touring shipyards, interviewing personnel and collecting data, the urgency of their task could not have been more appropriately underlined, when word of the Fairfield collapse made headline news. Derek Kimber, Deputy Managing Director, had the unenviable task of telling the secretary of the Shop Stewards Committee, at 4pm on 15 October 1965, that the Fairfield Shipbuilding & Engineering Co Ltd and its marine engineering subsidiary, Fairfield Rowan, were bankrupt.

This news was shocking and reverberated up and down Clydeside and beyond. How could the Fairfield yard, one of the most modern in Europe, and with a brilliant record of achievement going back to the early days of marine engineering and shipbuilding, find itself in this predicament?[86] It was well understood that the shipbuilding industry had been struggling to build profitably for several years, but this announcement elevated the seriousness of the situation to new heights.

The events that began in October 1965 following the collapse, marked the end of the 'old' Fairfield yard which, despite several bumps on the road, had represented continuity and competence not just on Clydeside but as part of the UK's industrial estate. From this point onwards, the narrative on Clydeside began to change from one of difficult times ahead to the outright recognition that it was now a struggle for survival. The Fairfield yard would continue to hit the headlines, but not so much for the ships it built but for the many crises it would face to stay in business.

At the time of the collapse, Fairfield employed about 5000, with 1500 of these in subsidiary Fairfield Rowan. The shipyard had a full order book valued at £34m, while Fairfield Rowan had 11 sets of marine engines on order, both steam and diesel, worth £11m.[87]

The ships on order at the time of collapse were:

> *Fife* and *Antrim*, for MoD. *Fife*, fitting out and due for delivery in May 1966 and *Antrim*, due to be laid down in December 1965.
> *Duhallow*, for Charter Shipping Co, a bulk carrier, due to be launched on 26 October 1965.
> *Wahine*, a ferry, launched in July that year and due

for completion in March 1966.
British Commodore, for BP, on the building berth and due to be launched in April 1966 (in the event not until 8 August 1966).
Atlantic City and *Indian City*, bulk carriers for Reardon Smith, were not due to be laid down until January 1966 and July 1966 respectively.

Fairfield and Fairfield Rowan were both members of the Lithgow group of companies. Group Chairman, Sir William Lithgow, when asked later about the collapse of Fairfield said that:

> in 1935, Fairfield was in trouble for reasons outwith the Company's control, and that his family had rescued it. I am asked why not again? And I must answer, because these are very different times. Against the background of national inflation and great uncertainty, the task of rescuing is not such as Lithgows could achieve alone.[88]

The Making of an Experiment

With speculation in the press rife about the future of the yard, on 18 October, at a mass meeting of Fairfield workers, the demand for a Government inquiry into events at the yard was made, with members urged not to find work elsewhere but to stand and fight to keep the yard going. Inevitably, a crisis of this proportion reached the highest levels of Government. On 28 October, the Chancellor of the Exchequer, James Callaghan, agreed to consider whether the Bank of England could advance funds to tide Fairfield over until appropriate action could be taken, possibly through the recommendations of the Geddes Report, which was then in preparation but not due to announce its findings until the following March. It was noted that the yard had been extensively modernised, had sufficient orders but had taken them at prices that barely covered costs to keep labour and machinery employed. The figure thought necessary to keep the company going in the short term was £1m.

At a meeting held in the Chancellor's office on 29 October, with Roy Mason, Board of Trade, and J Dickson Mabon from the Scottish Office, Sir William Watson of the Bank of Scotland, Fairfield's bankers, a report into the affairs of the Fairfield shipyard by the Receiver, Alexander I Mackenzie, was considered. Total liabilities of the company amounted to £5.5million while the amount due to creditors was £560k. The Chancellor's initial response was that, if the Government were to advance funds, there was a strong possibility that these might never be returned. Moreover, the ultimate conclusion might be in favour of closing Fairfield down. It was not the job of the Government to save a concern which had got into serious financial plight through mismanagement. Government assistance could only be justified if, without it, a serious social and economic loss would be entailed. The Chancellor was not convinced that this case had been made in relation to Fairfield. Moreover, only very scanty information was available on which to take decisions which might involve large sums of public money, at a time when Government expenditure was being strictly controlled. Once the government extended assistance to Fairfields, there would be growing pressure to continue, however poor the prospects, and this pressure might be extended to other shipbuilding firms.

Three main points arose from the meeting:

- The Geddes Report was due in February of the following year and it might be inappropriate to close Fairfield prior to that, particularly if the Report envisaged rationalisation, where Fairfield's physical assets would be of value. A merger might be possible and it was likely this would take two years, although other Clyde yards might not want to merge with Fairfield.
- Other Clyde yards would benefit from Fairfield's elimination, although this would be a 'damaging blow to the image of Scottish and Clydeside shipbuilding. The effect of this should not be underestimated.'
- Formidable political and psychological repercussions on Clydeside and throughout Scotland would be bound to result from the closure of Fairfield.

In conclusion, the Chancellor said that he and his colleagues would give urgent consideration to the position, but he felt bound to point out that this was the first request which had been made for Government assistance and, while recognising the urgency, it would clearly be impossible for the Government to produce a decision overnight.[89]

On 3 November, it was agreed that £1m should be advanced from the Bank of England to the Receiver to keep Fairfield going, although it was understood that a further £3 to £4m might be necessary to keep the yard in business until 31 March 1966, at which time a plan for viability, the Geddes Report, might be in place. The Chancellor made a statement to this effect in the House of Commons on the following day, while George Brown, First Secretary of State and Minister for Economic Affairs of the Labour Government, commented at the bottom of a report: 'I am strongly in favour of going the whole way. There's no point in shutting this one down before Geddes reports.'[90]

Many meetings ensued with a sense of urgency, not

least because of the possibility that Reardon Smith would cancel their order for two bulk carriers and this would likely seal the fate of the yard.

At the height of the crisis, industrialist Iain Maxwell Stewart, took an interest in the fate of the company. In addition to various directorships, Stewart was the Chairman of Hall Thermotank Ltd, and Managing Director of Govan based Thermotank Ltd, the leading UK manufacturer of air conditioning and ventilation systems. Over the previous five years, Stewart had been increasingly vocal over what he considered were serious failings in British industry, including low productivity, old methods, lack of investment and outdated management. Expressed through newspaper articles and public talks, he believed that these issues should be addressed head on; that new methods in management and industrial relations should be implemented.

On 19 November, Stewart wrote to George Brown:

Dear George,

The industrialist Iain Maxwell Stewart seen against the backdrop of the Fairfield yard. He persuaded the Labour Government of Harold Wilson to save the yard and operate it under new management and new ideas which gave rise to the 'Fairfield Experiment'.

> As you know, I have been deeply interested in the Fairfield situation . . .
>
> In my view, the situation presents a heaven sent opportunity to expose the deep problems of the industry and grasp the nettle of overmanning, demarcation and rank bad management.
>
> Politically, of course, the saving of Fairfields would have immense advantages, but quite apart from that, in my view it would be criminal at this stage to pass up such a golden opportunity to benefit not only shipbuilding, but industry across the board.
>
> ... if the Government would be prepared to back an experimental project ... I would undertake to produce a profit on the Fairfield balance sheet, certainly within 5 years and probably within 3.[91]

On 23 November, an evening meeting between George Brown, Jim Callaghan, the President of the Board of Trade (Douglas Jay) and the Minister of Defence for the Navy was convened to discuss the options: that is either to increase the loan from £1m to £3m, or for the Government to take over the yard. Only Jay wanted the first option. Of the second option, Callaghan wanted to close the yard, while Brown wanted to nationalise it.[92]

The following day, Iain Stewart wrote to the Chancellor:

> Dear Chancellor,
>
> I was very glad indeed to have the opportunity of meeting you again yesterday evening but was sorry to be so rude and divert your attention away from social relaxation to the more serious subject over affairs on Clydeside.
>
> In my view, the opportunity which now presents itself at Fairfields is not only a political godsend, but is also an invaluable catalyst towards bringing about the improved productivity in British industry as a whole. The question of productivity and its direct relationship to mobility and security has been the subject of my pleas since 1961 to the Tories, to several of your own ministerial colleagues and also to many leading members of the trade union movement both in London and in Scotland. I firmly believe that a calculated investment in the venture will be rewarded with handsome dividends in various forms.

On the 25 November, George Brown put all of the arguments for and against saving Fairfield together in a long memorandum, part of which included the following:

> The disadvantages over this course, alone or in

> partnership, are that we would be stepping in and acquiring a company which had failed, a company in an inefficient and contracting industry, an industry which must shortly be reorganised if it is to survive. On the other hand, it seems to me that this is a wonderful opportunity to get to grips with the situation and show what can be done by increased productivity. I have been impressed by what Iain Stewart has told me can be achieved in this industry in cooperation with the Unions. I am sure he would be prepared to take a principal part in putting these ideas into practice and he has in mind a competent managing director who is already achieving success on these lines in an adjacent yard and who is likely to be willing to act as chief executive. They would need to be supported by other directors with particular experience in industrial relations and finance and it would probably be helpful to have a member of the Scottish TUC on the Board. I would expect that in due course this experiment would be an example to the rest of the shipbuilding industry.

Brown ended by saying 'I hope my colleagues will agree that this, although the boldest, would be the best course in the long run.'[93]

In effect, Stewart, with enthusiastic support from Brown, had turned the crisis on its head and presented it as an opportunity not simply to rescue Fairfield but to address many issues that were known to beset British industry in general.

While all of this was in train, and while ways were explored by which the yard could be acquired, retaining ship orders became the focus of the trade unions, who recognised that, without swift action, the yard was doomed. At the same time, Reardon Smith were prepared to wait no longer that 2 December for confirmation that their two bulk carriers would be laid down and taken to completion.

On 30 November the Cabinet agreed to acquire the physical assets of Fairfield, preferably in co-operation with private capital. The Cabinet invited the First Secretary of State, the Chancellor, President of the Board of Trade, the Secretary of State for Scotland and the Attorney General to pursue the necessary negotiations to form a new company. This required a valuation of the company to be made which was carried out by James Barr & Son, chartered surveyors. The valuation, which said that the works were modern and in excellent condition, was valued at a break-up price of £750k. As a going concern, the figure was £1.855m. The Fairfield Rowan engineering works on the same site as the shipyard was valued separately at £2.6m.

On 3 December Iain Stewart met with the Scottish TUC and representatives of all the unions[94] to seek cooperation on improving productivity at Fairfield if the yard was to be kept open. According to Derek Palmar, an industrial advisor to the Department of Economic Affairs,[95] this 'went better than I could possibly have expected'. There was a general acceptance of the principle of flexibility. It was suggested that the First Secretary meet national officials to get their acceptance.[96]

On 6 December Derek Palmar reported that the situation at Fairfields was deteriorating. The Deputy Managing Director, Derek Kimber, was leaving to take up a job elsewhere, and men were drifting away. The attitude of creditors was getting harder. A decision had to be made before the creditors' meeting on 13 December.

On 7 December George Brown met with Iain Stewart, Roy Mason, Minister of State for Trade, plus representatives from the Treasury, Board of Trade and Ministry of Labour. Most importantly, trade union leaders and union representatives from Fairfield were in attendance. Brown explained where the Government was in the negotiations for the takeover of the yard. He said that labour flexibility and interchangeability were essential, and that the most progressive production techniques would be introduced. Subject to the completion of financial negotiations, ministers would report back to the Cabinet the following Thursday and a statement would be made to Parliament that afternoon.

However, as moves to finalise the arrangements to save the yard continued, a less than upbeat letter was delivered by hand to George Brown on 9 December from the Presidents (there were two) of the Shipbuilding Conference, the body that represented shipyard owners and management across the UK. They:

> keenly regret that they have not been taken into the confidence of Her Majesty's Government. If the proposals now under consideration materialise before the Geddes Committee has reported, then this firm will be placed in a privileged position compared with other shipbuilders.
>
> We have been led to believe that the industrial sponsor of the new scheme intends to carry out an experiment on labour-management relations at Fairfields. If our information is correct, this inevitably would have immediate and serious effects on wages in all shipyards and marine engineering works.

This letter was followed on 10 December, by another, this time to Douglas Jay, from Reay Geddes, echoing some of the points raised by the Shipbuilding Conference. Jay considered this letter unhelpful, given the advanced state of negotiations and the lack of time.

In answering questions in the House of Commons on

9 December, George Brown said: 'This is a matter of tremendous interest to a part of this nation. It involves the livelihoods of a lot of people. It is tremendously important to the Scottish economy. When it became clear that events controlled by banks and others over whom we had no control seemed about to bring the yard to a close, the Government had to move rather more quickly than we would have done otherwise.'[97]

9 December conclusions

At a meeting of the Cabinet on 9 December, various points were made, including:

> that it was considered to be less than advantageous for the Government to go it alone, in which case closure would follow. Too much attention should not be paid to the Shipbuilding Employers' Federation as they 'no doubt feared the introduction of a new and dynamic management into Fairfields'. Harold Wilson, the Prime Minister, summarised the situation, saying that the Government should still try to prevent closure, but that private capital should share with the Government in the new enterprise. If this could not be achieved, or before orders were withdrawn and the labour force dispersed, then closure would have to be accepted, no matter how much they might regret it. The Cabinet invited the First Secretary, Chancellor, President of the Board of Trade and Secretary of State for Scotland to continue negotiations.

On the following day, the President of the Board of Trade, Douglas Jay, wrote to George Brown to say that Reay Geddes had called on him to express concerns, namely:

> that it is imperative to bring the Shipbuilding Conference into the discussions. They will be able to provide an effective management team to run Fairfields meantime. It is undesirable to bring in outside interests at this stage in the reorganisation of Clyde shipbuilding. Existing shipbuilders might be ready to take Fairfield over, Sir John Hunter for example. A 50/50 arrangement is not good, better if Government does a 100% takeover until reorganised private enterprise takes over. Jay thought it would be impossible to convince shipbuilders that Fairfields with Government participation would not receive preferential treatment.

Also on 10 December, Fairfield workers had met with union delegates secretly and agreed to the principle of flexible working – in effect, to take Fairfield out of the rule book. Stewart needed to get the agreement of the men in the yard and proposed to meet them soon. At the same time, he was hopeful to win over the Shipbuilding Conference, most members of whom he considered to be friends.

The summary position by 10 December was that the Government, in partnership with private enterprise, was preparing to make an offer of £1.92m for the assets of the Fairfield Shipbuilding & Engineering Co and £2.6m for Fairfield Rowan. It had also been agreed that Iain Stewart should meet the men in the yard on 17 December.

Losses on the passenger ferry *Nili*, seen here on trials, pushed Fairfield into receivership in October 1965.

ERA THREE 1966–1999

Fairfield (Glasgow) Ltd

On 15 December 1965 the First Secretary, George Brown, met with the Shipbuilding Conference and Employers Federation, to tell them what the Government was endeavouring to do.

Derek Palmar provided him with briefing notes which included:

- The Employers complain of increased labour costs and the effect on long term fixed price contracts. This is precisely what the new Fairfield would try to reduce by better productivity.
- The new Fairfield will not be in a privileged position but be exactly the same as other yards. It has already been established that labour relations for the new Fairfield will not be used as a lever for general wage increases without comparable increases in productivity.
- The Shipbuilders [the Employers' Federation] are naturally pressing for the closure of Fairfield because it would ease the competitive position for skilled labour. It was clear that the Employers' solution would see the yard wound down.

Brown gave a robust reply to the points the Shipbuilders made, while, at much the same time, agreement was being reached elsewhere on transferring the assets and goodwill of the old Fairfield company to the new company, to be called Fairfield (Glasgow) Ltd.

The details of the deal to save Fairfield had largely been worked out and the shipyard saved, although not everyone agreed with the way it had been done. Political opposition to Government intervention was to be expected but the

The Board of Fairfield (Glasgow) Ltd in September 1966. left to right: Andrew Cunningham (Trade Unions), Hugh Stenhouse (private enterprise), Oliver Blanford (Managing Director), Sir William Carron (private enterprise), Iain Stewart (Chairman), Derek Palmar (Government), Reg Ingram (Financial Comptroller), Jim Houston (Productivity Services Manager), Ken Alexander (*University of Strathclyde*).

Shipbuilding Employers were less than enthusiastic, principally, because, with the contraction of the shipbuilding industry expected under the terms of the Geddes Report, Fairfield, through commercial failure, had selected itself as an obvious candidate for closure. Additionally, Government equity in Fairfield (Glasgow) Ltd was bound to lead to unfair advantages over other, struggling yards, while much later, the knowledge that management from outside the industry was being brought in to run the yard would be anathema.

21 December 1965

When asked by John Rankin, MP for Govan, about Fairfield, George Brown said that he was actively trying to find a solution. 'May I add that those who have charged the Government with indecision and delay should realise that the only decision which could have been taken promptly and effectively would have been to close the yard. It is because we have tried and are still trying so hard to keep it going, that I hope to be able to make a statement tomorrow.'

The marine engineers David Rowan & Co was merged with Fairfield's engine works under the title Fairfield Rowan in 1963.

On the following day, Brown told the Commons that arrangements had been made to safeguard the future of Fairfield shipyard.

On the 27 December Stewart addressed the workforce in the Lyceum cinema, just 500 metres along Govan Road from the shipyard. It was the first of five such meetings he would have. In a lengthy speech, he set out his case for taking the yard forward, asking for a vote of confidence which he duly received.[1]

On 7 January 1966, Fairfield (Glasgow) Ltd, in partnership with Government, private investors[2] and trade unions,[3] began trading. The new Board announced by Iain Stewart included Oliver Blanford (Managing Director), Richard Atkinson, James Lenaghan, formerly Managing Director of the old company and Derek Palmar representing the Government's interests.

Fairfield Rowan

The rescue of the shipyard did not extend to the engine works. Chartered accountant, Robert Courtney Smith, was appointed as Liquidator to take over the affairs of the company on Christmas Eve 1965. Despite interest from David Brown Ltd, manufacturers of gearing and Dutch diesel engine manufacturers, Stork Werkspoor NV, with whom Fairfield Rowan was a licensee, nothing came of it and Fairfield Rowan closed on completion of its last contract in April 1966, with the loss of approximately 1000 jobs.[4] George Brown had previously told the Fairfield Rowan workforce that he would come back to Glasgow to tell them whether or not their company was to be rescued and this he duly did, despite the distressing message. Closure was a significant, historical conclusion, as marine engineering lay at the foundation of the Fairfield shipyard when Randolph & Elder began the manufacture of compound steam engines at Centre Street in 1854 as a prequel to building ships in Govan. The subsequent fame of John Elder and the Fairfield Company owed much to innovation in steam marine engineering, the pioneering nature of which had been maintained by Alexander Kirk with the triple expansion engine.

Farewell to Chepstow

One further casualty of the October 1965 collapse was the Chepstow Works, which had gained momentum in recent years as an important structural steel company in the UK. Profitable, with a full order book and employing 350, the Liquidator soon found a purchaser for Chepstow in London based Mabey & Johnson Ltd, in July 1966. The London company established a new company, Fairfield-Mabey, probably retaining the Fairfield name for market continuity. Fairfield-Mabey went on to win major contracts for the cable-stayed bridge at Erskine on the Clyde, the

This poorly designed advert for Fairfield's Chepstow structural steel company dates from 1964. Following Fairfield's collapse in 1965, the profitable Chepstow works had to be sold.

Avonmouth Bridge and sections of the Humber Bridge among others.[5]

From the outset, the Fairfield Experiment, also referred to as a 'national proving ground', attracted huge attention from the media, not all of it positive. For Iain Stewart, who had relished the situation as a means of putting his often quoted theories about job security, redundancy and productivity into practice, the time had come to act. The Experiment sought to fundamentally rewrite the script of low productivity by the introduction of new management and working methods and to replace strife, by putting labour relations on a consultation and cooperation basis rather than confrontation. To do this, Stewart wanted a radical departure from existing shipbuilding management by introducing the latest techniques already in use in other industries. One of Stewart's conditions on taking on the new Fairfield was the ability to negotiate with unions in isolation, that is, that he would not be bound by national agreements or negotiating machinery. To assist in the effort to improve communications in the yard, a works newspaper, *Fairfield News*, was produced on a monthly basis. The commercial success of all of the above would be measured by making Fairfield profitable within five years, subsequently reduced to three years.

Fairfield (Glasgow) Ltd began trading on 7 February 1966, having been acquired from the Receiver for £1.4m. The shipbuilding programme, inherited from the old Fairfield company, had continued uninterruptedly, the only delay being stage payments which ship owners had held back in the event that the company would cease trading altogether.

Stewart's first task was to build the management team, reflecting the strategy that he had outlined to the Government, investors and the unions. First to be appointed was a new Managing Director, Oliver Blanford, who had previously been Engineering Manager at Alexander Stephen's shipyard, where he had successfully introduced Measured Daywork.[6] Next was James D Houston, appointed as a Director and Productivity Services Manager. Houston had previously been Director of Engineering at the Singer's sewing machine factory in Clydebank. He would be the person who would be largely responsible for the introduction of new methods. This involved setting up a department quite new to shipbuilding, staffed by expertise from outside the industry. This was, at once, in line with recommendations in the Geddes Report but at variance with views expressed by the shipbuilding bosses on the Clyde and elsewhere. With these key appointments in place, the rest of the management team was put together, partly from the old Fairfield and partly from new appointments over the next few months.

The Board of Directors would have to reflect the trinity between Government, trade unions and private enterprise which had made Fairfield (Glasgow) Ltd possible. Ken Alexander, Professor of Economics at Strathclyde University, joined the Board to undertake a live research project which would 'record in comprehensive detail the ways in which new ideas, methods and attitudes are introduced, accepted and used'. Representing the Government was Derek Palmar, the industrial adviser at the Department of Economic Affairs, who had done so much to bring Fairfield (Glasgow) Ltd into being.

The Geddes Report

Running in parallel with the many steps taken since October to get the new company established was the Geddes Report. Under the chairmanship of Reay Geddes, the Report looked into the future operation of the shipbuilding industry in the UK. In March 1966 the Geddes Report was finally published, a 200 page document describing the industry, together with an analysis of why it was failing. Most importantly, however, were its conclusions and recommendations. These ran to 101 points. One of the main findings was that shipbuilding groups should be formed, thus creating larger companies, giving them more influence, as well as reducing duplication and costs. The Fairfield Experiment was briefly mentioned in the Report, but Geddes was made to understand by the Government that Fairfield should be included in whatever Geddes and his committee prescribed, irrespective of whether the Experiment had reached conclusions or not. Equally, the authors of the report appeared to be oblivious to the significance of events at Fairfield or that any important findings that might arise from the Experiment should be taken into account in the future.

Soon after the report was published, and in response to its findings, talks about merger discussions between the two largest upper Clyde shipyards, Fairfield and John Brown were proposed.

An Industrial Engineering Department was established at Fairfields in March. The main objectives were to introduce work study and eventually, a Measured Daywork Scheme. In April, an interchangeability agreement was signed with boilermakers, enabling relaxation of working practices for an increase of nine pence per hour and an agreed 24 months' work guarantee.

Embedding the new management culture would take time and involve training in new techniques. In May, the first such exercise began when 26 shop stewards went on work study courses at the Electrical Trade Union's training college at Esher in Surrey. Courses in work study and analytical estimating were run at the local Pearce Institute in Govan.

On 2 June negotiations for what was termed the Procedure Agreement were completed between management and the 13 unions[7] in the yard for a no strike, no overtime ban and no limitation of production. Under the terms of this agreement, a Central Joint Council was established. This comprised one representative from each of the 13 unions and three members from management. The point of this Council was the speedier resolution of disputes through joint consultation, in order to ensure that the workforce was 'well informed about the company, its prospects, trade conditions and policies'. One element of this agreement, described at the time as 'revolutionary,' was the creation, for the first time, of a full-time yard convener and an office in the yard where shop stewards could meet to discuss issues.

In July, the Shadow Minister of Labour, Sir Keith Joseph, visited the yard with Nicolas Ridley MP. On that occasion, Joseph noted 'I would hope that, if this yard can show that it can win orders and make profits and pay higher earnings because of the new air of cooperation, then management and labour elsewhere in the industry will be bound to want to copy.'

That month, the editor of the journal *Shipping* wrote, with reference to Iain Stewart:

> It is our blessing for the whole of British shipbuilding that he had the will, the stamina and absolute determination to get his new methods accepted by yard workers (stiff with prejudice) and staff (almost as stiff), with the additional burden of a Government department as a partner.

In September, there were three new additions to the Board which now comprised:

> Andrew Cunningham (Trade Union), Hugh Stenhouse (Private Enterprise), Oliver Blandford (Managing Director), Sir William Carron (Private Enterprise), Iain Stewart (Chairman), Derek Palmar (Government), Reg Ingram, (Financial Comptroller), James Houston (Productivity Services Manager), Ken Alexander (Research).

In August, in line with Geddes recommendations, a training centre with a full time training manager was established, the objective being, to raise standards in apprentice and journeyman training.

In October it was announced that Work Study had been introduced in the yard. To explain what this was, the works newspaper, *Fairfield News*, carried a question and answer article where the Shop Stewards' Convener, Alex McGuinness, asked the questions and Chief Industrial Engineer, Alistair Lowther, answered them. Among many questions asked, the Convener was anxious to know if Work Study would result in reductions in the labour force. He was told that anyone affected would be redeployed or retrained and that the Company would simply stop taking people on as a means of controlling numbers.

Over the months of August, through to November 1966, meetings took place between senior John Brown and Fairfield managers on the subject of a potential merger. While both sides appeared open minded about how this might develop, it was recognised that there would be a different managerial approach to modern management techniques and in particular, work study. On 8 November Lord Aberconway, Chairman of the John Brown group of companies, and Iain Stewart, Chairman of Fairfields met to discuss the proposal further. While both were favourably disposed towards a merger, Aberconway said that Fairfields still had to prove itself as an experiment. Stewart replied that it would probably be the end of 1967 before he could say definitely that the experiment had been a success or not. On that basis, Aberconway thought it best to wait until it had been clearly established that ideas about the application of new management techniques on shipbuilding had been clearly seen to be practical and profitable. There the issue lay.

On 11 November 1966 Stewart met with the workforce at the Lyceum cinema for the third time to tell them that the 'fight for higher productivity starts now'. He said that, because progress has been good, the firm would be able to cut one year off the proving ground period which could come to a close at the end of 1967, by which time he hoped the firm would be producing ships on time, at the right price, with a reasonable profit.

To remind the workforce of the pledge they had given him one year earlier, he reiterated that those who did not agree with the pledges that were made then 'were free to go elsewhere'. As usual his language was direct:

> As most of you know, this is the third time we have met in this cinema. On the first occasion, December 27th, we all made the decision to start a fight, a fight to save Fairfields and reconstruct it as a model shipyard. The preparation work for that fight has nearly finished and now the real battle, that is, the battle for efficient performance and measured productivity, is about to begin.

Stewart chose to remind everyone what George Brown had read out in the House of Commons the previous December by reading it out again:

> in setting up this tremendous new format in British business, all the partners, the Government, private enterprise and the headquarters of the unions are relying on the men in the yard to unreservedly co-operate in working the yard as efficiently as possible and in particular, achieving the flexible manning arrangements and interchangeability of workers which are essential. If this cooperation were not forthcoming, this whole scheme, involving the combined support of Government, trade unions and private enterprise, would fall to the ground and the shipyard would have to close.

Iain Stewart told the several thousand persons assembled, that the contract for a £13m Type 82 destroyer [HMS *Bristol*] which the company had tendered for, had gone elsewhere [to Swan Hunter].

> We could have had it if we had been prepared to guarantee a delivery of 54 months. But, with productivity at its present level, and even with an optimistic assessment of future productivity, we could not promise to complete the ship in under 60 months and said so to the Ministry of Defence. The Ministry, quite rightly, accepted the 54 months guarantee [from Swan Hunter] and so the vacuum in our order book still remains.

Stewart was doing everything in his power to ensure that all the workforce were reminded of the gravity of the situation, that one false step could undo everything so far achieved and of the commitment they had made. He now had the company structure that he wanted in place, management, communication processes, agreements and

The US Navy survey ship *Harkness* running trials on the Firth of Clyde in January 1971. Sister ship *Chauvenet* was also built by Fairfield.

good relations with unions etc. Now the next phase would begin, the battle to improve productivity by reducing man-hours per ship.

Not everything went to plan however. One hundred and thirty apprentice boilermakers went on strike for an increase in pay in line with that awarded to journeymen. Fortunately for the Experiment, the journeymen chose not to join them. This dispute was resolved through what was termed the Apprentice Charter, created to give apprentices better training and clarify wage rates. In resolving this dispute, it was the first instance where the Central Joint Council could perform the function for which it was intended. The new Apprentice Charter cut the time served by apprentices from five to four years.

Another timely reminder of just how bad the situation for UK shipbuilders was came when Alexander Stephen & Sons, at the adjoining Linthouse shipyard, disclosed a loss of £791,764 for the year, with £1,219,000 set aside for contracts in progress. In August, the Blyth Shipbuilding & Dry Dock Co on the North East Coast of England called in the Receiver.

November was also the month when Anthony Hepper visited the yard in his role as a member of the newly established Shipbuilding Industry Board. On 25 November, Ted Heath, Leader of the Opposition, toured the yard. Veteran Labour MP Emanuel Shinwell, Chairman of the Labour party, also visited the yard, after which he made the somewhat downbeat but prescient comment, 'If Fairfields does not succeed, it is very doubtful that even with the implementation of Geddes, that British shipbuilding can be put back on its feet.'

And then there were the ships. In February, Sophie Brown, wife of George Brown, laid the keel of the guided missile destroyer HMS *Antrim*. The ferry *Wahine*, built for the Union Steamship Co of New Zealand, was handed over to her owners on 19 June, while on 21 June, *Antrim*'s sister ship HMS *Fife* was commissioned in the fitting-out basin. In August, following the launch of the 67,000dwt BP tanker *British Commodore* by Peggy Jay, wife of Douglas Jay, President of the Board of Trade, the new company was able to announce its first order which was for two survey ships for the US Navy. This order was worth £6m. The order brought with it various visits by US Navy personnel to the yard including the Assistant Secretary of the US Navy, Graham Bannerman. To complete the year, Cardiff-based Reardon Smith ordered two 15,000dwt cargo liners.

The year 1967 began in an optimistic mood, captured in an article by industrial journalist John Hossack headlined

The container ship *Jervis Bay* at Govan in the final stages of completion. She is being fuelled by a John H Whitaker bunkering tanker.

'The Fairfield Experiment Looks Like Succeeding'.[8] According to Hossack, productivity was improving. Where 88,000 man-hours had been lost in the last nine months of trading by the old Fairfield company, the figure for the new in the equivalent period for 1966 was just 9000.

The next step along the road to greater efficiency was the introduction of a job evaluation programme, developed in consultation with P A Management Consultants. One of the objectives behind this was simplifying a complex wage structure, which Iain Stewart had described as a 'dog's dinner'. Whatever misgivings the unions might have had over concepts such as Work Study, Measured Day Work and Job Evaluation, these were accepted as part of the survival plan. To reduce the costs of components, the concept of Value Engineering was introduced, whereby function was optimised without compromising performance. The aim of Value Engineering was to generate savings equivalent to five times the cost of running the Value Engineering Department.[9]

On 10 February 1967 the 45,000dwt bulk carrier *Atlantic City* was launched. She was the first ship to be laid down and completed by Fairfield (Glasgow) Ltd and the first, in a number of other significant ways, demonstrating the changes that were being made. She was the first to be built from the stern forward on the building berth to allow an early start by fitting-out trades. She was 'programmed' from the start, although Blanford admitted that they still had to learn a lot about making programmes and, even more, to learn about abiding by them, particularly as *Atlantic City* followed ships that were unprogrammed and very late. She was also the first attempt at quality control in a formal sense and the first, at Fairfield, built under a Ship Contracts Manager, responsible for cost, method, adhering to time and coordination of trades. *Atlantic City* was launched on time and delivered on time two months later. The ship was built for Reardon Smith, who had stuck by the yard in the very difficult days at the end of 1965 when the orders for *Australian City* and *Eastern City* effectively kept the yard in business. At the launch, Iain Stewart, in a lengthy speech, said 'through this critical year of 1967, we hope to acquire a full orderbook and so attain a continuing rise in productivity which will not only enable us to maintain the promised quality and delivery, but will also reveal, perhaps as early as the end of 1968, if all goes well, that the capital investment by Government, unions and private enterprise has been profitable.[10]

In February, the Company was able to announce the order for another vessel, the container ship *Jervis Bay* for OCL, booked at a price £1m below other British yards and with a delivery date of May 1969. One of a class of six ships, the five others went to German yards. This order placed Fairfield favourably at the beginning of what would be a growing market for ships of this type.

The Bowler and the Bunnet

In April 1967 Scottish Television, of which Iain Stewart was a Director, had decided to make a documentary highlighting the new methods being introduced at Fairfields. The script was written by journalist, author and broadcaster Cliff Hanley, while the director was Sean Connery, who also narrated and appeared in the film. The coup of getting Connery, then at the height of his success playing James Bond, to make this one hour documentary, came about because of his interest in the subject and his meeting with Iain Stewart, at which the latter impressed Connery with his vision for the future of shipbuilding on the Clyde and elsewhere in Britain. Hanley commented that 'This Fairfield film that Sean Connery dreamed up is the most ambitious piece of industrial enquiry ever shot inside a fortnight. It is trying to take an absolutely honest look at the Fairfield Experiment, with no punches pulled and no favours, but no distortion either.' The documentary, shown in July of that year, was titled *The Bowler and the Bunnet*, a reference to the divide between management and workers symbolised by their head gear.

June 15th saw the introduction of the Measured Daywork Scheme across the yard. According to Managing Director, Oliver Blanford, this scheme was the single most important activity in the new Fairfields, along with the Procedure Agreement, the Apprentice Charter and the Training Scheme.

> No one should need reminding that, without substantial reduction in costs, neither Fairfields nor many British yards building merchant ships will survive. It follows that we must get a greater increase in output per man-hour than the increase in wages and that, without this increase in output per man-hour, there will be no merchant shipbuilding industry, no jobs and no wages. The essence of the Fairfield Experiment is improved output per man-hour, increased wages, and decreased costs.[11]

Under the scheme, each worker was given a job card and each foreman a route card for all jobs and the target time to be taken: *ie*, the work to be done, the tools and materials required and the man-hours to be taken.

One visual outcome of new management methods at Fairfields (Glasgow) was the disappearance of the traditional bowler hat worn by management, to be replaced with a hard hat. This was subsequently further developed when it was decided to colour code the helmets. Thus, the appearance of various coloured helmets denoting a particular group – Executive Managers, green helmets; Managers, white; Foremen, red; Productivity Services, blue; Quality Control, orange; Steelworkers, yellow and Outfitting Trades, black.

This moved one shop steward working on a ship to refer to the scene as Smarties on the move.

At the same time, the Company employed the Woudhuysen Design Group to design a logo, the inspiration for which was the three investment groups the Government, trade unions and private enterprise combined with the letter F.

In August Blanford published an article in *Fairfield News* under the heading, borrowed from Mao Zedong, if slightly modified, 'The Great Surge Forward'.[12] He was referring to the timely launch of the next Reardon Smith bulk carrier, *Indian City,* and improved productivity because of the Measured Daywork Scheme. Moreover, building on the introduction of new methods and practices established on *Atlantic City*, *Indian City* became the first ship subject to Budgetary Control, and the introduction of Value Engineering. It was also the ship where, under the terms of the Flexibility Agreement, 'integrated team-working' was employed, whereby one foreman was in charge of various steel trades, welders, caulkers, platers etc, working in one unit or area where, previously, one foreman for each trade would have been present.

Throughout the first half of 1967, while these events were unfolding, the Shipbuilding Industry Bill was making its way through the House of Commons, finally passing as the Shipbuilding Industry Act on 28 June. This Act paved the way for the creation of the Shipbuilding Industry Board that would implement the findings of the Geddes Report. It would also mean, in due course, the end of not just the Fairfield Experiment but of Fairfield (Glasgow) Ltd, and the consignment to history of the Fairfield name.

In debating the Shipbuilding Industry Bill in the House of Commons in March 1967, John Rankin, Labour MP for Govan, said with reference to the proposed Council which would inform the implementation of Geddes:

> When so much that is new is being tried out at Govan and is succeeding, it seems strange that the Council, which is to influence the future of shipbuilding in Britain, is not to have among its members someone from Fairfields. Does this mean that the Council will be tenanted only by individuals who are representative of the older ways of thinking in shipbuilding?[13]

It was a point well made, given that Fairfields was a Government-backed, national proving ground, tackling the issues of low productivity and bad labour relations which had contributed greatly to the plight of the shipbuilding industry. The *Financial Times* made a similar point. 'The re-organisation of upper Clyde shipbuilding certainly provides an opportunity for radical departures from traditional shipyard management methods, practices and philosophy. Whether it will be taken is largely up to the half dozen men who, under Mr A E Hepper, form the Steering Committee.'[14]

On 31 October 1967 the first – and only as it would turn out – Annual General Meeting of Fairfield (Glasgow) Ltd took place. The results for the thirteen months to 31 December 1966 showed a loss of £220,017 to which a provision of £368,000 had been made for losses against some contracts, making an eventual loss of £588,017.[15]

However, as 1967 ran its course, Iain Stewart and Oliver Blanford reflected on the achievements of the last two years and their hopes for the future going forward. The success of the Experiment was difficult to measure because it was terminated after just two years. Moreover, the new company had spent £0.5m setting up the Productivity Services Department under Jim Houston which employed 225 personnel with an annual running cost of £300,000. In response to criticisms made of this burden on overheads, Oliver Blandford said that this administrative cost of the Department [£300,000] was equivalent to £5 per ton of steel worked through the yard, and that this was relatively insignificant against the hoped for reduction of 75 man-hours per ton of steel, reduced to 25 or 35 per ton over the life of the Experiment. He added that the investment in overheads was considerably higher in Japan.

Despite the premature passing of Iain Stewart's great shipbuilding experiment, it had been, among other things, a serious attempt at rewriting industrial relations. Although not in existence long enough to test his theory, Stewart's idea of continuity of employment would most likely have reduced the need for restrictive practices. The key achievement was a whole new concept of industrial relations, one of consultation, cooperation and negotiation rather than confrontation. They tried to introduce a sense of partnership.

A report for the 1967 AGM prepared by Iain Stewart listed the ships completed since January 1966, together with their price as:

HMS *Fife*	£4,689,000
Duhallow	1,873,000
Wahine	3,085,000
British Commodore	3,599,000
Atlantic City	1,868,000
Total	15,114,000
Contracts in hand	
Indian City	2,028,000
HMS *Antrim*	7,015,000
2 'Glen' Liners	2,790,000
2 US Navy vessels	5,924,000
Jervis Bay	4,950,000
Total	22,707,000[16]

As the Fairfield Experiment drew to a close, Iain Stewart met with the workforce for a fifth and last time in the Lyceum cinema on Govan Road. Characteristically, he spoke in direct language. The last part of his long farewell as transcribed read:

> As I have said before, the proof of the Fairfield pudding will be in the eating of it, and I know that the new Board do not intend to waste any time in ensuring that the best ingredients of that pudding will soon be absorbed and consumed in the Upper Clyde area. You, yourselves, can do more than anyone to ensure that this will happen, by transferring your unreserved co-operation to the new management, for you have no other choice if you want to follow the only road from threatened bankruptcy to a healthy industrial livelihood. Before closing, and in thanking you for your pledge and for the way you have stood by it, I would like to stress three points.
>
> One. The future of the upper Clyde depends on the rapidity with which the various companies can be galvanised into one efficient trading group.
>
> Two. The Fairfield unit, now called Govan Division of UCS, is already earmarked as a key factor in the grouping so that much responsibility falls on your shoulders to ensure that none of the valuable ground which has been gained will be lost.
>
> Three. You can move into this new group with confidence and with genuine pride in what has been achieved to date.
>
> These three objectives may not have been reached in full and we may not have hit the jackpot in any of them, but, when the records are written, as they have been, are being, and will be, they will show that we have made a hell of a lot of real progress in umpteen different ways and that you have no need to apologise for any aspect of Fairfield's performance during our two years' experiment. Good luck to you all, Iain M Stewart.[17]

The Experiment had been roundly attacked, mostly by those who believed the State had no role interfering in private industry, and others, because it employed managers from outside shipbuilding. Despite this, there can be no doubt that, what happened at Fairfield during Stewart's tenure was a remarkable attempt to tackle deep seated issues in a hard bitten industry set in its ways.

On the key question of profitability, was the Experiment on its way to being profitable? The trading results for Fairfields (Glasgow) Ltd to the year ending 31 December 1967 were left to be prepared and registered by managers at UCS Ltd. Of these, a draft announcement made by Fairfields (Glasgow) Ltd stated that:

> The former directors of Fairfields (Glasgow) Ltd, were confident that, under the former Senior Management of the Company (the majority of whom subsequently resigned) and the methods of systematic control which were exercised by that Management, the Company as then constituted was trading profitably by the end of 1967, and was expected to make a reasonable profit for 1968.

This position was confirmed by management consultants appointed by UCS who, based on expected productivity improvements, projected a profit of £296k.

The Three Wise Men

The reorganisation of the shipbuilding industry in line with the recommendations of the Geddes Committee was managed through the creation of the Shipbuilding Industry Board (SIB) made up of three people: William Swallow, industrialist, Joe Gormley, National Union of Mineworkers and Anthony Hepper, formerly Managing Director of the hosiery business, Pretty Polly. The SIB's main function was to direct the creation of shipbuilding groups across the UK and to distribute grants to assist in the organisation of groupings. To explore the possible grouping of yards on the upper Clyde, a Working Party was set up on 2 April 1967.[18] The Working Party comprised three men, referred to at the time as the Three Wise Men: Anthony Hepper, Chairman and member of the Shipbuilding Industry Board; A W Giles, Managing Director of Baring Brothers & Co Ltd, and J F H Macmichael, a Director of PE Consulting Group Ltd.[19]

The working party's main recommendations included the following points.

- Sales of the new shipbuilding group should be £36m for merchant ships and £15m for naval vessels.
- The Stephen and Connell shipyards should be phased out over four years and the yards closed.
- There should be three Hull Production Divisions (one at John Brown's West Yard, one at Fairfield and the third at Barclay Curle's Clydeholm yard),[20] and one Fitting-out Division (at John Brown's East Yard).
- A two-shift labour scheme for the steelworking shops should be introduced.
- Productivity would need to rise by 30 to 70%.
- Capital expenditure should rise over five years from £7.26m to £10.97m.
- Profits should begin by 1972. This would be in

comparison to the average loss of £2.4m made by the five yards over the last three years.

The Report noted that the five yards had built an average of 210,000dwt of merchant ships per annum since 1960, valued at £23m per annum. This was equivalent to 2% of the world market open to UK shipbuilders. Tankers and bulk carriers accounted for the greater part of the tonnage, but only a quarter of the value. Most of these vessels were under 40,000dwt, while the largest (a BP tanker) was 71,000dwt. Over the previous ten years, the yards had built 33 naval vessels including two cruisers, two guided missile destroyers and 18 frigates.

The final group to be merged comprised John Brown, Connell, Fairfields, Stephens and Yarrow, all well known and long established companies with distinguished histories. The name chosen for the grouping was Upper Clyde Shipbuilders Ltd.

The management structure, with Antony Hepper, one of the 'three wise men', in the Chair, had three deputy chairmen, Sir Charles Connell, Sir Eric Yarrow and Iain Stewart. The three directors were Jim Stephen (of Stephens), Tom Burleigh (of John Browns) and M Hughes. Altogether, the five yards employed 13,000 people, had 26 building berths and covered an area of 203 acres. On formation in February 1968, the new company had 18 ships on order, valued at £87m. These ships were close to reflecting what the Clyde was best known for, including passenger liners, cargo liners, warships and 'one-off' types. Almost all of these contracts would be loss makers.

The unenviable task facing the new management was to bring together the five yards, each with their own working practices, markets, management styles and a very deep sense of tradition, and to forge them into a single entity. Only Yarrow was profitable, while Fairfields was part way through its great experiment in new management techniques and industrial relations, with many questions yet unanswered. Upper Clyde Shipbuilders did, however, have the most significant and prestigious shipbuilding contract in the world for the Cunarder QE2.

Upper Clyde Shipbuilders Ltd

In October 1967 Anthony Hepper was appointed Chairman of Upper Clyde Shipbuilders Ltd (UCS). It was not until 7 February 1968 that UCS began trading with Chairman Anthony Hepper announcing a £2.5m contract with the Ben Line. The hope for the new enterprise was that, given the lossmaking contracts previously booked, and the costs involved in setting up such a complex business, UCS would hope to be in profit after three years, but that it could take up to five. With the order book thinning out, the decision was made to take orders at a loss if necessary, in order to keep production going and labour employed while the company settled down and began the hunt for profitable work. The transitional period between loss making and profitability would be covered by loans and grants from the Shipbuilding Industry Board (SIB).

Top two at Upper Clyde Shipbuilders. Ken Douglas at left was the Managing Director, appointed in August 1969, and Anthony Hepper, Chairman.

Smoked Out

However, no sooner had the new Company begun trading than unresolved issues regarding the integration of the Fairfield yard into UCS surfaced. At one of the last meetings of the Fairfields (Glasgow) Ltd Board, held in February 1968, it was noted that no Fairfields personnel were represented on the UCS Board. Hugh Stenhouse said that the composition of the UCS Board gave 'little recognition to Fairfields achievements, nor to the fact that our shareholders were receiving the largest single block of shares in UCS'. He was concerned about the position of Fairfields' managers and personnel who might suffer because of the prejudice which had accumulated against Fairfields.[21] This issue drove Iain Stewart, in his position as a Deputy Chairman of UCS, but now isolated, to write to Hepper on 12 February:

> . . . they [Fairfields managers] are restless, unsettled and highly suspicious of the new Board's intentions, in spite of anything that may have been said at

> private interviews. They have been waiting for some private statement, other than my own, which will give tangible recognition from UCS that the Fairfields principles will find a place in the sun and will not be smoked out. As you could no doubt sense, however, the present list of appointments suggests to them that the latter is more likely.

For Iain Stewart, the situation he found himself in was intolerable and on 11 March 1968, he resigned from the Fairfields (Glasgow) Ltd Board and also as Deputy Chairman of UCS.

By April 1968 Fairfields (Glasgow) Ltd had been fully incorporated into UCS Ltd. Stewart later referred to those who decried his attempts to pioneer a new way forward at Fairfields in the following terms: 'Not unnaturally, short-sightedness, prejudice, parochial thinking and perhaps sheer bloody mindedness still linger in the Halls of Tradition.' Regardless of the lack of recognition given to Iain Stewart's 'Experiment' in the formation of the Board and the management structure at UCS, he was knighted in the 1968 Queen's Birthday Honours List for his role in saving the Fairfield yard.[22]

In October 1968 a group of UCS management and trade union officials organised a fact finding visit to Swedish shipyards, then among the most productive in the world. The objective was to look at man-management and the organisation of manual labour in one union. The three yards were Eriksbergs at Gothenburg, Gotaverken Arendal also at Gothenburg and Kockums at Malmo. A report written in December noted that very high levels of productivity were driven by the need to extract the maximum output from what is perhaps the highest cost shipbuilding labour force in the world. Such high levels of efficiency were evidenced by general work tidiness, mechanisation of routine handling operations, a clear cut organisation with defined responsibilities and good control systems.[23]

In comparison to the profitable Swedish yards, UCS would lurch from one financial crisis to another. The first came to a head on 7 February 1969 when UCS, on the verge of bankruptcy, submitted an urgent request to the SIB for £6m. Hepper attributed the deficit to:

- a shortage of working capital from the outset
- original profit forecasts were not met
- formation expenses incurred
- Cunard stopping payments for the QE2

To this he might have added that inherited loss making contracts were significantly worse than forecast. The late delivery and subsequent turbine blade failures of the QE2 brought with it a host of difficulties, leading to a fallout with Cunard and dire publicity. The public relations triumph that might have been expected for having built the world's most prestigious ship failed to materialise. The saga that took place between UCS, the SIB and Tony Benn at the Ministry of Technology continued throughout much of 1969. The recommendations of the independent reports into the business, suggesting the capital necessary to put the company on a sound footing, were only met in part, which ensured that further crises were inevitable. In November 1969 the UCS board consulted their lawyers to confirm, or otherwise, the legality of their trading position, given the poor financial state of the Company. By January of the following year, the situation had deteriorated, forcing Hepper to cable the Paymaster General (Harold Lever): 'Extremely concerned at non-arrival of financial support. Situation becoming critical due to increasing pressure from creditors. Company prospects now at risk. Directors also concerned about personal integrity and reputations.'[24] When the money did arrive, it was £3.3m of the £4m requested. A new management structure was put in place and the post of Managing Director was and filled by Ken Douglas, previously the Managing Director at Austin & Pickersgill Ltd on Wearside.

Notable completions during the UCS period were the guided missile destroyer *Fife*, the two survey vessels *Harkness* and *Chauvenet* for the US Navy and a series of 'Cardiff' class bulk carriers. One contract would cause serious problems for the Govan yard and result in a significant loss and bad publicity. This was the contract for the container ship *Jervis Bay*, contracted for during the Fairfields (Glasgow) Ltd days and laid down at the very end of that Company's brief existence.

Jervis Bay

In August 1965 a group of British shipping companies established Overseas Containers Ltd (OCL) to exploit the economies that containerisation, then on the point of being widely adopted, promised. In 1966 the design of a class of containership was undertaken by Blue Funnel at their Liverpool offices by naval architect Marshall Meek. The following year, contracts were placed for six ships of the *Encounter Bay* class. Five were awarded to a consortium of Hamburg based shipbuilders, Blohm & Voss, Deutsche Werft and Howaldtswerk. The sixth was awarded to Fairfields (Glasgow) Ltd in February 1967. All vessels were scheduled for completion between February and May 1969. In order to build *Jervis Bay*, Fairfields had to work with the German consortium acting as lead yard. In April 1967 agreement was reached for the delivery of all technical information and drawings, with the request that the first steelwork drawings and 1/10th scale burning prints should be supplied in June, with subsequent drawings thereafter,

Cutting a dash one year late. *Jervis Bay* during turning trials on the Firth of Clyde May 1970.

along with steel requisition lists, machinery and equipment ordering lists. Later however, the consortium said that they would not be able to begin the supply of drawings until mid-August, although this was brought forward to July. Consequently, Fairfields drew up a construction programme which envisaged a start on steel preparation on 16 October 1967, with the first units laid down in December, launch in mid-October 1968 and completion date as per contract on 27 May 1969. The penalty clause for late delivery was £1000 per day after four weeks. An additional issue for Fairfields was that the drawings were metric, whereas Fairfields was still working to the Imperial system, with the UK making a slow switch over to metric during this period.

By February 1968, with the Govan yard now part of UCS, the contract had already gone badly off programme by three months because drawings eventually received from the consortium were up to 19 weeks late.

A lengthy report made by Fairfields[25] into the problems facing the construction of *Jervis Bay* describes the late delivery of drawings, showing that, for example, double bottom drawings promised in July 1967 were not delivered until December 1967 and January 1968, that is between 15 and 22 weeks late. Engine foundation drawings were 12 weeks late, while certain transverse watertight bulkheads were 24 weeks late. It was noted additionally that some of the drawings supplied by the consortium were early versions which had not been updated when errors were found by the consortium in Hamburg. Where it was initially hoped to have 1000 tons of steel at various stages of preparation by December 1967, only 300 tons had been worked. It became clear that, under these circumstances, if the contract completion date was to be maintained, the ship would have to be launched no later than the first week in December 1968.

On 17 January 1968, a letter was written to Marshall Meek, Naval Architect with the Blue Funnel Line and who was by this time a Director of Ocean Fleets Ltd, that 'the late build-up of steel throughput has seriously disturbed our overall steel production programme and we have been obliged to subcontract steelwork for other vessels to local shipbuilders in an effort to maintain the programmed launch date for Ship No 839 [*Jervis Bay*]'. At the same date, the shipyard wrote to Shaw, Savill & Albion Co Ltd who would operate the ship, telling them that matters beyond the control of the shipyard may affect delivery of the vessel as covered by Clause 9 of the shipbuilding contract. *Jervis Bay* was not launched until 3 April 1969, five and half months later than the original programme date and six weeks before the original delivery date. Penalties of £1000 per day began in June. It was another public relations disaster.

By the end of 1969 UCS had received £20.2m from the SIB plus £4.3m for capital investment. The order book was worth £100m, with work through to 1972 and a 20% increase in productivity.[26] The completion of the container ship, which would take another year, was against a background of labour disputes, with 3500 redundancies to be carried out over the year, coupled to another cash crisis for UCS in March through to June 1969.

Training for all shipyard employees from professional design and technical staff to apprentices was an important recommendation of the Geddes Report. On 1 April 1970 Upper Clyde Shipbuilders Training Ltd, a subsidiary of Upper Clyde Shipbuilders Ltd, came into existence to train apprentices and men whose skills had been overtaken by incoming technology. The new company had offices in the former Alexander Stephen office block and used the refurbished former West Yard plater's shed as the centre of training activities.

At the beginning of 1970 a further crisis resulted in the Company receiving £7m as a final payment from the SIB. In February, with UCS now two years old, Hepper gave an upbeat spin on events, commenting 'we have a substantial order book, we have the necessary financial support, morale is high and we have the indications of rising productivity which is essential to make our company a success.'[27]

Despite this appraisal, in April 1200 boilermakers went on strike at Govan over a bonus payment dispute. With *Jervis Bay* in the final stages of completion and delivery now threatened, Hepper felt obliged to fly to Hamburg to arrange for the completion of the ship there if need be, with 14 May fixed as the delivery date. A return to work made this drastic move unnecessary and the ship was handed over on 14 May, one year late. Penalty payments were just short of £250,000.

In April 1970 the image of UCS was further dented when Yarrow left the Group, suggesting that they were leaving a sinking ship. Despite the difficulties the Group was going through, there were positive signs of improvement, particularly with the completion of troublesome legacy contracts such as *Jervis Bay*. Over September and October, orders for six standard ships were booked bringing the total for UCS's three standard designs to 30 ships worth £94m.[28] This reflected the company's aim to achieve stability through series production, or, as Ken Douglas put it, 'In the last eight months, we have been successful in achieving a balanced intake of orders of mainly standard ships, in order to provide a stabilised production schedule for the next two years. With this solid background behind us, we now feel that we can aim at widening the band of ships we can offer.'[29] As well as the three existing ship designs on offer, UCS was also developing designs for Panamax-sized bulk carriers of between 67,000 and 75,000 tons.

The 25,000dwt 'Cardiff' class bulk carrier *Irish Larch* on the day of her launch, 3 April 1973.

One notable achievement made during the UCS years was the creation of teams under technical director John Starks to look at new designs for vessels: cargo liners at Scotstoun, container ships at Linthouse and bulk carriers at Clydebank. The Clydebank team, led by naval architect James Shand came up with the successful 'Clyde' class cargo ship and the 'Cardiff' class bulk carrier.

In July 1970 Hepper told the workers, through the medium of the company newspaper, *Upper Clyde News*, that by the autumn the majority of loss making contracts will have been completed and that series-produced standard ships would result in break even for 1970/71 and profit by the end of 1972. The paper reported that orders for 11 ships had been booked worth £30m. Nine of these were for the 26,000-ton standard bulk carriers and two for the new 18,500-ton multi-purpose 'Clyde' class. Together with existing contracts, this meant that UCS had series-production of 15 bulk carriers and 9 'Clyde' Class, valued in total at £60m taken at realistic prices, with deliveries into mid 1972.[30]

As late as March 1971 Ken Douglas said he believed that 'UCS was now the best placed shipbuilding company in the country . . . recently booked contracts were healthy and realistically matched to our capacity and work force. Productivity, for long at a low ebb, is showing an encouraging increase. We are now entirely on our own. The last Government made it clear that we could expect no further aid and the present Government have made it equally clear.'[31]

By the early summer of 1971, UCS was once more facing the all too familiar cash crisis and was forced to ask the Government for a £6m loan. The new Conservative Government, elected in June 1970, took a different view of state bail-outs to industry than their predecessors and, in the wake of its rescue of Rolls Royce following the collapse of the latter in February 1971, they said that no further State aid would be provided for 'lame duck'[32] industries.

On 8 June 1971 Hepper visited the SIB to ask for a further £5m to maintain the Company's position. While the SIB was 'sympathetic and recognised the tragic nature of the situation in view of the recent improving trend which they had noted with approval' they were nevertheless unable to provide more money.[33]

With the refusal of the Government to provide the working capital requested, the directors, despite 'the certainty of a viable future' had no alternative but to file for the appointment of a provisional liquidator. This they did on 14 June. The liquidator was Robert Courtney Smith CA, who had previously 'wound-up' Dumbarton shipbuilders Wm Denny and Co and, most recently, Fairfield Rowan Ltd.

Despite the upward trend in productivity, in the end it was a shortage of working capital that forced the Company into liquidation. The refusal to accept the collapse of UCS brought on one of the trade union movement's finest hours.

The UCS Work-in

While the Government saw a way forward for shipbuilding on the upper Clyde, the proposal put forward by John Davies, Secretary of State for Trade and Industry, was to form a new company, comprising the Govan and Linthouse yards only, which would reduce the workforce from 8400 to 2500. Since the collapse of Fairfield in October 1965, the fight for survival against the odds had engendered a justifiable defiance among the workforce which the Government's plan brought to a head. Job losses on this scale were completely unacceptable and resulted in shop stewards forming a Coordinating Committee under the chairmanship of Jimmy Airlie, the Convener at the Govan yard. The stewards formulated a campaign which did the opposite of what many expected. They decided not to strike but to continue to build the ships on order, despite the fact that UCS was bankrupt. They called this, the 'Work-in', which immediately transformed perception of the unions. The success of this campaign and the public support it created was very much to do with the way in which Clydebank shop steward Jimmy Reid so eloquently and passionately articulated the objectives of the Work-in, capturing public imagination and wrong footing the Government of Ted Heath. The mantra of the shop stewards, 'the right to work', was supported by a Fighting Fund, made possible by contributions from individuals and organisations from all over the world. It was a master class in public relations. This Fund enabled workers made redundant to remain in their jobs with full pay. The sympathetic approach taken by the Liquidator, Robert Courtney Smith, ensured good communication and understanding with the shop stewards.

Nevertheless, the Government continued with its strategy and, on 21 September 1971, a new company, Govan Shipbuilders Ltd, was incorporated, based on the Govan yard and fabrication facilities at Linthouse. Hugh Stenhouse, who had developed a major insurance broking business based in Glasgow, was appointed as Chairman, Ken Douglas, who had been Managing Director of UCS, was Deputy Chairman and engineer Archie Gilchrist, most recently in charge of Edinburgh based Brown Brothers Ltd, was Managing Director. However, there could be no meaningful progress without union support and the resulting stand-off between John Davies at the Department of Trade and Industry (DTI) and the UCS shop stewards was played out against a dwindling order book. Contracts booked by UCS for the Irish Shipping Co, but not yet laid down, stalled, as the ship owners understandably wanted

Jimmy Reid addresses a mass meeting: 'The workers of Britain are getting off their knees, getting on their feet and asserting their dignity . . .' Part of the speech delivered by Reid in the engine shops at the Linthouse Division of UCS in August 1971.

completion guarantees, which the new company could not give without union support. The 'four yards or no yards' stance taken by the stewards held firm, supported by virtually all of the workforce.

While UCS workers campaigned outside the Conservative Party annual conference in Brighton in May 1971, the Labour Party conference, also held in Brighton that month, passed a resolution that the shipbuilding industry should be taken into public ownership. This was the beginning of a process that would lead to a manifesto commitment and, finally, an Act of Parliament in 1977.

The first sign that the Government was reconsidering the formation of a new company occurred when, subject to a feasibility study, it agreed to consider including the Scotstoun yard in the new grouping. By mid-October, intense discussions between the shop stewards, Govan Shipbuilders, the Liquidator and the DTI, resulted in union cooperation. This, in turn, allowed the Government to provide guarantees for the Irish Shipping Company's £13m order for the four bulk carriers. The shop stewards still insisted that the Clydebank yard had to be included or at least saved as an independent facility. The first indications that this might happen occurred when several American based companies expressed interest in the formerly iconic John Brown yard. On 8 August 1972, the last of the shop stewards' demands fell into place, when the Marathon Manufacturing Co of Houston, Texas, acquired the Clydebank yard, not to build ships but to build oil rigs. This signalled the end of the 15 month long Work-in[34] and brought the short life of UCS to a close. These events, which at times seemed doomed to failure, were played out against a blaze of press and television publicity.

Perhaps the simplest and certainly the most concise

summary of the UCS debacle comes from its Liquidator, Sir Robert Courtney Smith, who said: 'The government had control of the financial destiny of the company, it encouraged it to continue, removed the support and then denied moral responsibility.'[35]

Govan Shipbuilders Ltd

At a meeting in London on 16 December 1971, while the Work-in was still in full swing, the SIB/DTI asked Consultants Hill Samuel/Maynard PRC to examine a viable way forward and an appropriate investment strategy for the production facilities of the failed group. The work of formulating this strategy was undertaken by the Swedish engineering consultants AB Svenska AG, of Goteborg. One month later, on 19 January 1972, at a further meeting in London, an initial set of proposals was put forward in response to the objectives that had been tabled at the December meeting, namely 'to recommend improvements to make profitable production in the future, and at the same time, keep the total workforce occupied'. This report was followed by a second, more detailed study, tabled in March 1972. Meanwhile, numbers employed continued to fall. While employment in June 1971 stood at 2932 in Govan, 410 at Linthouse and 1225 in Scotstoun, by January 1972 figures had fallen to 2695 at Govan and Linthouse combined and 1057 at Scotstoun.

The measure used for comparison purposes in determining new facilities for the proposed Govan Shipbuilders was described as a European Competitive Yard (ECY), that being a typical modern European yard of the same type as the Clyde yards. Against this, the Govan yard, although comprehensively modernised ten years earlier, was some way behind the ECY, and the Scotstoun yard significantly behind. The initial proposals demonstrated some blue sky thinking and significant upgrading of yard layouts and facilities:

Govan/Linthouse
Alternative 1A
This totalled over £15m of investment across the integrated Govan and Linthouse yards covering approximately 150 acres. The shipbuilding berths at Linthouse were dispensed with, while the Govan yard was reduced to two berths, capable of accepting hulls up to 45,000 tons. The main feature of the proposal was a 300m x 56m building dock of 150,000dwt capacity on the former Fairfield West yard site, with large 'build-up' shops to the west and an outfitting terminal cut into the river at Linthouse.
Alternative 1B
Linthouse. The total spend at the former Stephens yard totalled £10.1m and included a conventional building berth, capable of accommodating ships of 150,000dwt.

At Govan only
Alternative 2
This proposal was concentrated solely on the Govan yard, and proposed the reduction of berths from five

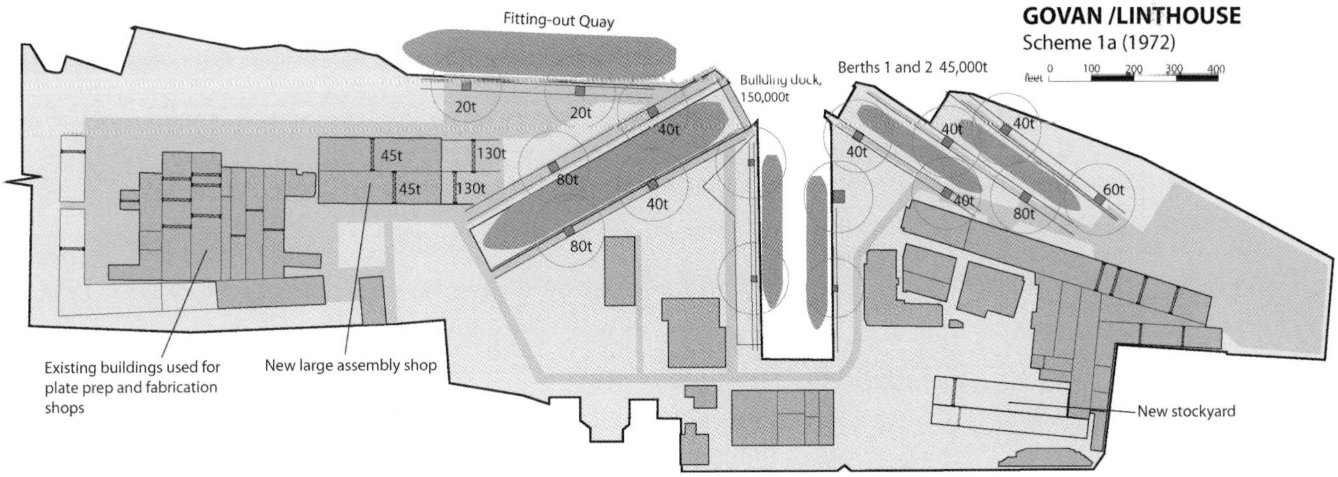

What might have been. This is the most ambitious of several proposal for the Govan, Linthouse and Scotstoun yards. The schemes were produced by AB Svenska Mec of Gothenburg for Hill Samuel & Co, who had been commissioned by the Government to find a way forward following the collapse of UCS. In this scheme the Govan yard was largely unchanged with the exception of the large building dock where the West Yard had been, the demolition of the old engine works and the re-siting of the stockyard. The large assembly shop, larger than the subsequent Tank Assembly Shop, was located on ground between the Linthouse and Govan yards.

to three, each capable of building up to 45,000dwt. This scheme was costed at £5.3m.

At Scotstoun only
Alternative 1
Two new berths were proposed in this scheme, one for vessels up to 45,000dwt and the other for 30,000dwt. The total proposed investment here was £3.8m, including new berth cranes and fitting out wharf.
Alternative 2
Total spend here was £4.5m, including one new berth for hulls up to 125,000dwt.

The outcome, however, was much less ambitious, envisaging the eventual elimination of the Linthouse yard and the upgrading of both the Govan and Scotstoun yards, to enable a total of 70,000 tons of steel to be worked annually, 45,000 at Govan and 25,000 at Scotstoun. There was no major increase in sizes to be built, with vessels up to 80,000dwt at Govan and up to 62,000dwt at Scotstoun.

The general trend was to build more of the ship under cover and, where up to 40 tons had been the maximum lift of berth cranes in the Fairfield modernisation of ten years earlier, new berth cranes of 60 to 80 tons were to be the standard. At Scotstoun, a new fitting-out quay with craneage was cut into the river bank. The plan developed for both sites envisaged productivity improvements across the board, organisation, systems, procedures and methods.[36]

In February 1972, with the feasibility report available, the Government confirmed investment of £35m for Govan Shipbuilders, somewhat more than expected, of which £17m went to meeting losses over the first three years and £18m for investment in facilities and working capital.

The Govan and Linthouse yards were rebranded as Govan Shipbuilders and the former Connell's yard became Scotstoun Marine, a wholly owned subsidiary of Govan Shipbuilders. Although Linthouse had ceased ship construction at the beginning of the UCS period, its fabrication facilities were retained for a time, and the site was not sold off until 1987.

On 1 July 1972 Govan Shipbuilders Ltd, wholly owned by the British Government, began trading. Secretary for Trade and Industry John Davies announced that the Government would support Govan Shipbuilders for a period of five years or until it was viable. A few weeks later, the new Company was able to announce that a four ship order for the Kuwait Shipping Co had been booked. In September, a cheque for £2.1m was paid by the new company to the UCS Liquidator for the facilities and goodwill of the old company.

Archie Gilchrist addressing an audience in the Boardroom at Fairfield.

As early as September 1971, Davies had asked Archie Gilchrist, the Managing Director of Brown Brothers, manufacturers of stabilisers and other marine gear, if he would manage the new company. Gilchrist had served an apprenticeship as a marine engineer at Alexander Stephen's shipyard in Linthouse, Glasgow. After a period at sea, he worked at the Barclay Curle shipyard, where his father was Chairman. Gilchrist subsequently became managing director of Barclay Curle, as well as a director of parent company, Swan Hunter. In 1964 he was Managing Director of Brown Brothers, until he took on the role of Managing Director at Govan Shipbuilders Ltd.

The order book at the newly formed Govan Shipbuilders Ltd (GSL) comprised a series of 'Cardiff' class bulk carriers, four for the Irish Shipping Co, which had been ordered under UCS. The second of these, *Irish Maple,* became GSL's first launch. With the support of Government subsidies, the new company under Archie Gilchrist had no difficulty winning orders for 'Cardiff 'and 'Kuwait' class ships. The main issue facing Gilchrist was to make significant improvements in productivity, thus paving the way to a profitable company.

The promised capital investment materialised in early

1974, secured against an order book which then stood at £100m. Commenting on this, Gilchrist said:

> We have the go ahead from the Government to press ahead with a £15m modernisation plan, £10m at Govan and £5m at Scotstoun Marine. Broadly, our plan is aimed at reducing the number of berths at Govan but, by better use of the present labour force, doubling production, to give a theoretical launch capability of 15 vessels per year.

If investment was one side of the coin leading to higher productivity, then working practices and labour relations was the other. Gilchrist commented: 'We now have a situation where, if the workers feel they have a grievance, they can come and talk about it instead of going on strike. We have had several situations which, in the past, would almost certainly have led to a walk out. But with this, a feeling of mutual confidence, problems are ironed out the sensible way, round the table.' Commenting further, he said that management had brought workers in on planning future strategy. 'Workers have a lot at stake and a lot to offer. If they are in on the ground floor, it saves problems developing later.'[37]

However, higher productivity levels were proving elusive and after three years of trading Chairman Ken Alexander, who had succeeded Lord Strathalmond, used the occasion of the launch of *Ibn Hayyan* for the Kuwait Shipping Co in February 1975 to say, 'We have not yet succeeded in achieving and sustaining the very substantial improvements in efficiency and productivity which were set for us at the establishment of this company.'[38] Meanwhile, the Labour Government of Harold Wilson responded to the demand for nationalisation of the shipbuilding industry by preparing a Bill, which was not passed until 1977 as The Aircraft and Shipbuilding Industries Act.

Winning and Losing

The events that unfolded in March and April 1976 demonstrated just how difficult the market for new orders had become. What amounted to a bidding war between GSL and Hyundai Heavy Industries (HHI) at Ulsan took place in the race to secure six 'Kuwait' class ships. Ironically, Govan Shipbuilder's design of the 'Kuwait' class ships in question had been sold to the South Korean yard in November 1974. HHI paid royalties for the first 15 ships built under the terms of agreement, after which they could build the ships royalty-free. Govan was running out of work, which made winning this order vital. With the deadline for bids approaching and, aware that they were up against the Korean yard, a price of £8.25m per ship was initially submitted. This, however, was not sufficient to win the contract and neither was a subsequent bid of £8.2m. At the very last minute, it was agreed to submit a loss-making bid of £7.85m per ship, for which Cabinet approval was required. This was given on 16 March, the day that Harold Wilson resigned as prime minister. The loss making bid was too late, as the deal had already been done with HHI. In the aftermath of losing the contract, the future shipbuilding programme was reworked, which meant transferring a contract from Scotstoun to Govan. This stop-gap solution meant that large scale redundancies among GSL's 5600 employees had been averted for that year but were certain to take place in the following year unless new orders could be found.

Small compensation for the loss of vessels to HHI lay in productivity, which had improved when the 26,000-ton bulk carrier *Baron Pentland* was launched in April 1976. This ship had taken just 15 weeks to the launching stage since being laid down, where previously 25 weeks or more had been typical.[39]

Koreans on the Clyde

In the 1970s, shipbuilding in South Korea began its remarkable rise and, like the Japanese 60 years earlier, they visited Clyde yards to see how it was done. Jim Thomson, a manager at Scotstoun Marine, describes his encounters with the Koreans from HHI in the mid 1970s.

> Scotstoun Marine was part of Govan Shipbuilders and that is where I first met some of the Koreans, who came to Glasgow after Govan Shipbuilders sold HHI the design of the 'Kuwait' class vessels. I was instructed to explain what I did and have them tag along with me during the building of the vessel I was working on. I explained to the HHI personnel that I was called the berth manager and responsible for the vessel from laying the keel to launch, after which I handed the vessel over to the outfit manager who then completed the vessel and delivered it to the owner. I then explained the way we would build the vessel on the berth and the work we would do covering dry surveys, structural testing and the preparations for painting the vessel.
>
> As you can imagine, there were many questions by the HHI lads and this was very demanding for me, as I still had to undertake my normal duties on the berth. I introduced them to the different heads of departments – Shipwrights, Platers, Welders, Caulker/burners and Pipework etc, and had them explain what their work involved for building the vessel on the berth.
>
> I also tried to explain that I worked closely with the technical departments such as lofting during the

> build of the vessel. The final part was to introduce the HHI people to the classification society and finally the owner, who gave their view on the process of shipbuilding. I don't remember the number of HHI groups that I had to work with but it was several, going through the same process each time.

However, the high drama generated over the loss of the six-ship order to HHI was soon trumped just two months later when another six-ship contract for Kuwait was won by the Govan yard. This guaranteed work for another 18 months.

Meanwhile, the Public Accounts Committee at the House of Commons, in response to the yards' inability to meet productivity targets in 1975, coupled to greater than expected losses against increased financial aid given to Govan Shipbuilders, posed the question whether or not the Government should continue to bail out the yard or meet the social cost of closure.[40] The foundation of the case for continued public money was a future viable industry. The answer from the Department of Industry was that, once the major programme of modernisation was complete, now estimated at £25m, the Govan and Scotstoun yards would be equal to the best yards in Europe, provided its facilities were used to the maximum.

'I Saved this Bloody Yard'

During the 1970s the Govan yard proved to be a popular place for politicians and personalities to visit. In June 1975 Dennis Healy, then Chancellor of The Exchequer, visited the yard to see improvements taking place. One year later, in July 1976, Lord George Brown, who had done so much to save Fairfields with Sir Iain Maxwell Stewart ten years earlier, arrived to launch the bulk carrier *Dona Hortencia II*, saying good humouredly 'I saved this bloody yard.' In November of that year, it was Barbara Cartland who came to launch *Dona Magdalena*. In January 1977 Lord Mountbatten launched the cargo ship *Dona Paz,* using the opportunity to promote a potential £80m contract from shipping chairman Sirdar Aly Aziz. Yard convener, Jimmy Airlie, gave him the assurance that, if the ships were ordered, they would be delivered on time.

Most importantly, in January 1977, at a mass meeting in the Lyceum cinema, boilermakers, knowing that further contraction of the shipbuilding industry was a distinct possibility, voted to end demarcation, signalling that they were doing their utmost to save their yard despite past intransigence.

In December 1977 *Marine Week* published a major article on Govan Shipbuilders which was then celebrating its 5th anniversary. The essence of the article was that building repeat orders of a standard ship type, coupled to good industrial relations and improved facilities had resulted in greatly reduced building times. According to the article, man-hours expended in building the first 'Cardiff' class bulk carrier had dropped from 850,000 in 1971 to 600,000 in 1977. The target set for 1979 was to build the ship in 400,000 hours. At Scotstoun, the most recently completed ship was built in 740,000 hours against a target of 708,000. So far, 27 'Cardiff' class bulk carriers and 19 of the 'Kuwait' cargo class had been completed.

Govan Shipbuilders' Managing Director Archie Gilchrist, left, instructing Lord George Brown on how to launch a ship, in this instance *Dona Hortencia II* on 14 July 1976. Brown had been the political power behind the rescue of Fairfield after the October 1965 crash.

British Shipbuilders

In March 1977, in the midst of a major downturn in shipbuilding world-wide, a manifesto commitment by the Labour Government to nationalise the shipbuilding industry became a reality with the passing of the Aircraft and Shipbuilding Industries Act, which, from 1 July 1977, saw British Shipbuilders come into being, with 27 shipyards, both merchant and warship, five marine engine builders and three training companies. Its principal objective was, in the first instance, to stem losses, by reducing the size of the industry in line with what the shipyards were expected to win in a market that suffered from over capacity. Thereafter, it hoped to create a viable industry once trading conditions improved. The reality was however, that merchant shipbuilding in the UK and much of Europe was an endemic loss maker, retained for social and cultural reasons, by the public purse and political will. State assistance was vital if the industry was to survive and thus, in parallel with the creation of British Shipbuilders, the Government established the Shipbuilding Intervention Fund[41] to assist the industry in competing against severe competition, especially from non-EEC countries. Initially, this fund provided up to 30% of the contract price of a ship, but this was progressively scaled down over the years. Further incentive to buy British was also provided to ship owners through 'soft' credit schemes, typically 8.5 year loans at 7.5% interest for 80% of contract price.

The bulk carrier *Al Muharraq* on 25 April 1978, the day of her launch at Scotstoun Marine, a wholly owned subsidiary of Govan Shipbuilders and formerly known as Connells. Ships of the 'Kuwait' and 'Cardiff' classes were the backbone of production during the 1970s.

Headed by CEO Michael Casey, British Shipbuilders began formulating a plan for the industry which, it was understood, would mean a reduction in capacity. Meanwhile, with closures hanging in the air, the hunt for orders began. The hunt led directly to Poland.

The Polish Contract

This highly controversial contract, uneconomical from the outset, was taken as a stop-gap to keep British shipyards on the Tyne, Tees, Clyde and elsewhere, in work, without which significant redundancies would have been necessary. In November 1977 discussions between the British and Polish Governments resulted in British Shipbuilders signing a contract worth £115m with the newly formed Anglo-Polish Shipping Venture Ltd. The contract was to build 22 bulk carriers of various sizes and two crane barges at yards including Swan Hunter and Govan Shipbuilders.

When labour issues caused Swan Hunter to decline its share of the contract, the three bulk carriers provisionally allocated there were re-allocated to Govan. This action was in itself a matter of some contention in trade union circles. However, although involved in the design of the smaller Polish vessels, Govan had no direct role in pricing the ships, the negotiations for which had been conducted at British Shipbuilders level.

From the start, management at Govan knew that the contracts would result in a significant loss, despite the Intervention Fund subsidy of £28m. Additionally, the Polish ships were smaller than was economical for a yard of Govan's size. Moreover, with these vessels occupying production time of several years, it meant declining contracts for the more suitable ships of the successful 'Cardiff' and 'Kuwait' classes. This occurred when a potential order for four 'Kuwait' class bulk carriers for United Arab Shipping Co had to be turned down.

Milton Friedman, the right wing American economist in a visit to Glasgow, to make the film *Free to Choose* used Govan Shipbuilders' Polish contract as a case study. This was one of the influences on Margaret Thatcher's eventual decision to privatise British shipbuilders.

Perhaps because Govan Shipbuilders had stepped into the breach by taking on the Swan Hunter contracts, Michael Casey, Chief Executive of British Shipbuilders, said of them in a press announcement: 'This yard has had a chequered history but has shown very considerable progress in recent years with a substantial improvement in productivity and remarkable freedom from industrial strife. I view Govan as being a very promising yard indeed, particularly if current progress is sustained.'[42]

Nevertheless, in November 1978, British Shipbuilders issued a press release stating a loss of £108m for its first 9 months of trading. This was attributed to 'inherited losses, political delay and market conditions'. The release also noted that 'it was the Corporation's immediate task to bring under control the rapidly deteriorating situation with which it was presented.' British Shipbuilders' basic function was a restructuring of the industry by the reduction of capacity, particularly the merchant division, in line with market realities and the minimising of losses by raising productivity.[43] Yet again, uncertainty swept through the yards, something which Govan was well used to.

In January 1979, as the Polish contract was working its way through Govan and Scotstoun,[44] Archie Gilchrist was appointed Chairman of Govan Shipbuilders in addition to his role as Managing Director. On this occasion, he acknowledged the great uncertainty about the future and that a world-wide shrinkage in shipbuilding capacity was inevitable, a process that had already begun in Japan and Sweden.

In June 1979 an important point of departure was reached for Govan when yard convener Jimmy Airlie left to become a full time AUEW official. Airlie had contributed significantly to greatly improved labour relations and flexible working arrangements, in the knowledge that the survival of merchant shipbuilding was hanging by a thread and any intransigence on the part of labour or old style disputes would hasten the end of the industry.

Under Airlie's guidance, the vehicle for great strides in industrial relations was the Shop Stewards' Coordinating Committee, first established during the UCS work-in. This remained throughout the Govan Shipbuilders era and, re-styled as the Shop Stewards Negotiating Committee,

Jimmy Airlie signing off as yard convener and handing over to his successor Sammy Gilmore in June 1979.

ensured that all issues in the yard were dealt with through this unified body which represented all the unions on site. This was a major element in the road to greatly improved working practices and higher productivity and finally, the abandonment of demarcation altogether, a stage reached earlier in 1977. His place as yard convener was taken by Sammy Gilmore who, with Sammy Barr, was joint Chair of the Negotiating Committee.

In August 1979 British Shipbuilders' long awaited corporate plan was announced. As expected, major closures were part of the plan in line with the decision to reduce merchant shipbuilding capacity by one third. With 10,000 jobs to go across the UK over a period of 18 months, the Clyde was hit with the closure of Govan Shipbuilder's subsidiary, Scotstoun Marine, and Scott Lithgow's yards at Cartsdyke and Bowling. With no compulsory redundancies at Scotstoun, the yard would be placed on a care and maintenance basis and the site tentatively offered to neighbour, Yarrow Shipbuilders Ltd. The Govan yard did not escape the cuts, with 850 jobs to go, 350 staff and 500 hourly paid workers. It was a bitter blow for the Clyde, but Govan had dodged the bullet yet again.[45] The same could not be said for Chairman and Managing Director, Archie Gilchrist.

Ruthless and Dictatorial

If senior management can characterise a particular era, Govan Shipbuilders first era came to an abrupt end on 11 December 1979. On that day, Archie Gilchrist flew to London to meet with British Shipbuilders' Chairman, Sir Anthony Griffin, and Chief Executive, Michael Casey, ostensibly to discuss the final stage in the contract for two 'Cardiff' class bulk carriers for Bishopsgate Marine. Or so Gilchrist thought. In what one journalist subsequently described as 'a deplorable case of bad manners on the part of British Shipbuilders',[46] Gilchrist was told that the orders had been secured, but that an unwritten condition was that he should either resign, now, or be sacked. The manner of his sacking caused a furore among politicians and others in Scotland. While the reasons for Gilchrist's sacking were never made clear, Casey subsequently said that, with reference to the Polish contract, promised improvements in productivity had not materialised. Gilchrist, who felt that he had been made a scapegoat for losses on a contract he had not negotiated, had achieved much in his time as Managing Director. Labour relations were greatly improved, productivity was rising (largely through repeats of the 'Cardiff' and 'Kuwait' class vessels) and a scheme of modernisation had been put through.[47]

On the day of Gilchrist's sacking, senior managers at British Shipbuilders compounded their earlier atrocious behaviour by, bizarrely, having his successor, Eric Dermot

Eric Mackie who famously said that he came to the Govan yard as Managing Director to close it down but then decided he would do everything he could to keep it going.

Mackie, arrive at Govan to take over Gilchrist's role. Mackie was a Harland & Wolff trained marine engineer who was British Shipbuilder's Director in charge of ship repair. His remit from the Board of British Shipbuilders was to assess the situation at Govan and close the yard if necessary.

Earlier, in October 1979, to sweeten the pill of industrial action likely to arise from redundancies caused by the closure of Scotstoun Marine, British Shipbuilders concocted a plan whereby a contract for two 'Cardiff' class bulk carriers would be placed at Govan for Greek owners, Liberty Maritime, on a hire purchase basis. This contract was considered unique, as the builder, British Shipbuilders, had the right to cancel at any time up until delivery, without penalty. This clause in the contract, which was never made known to the workforce, was there to enable British Shipbuilders to close Govan once the Polish contract was completed and if the performance of the Govan yard in building both bulkers was not satisfactory. Despite these threatening undertones, a report written after the ships had been delivered said:

> Govan's workforce more than justified the confidence placed in them. They achieved a breakeven figure better than expected, which lead to the repayment of approximately £0.5m of the total Intervention fund subsidy back to the Government. Moreover, the first ship was delivered one day ahead of time and the second vessel six weeks ahead of time. The Liberty Maritime deal achieved its objective of creating the basis for a future viable shipbuilding facility at Govan.[48]

When British Shipbuilders published their results for 1979/80, it perhaps goes someway to explaining their peremptory action in sacking Gilchrist, if not their cack-handedness. Total group losses were £109.9M. Warship building yards were profitable, while merchant yards were the loss makers and, of these, Govan Shipbuilders with its Scotstoun Marine subsidiary headed the list at £17.5m on a turnover of £25.75m.[49] Employment in the merchant shipbuilding sector stood at 20,500, of which 5000 were employed at Govan Shipbuilders and Scotstoun Marine, the latter already scheduled to close. Against this background, it is no surprise that when Robert Atkinson asked Eric Mackie, then British Shipbuilders Director in charge of ship repair, to take up the post of Managing Director at Govan, it was to make an assessment whether or not the yard should close.

With Scotstoun Marine being run down, the total workforce at Govan Shipbuilders fell from 5500 to 2300. By February 1980, 2000 people had volunteered for redundancy, of whom 1000 had already left. The Polish contract came to an end when *Kopalina Miechowice* left Govan in April 1980 and *Warka* left Scotstoun in June 1980.

While these events were underway, Eric Mackie began to form a solid relationship with the workforce at Govan. At a series of meetings with senior shop stewards and subsequently, in mass meetings held in the Lyceum cinema, his blunt but straight forward approach was accepted by the stewards and, in particular, by what Deputy Managing Director, Bill Gallagher, describes as an 'enlightened group of senior shop stewards' headed by convener Sammy Gilmour.[50] By early 1980, Mackie had decided that the Govan yard was worth saving, a view that he communicated to the Board at British Shipbuilders. As if to test the resolve of both Mackie and the workers at the yard, the first contract taken under Mackie's control, was the re-engining of two container ships. The subsidy went not to the yard but to the ship owner, P&O. The ships concerned were the *Encounter Bay* class container ships *Moreton Bay* and *Discovery Bay,* German built sister ships to *Jervis Bay*. The work scope included removal of all asbestos, dismantling the steam plant and re-fitting with diesels. Additionally, the container guides had to be altered to accept 40ft containers. The refits, which lasted from August 1980 until March 1981, were delivered on time and to cost.

In March 1981 Robert Atkinson, the new head of British Shipbuilders since June the previous year, announced a £109m loss for the year. While the Warship Division made a profit of £32m, up £5.5m from the previous year, the merchant division lost £44m, an improvement on the £78m loss the previous year. He reiterated that the 'principal and cardinal objective of our policy is essentially the restoration and increase in productivity leading to a reduction in unit cost.' Later in the same narrative he said, 'failure by individual companies to improve their productive performance will inevitably lead to stable employment prospects being jeopardised.'[51]

The market for merchant shipping remained severely depressed, forcing many Western European yards into bankruptcy. In Japan, the world's leading shipbuilder, capacity was cut by 35%.

At Govan, the arrival of Mackie was like a fresh start and, following the successful contract for the re-engining and engineering work on the 'Bay' class container ships, the yard was awarded two further 'Cardiff' class bulk carriers, *Lord Byron* and *Lord Curzon,* for Bishopsgate Marine. This contract was largely down to the support given to Govan by John Parker, Depute Chief Executive of British Shipbuilders under Atkinson. These vessels would be the last but one of a long line of highly successful designs which, with the 'Kuwait' class, had been the backbone of output for the previous ten years. Last of the line of the 'Cardiff' class bulk carriers was *El Conqueror* for Panamanian owners, launched in September 1982. Thereafter, the order book began to change, with larger and more complex ship types. First of these, with a significant jump in size, were the 66,000dwt Panamax bulk carriers *Pacific Peace* and *Pacific Patriot* for CY Tung of Hong Kong and *Sir John Fisher* for Barrow-in Furness based James Fisher & Sons.

These contracts were followed by a £50m contract to build three specialised grain carriers, two for the Misener Transportation Co (*Selkirk Settler* and *Canada Marquis*) and one for Pioneer Shipping Ltd (*Saskatchewan Pioneer*), both of Canada. Of 35,000dwt, these dual purpose vessels were designed at Govan to work the Great Lakes during summer months and, during winter months, when ports on the Great Lakes were icebound, carry out deep-sea trading. The Misener ships demonstrated the strength of GSL's design expertise and quick response to shipowners' requirements. This, as much as price, tipped the balance in Govan's favour, particularly with a deep economic

Pacific Peace is launched on 15 October 1981, the first of three 66,000dwt Panamax bulk carriers. This was the largest ship launched at Govan since the BP tanker *British Commodore* in 1966.

recession underway, coupled to ongoing severe competition from the Far East.

As part of British Shipbuilders' corporate plan, Govan Shipbuilders, like other yards in the corporation, had to produce what was in effect a strategy for survival. This strategy recognised that Govan was 35% more expensive than their Japanese rivals and that man-hours per ship had to be reduced by 47% over the next four years. It was equally clear that financial support from British Shipbuilders would only be given to yards that were showing improvements in productivity. In response, an action plan was drawn up at Govan which included 'PIP',

Canada Marquis, one of three bulk carriers built for service on the Great Lakes.

An aerial view taken in 1982 showing the Panamax bulk carrier *Sir John Fisher* and the 'Cardiff' class bulk carrier *El Conqueror* in the fitting out basin, and the Great Lakes bulk carriers *Selkirk Settler* and *Canada Marquis* on Nos 1 and 3 berths respectively. The first units of *Lakenes* have been laid down on No 2 berth. Treelined Elder Park is at right while the widespread demolition of tenements in Govan is evident at top centre.

Productivity Improvement Projects, aimed at increasing productivity. There were 20 of these in total, covering various aspects of production.

Winning any contract was a significant achievement. Part of the struggle to achieve competitiveness lay in developing new production techniques and this period at Govan is characterised by the development of advanced pre-outfitting of units and modules. Where, previously, individual fabricated units weighing anything from 40 to 100 tons, had been mostly bare steel, certain units were now pre-fitted with significant amounts of pipe work and other items of equipment. Advances in CAD and technical planning had made pre-outfitting of modules possible and, in the case of pipe work, by investment in a new equipment,

Launch of the 45,000dwt bulk carrier *Loftnes* from No 3 berth on 24 March 1984, the second to be built for Jebsens (UK) Ltd.

the pipe shop became the most advanced of its kind not just in the UK but among the best in Europe.

With the Misener ships scheduled for construction with delivery dates from April to September 1983, the hunt for new orders to avoid a gap in production was on. Two 45,000dwt bulk carriers, *Lakenes* and *Loftnes* for the Norwegian Jebsen Group, were won. Completed in March and July 1984, they were, for a time, the only orders in the yard.

The unprecedented dearth of orders generally and overcapacity in shipyards worldwide was compounded by South Korea's State-supported entry into shipbuilding with several vast new shipyards.

Privatisation

On 25 July 1984, the Secretary for Trade and Industry (Norman Tebbit), reported that British Shipbuilders had made a loss of £161m in 1983/4, of which almost half was attributed to Scott Lithgow on the Lower Clyde. He stated that it was the Government's intention to privatise the warships yards such as Yarrow and Vospers by March 1986, and to privatise the remaining seven merchant yards, all of which were loss makers, later. He pointed out that, up to March 1985, assistance of £1.25m had been provided by British Shipbuilders to the shipbuilding industry, much of it in the form of redundancy payments.[52] Through 1985 and early 1986, all British Shipbuilders' warship yards were sold off.[53] The pressure on Eric Mackie, in the hunt to keep his workers employed, already considerable, now had the added complication of an imminent return to private ownership.

Eric Mackie with Norman Tebbit (Secretary of State for Trade and Industry 1983–85) and Graham Day (Chairman and Chief Executive of British Shipbuilders 1983–86) during a visit to the yard in 1984.

Early in 1984, the Central Electricity Generating Board issued invitations to tender for three 19,000dwt colliers to take coal from the River Tyne to power stations on the Thames. When the £28m contract was awarded to Govan Shipbuilders in April 1984, the press release said it was won against intense competition from the Far East and also from within British Shipbuilders. As if to correct any impression that this had been an unfair contest, CEGB Chairman Sir Walter Marshall said that, while Government Ministers made it clear they would be pleased if the order went to a British yard, there was no pressure from within the CEGB to choose a British yard. Graham Day, British Shipbuilders' Chairman since 1983, said that Govan had not been nominated by British Shipbuilders, but that Govan had produced the most competitive price after an internal competition for the order involving three yards.

These events unfolding in the UK were taking place against significant reductions in shipbuilding worldwide, with yards in Japan, France, Germany and the Netherlands going out of business. In Sweden, a country which had arrived late to large scale merchant shipbuilding but had modern and efficient yards, the decision was made to abandon the industry completely.

Norsea

In April 1985 the Govan yard pulled off one of the greatest successes of the British Shipbuilders era by winning the contract to build the 31,000-ton cruise ferry *Norsea* for P&O/Nedlloyd, for service between Hull and Rotterdam.[54] This was billed as the first major passenger vessel built on the Clyde since the *QE2* had been completed at Clydebank

Built for North Sea Ferries Ltd and the Hull/Rotterdam service, *Norsea* was the first passenger ship to be built at Govan since *Empress of Britain* in 1956 and indicated the passenger ferry/cruise market direction that Eric Mackie wanted to take the Govan yard.

in 1969. For the Govan yard, under its famed Fairfield name, this was the first passenger ship since the *Empress of Britain* in 1956. Worth £40m, the contract promised new jobs and a delivery date of March 1987. J Graham Day, the new Head of British Shipbuilders, said it was probably the most important order placed with British Shipbuilders in recent years, as it represented a breakthrough into an important sector. Eric Mackie's comment was much more direct:

> By winning the order for the cruise ferry, we are putting our reputation and future on the line. We must deliver within time, quality and budget and produce a passenger ship which is better than anything similar built anywhere else in the world. I have every faith this will be so, but I can't emphasise too strongly just how much the future for all of us depends on our performance.[55]

One of the reasons why Govan Shipbuilders' tender was so attractive was the Cabin Module Manufacturing Facility which had been set up at Govan a short time previously. A licence to utilise the technology and process to manufacture cabins was acquired from Finnish shipyard Wartsila, who had developed it for building cruise liners. Two teams of 20, including joiners, sheet ironworkers, plumbers and electricians, were trained in Finland, with total interchangeability between trades.

On 8 September 1986 *Norsea* was launched by Her Majesty the Queen Mother with considerable fanfare. A little over eight months later, the ship was delivered to P&O on 7 May 1987. *Norsea* was a significant achievement for Eric Mackie and the Govan yard. When interviewed about its importance, Mackie said, in reference to Govan's quality:

> I believe that *Norsea* – on time, within the price window asked for, to the standard and quality,

NORSEA HULL
FLYING SPINDRIFT

asserts that quality on the world market. … we've got the skills of generations, their fathers were shipbuilders, their fathers before them were shipbuilders, they've got that spirit, they've got that ability and that quality that I would find very difficult to match throughout the world.

Contracts were being pursued for passenger and container ships in what was still an exceptionally difficult time of low demand and over-capacity. The reality of the situation which the yard found itself in was that, after the departure of *Norsea*, there was no work at all and, unless this was found quickly, it would mean laying off over 740 people. Employment at the yard now stood at 1975, of which 1297 were direct and 678 were subcontractors. The shortage of work contributed significantly to a loss of £28m made by Govan Shipbuilders for the year 1986/7, of which £16m was down to under-recovered labour. Mackie reiterated that the yard was facing a crucial time and may have to face difficult decisions. Better news was received in May however, when Mackie was able to announce that they had booked an order from China Ocean Shipping Corporation (COSCO) of Shanghai for two container ships, with an option on a third.

The hoped-for entry into the passenger ship market made possible by *Norsea*, saw encouraging progress in discussions with Brittany Ferries for a 25,000-ton cruise ferry, subsequently the MV *Bretagne*. Months of design work were undertaken and even tank testing on what looked like a firm proposition. However, by June 1987, it was clear that the order for the French ship would stay in France, because the French Government, prepared to break EEC rules on competition, gave the French shipbuilder an additional subsidy.[36] In the event, the contract was placed with Chantiers de l'Atlantique in Saint-Nazaire and Govan was left with a major gap in its building programme.

Nevertheless, working-up designs for cruise ships for 250 to 600 passengers was seen as a major project for design staff and by June, company minutes noted that indications on the passenger vessel front were encouraging, in that there was a discernible trend towards higher price levels and longer deliveries. Meanwhile, the scramble for orders continued, with each enquiry taking up significant amounts of time and effort. By June 1987, the following contracts were being pursued:

Launched by HM Queen Elizabeth the Queen Mother on 9 September 1986. *Norsea* is seen here moments after entering the water. She is attended by the tug *Flying Spindrift*.

3 x Container ships (2700 teu) for Marine Computing Ltd (Zim) of Haifa.
2 x large passenger ro-ro ferries for Stena (in conjunction with North East Shipbuilders Ltd).
Small cargo liner, *St Helena*, which subsequently went to Hall Russell in Aberdeen.
2 x 2100 teu containerships for Hapag Lloyd.

At the same time, discussions were underway with:

Taylor Woodrow, for a floating dock at Coulport.
Holland-America Line, for a cruise ship.
The Institute of Production Control, for a cruise exhibition ship.

Despite market difficulties and the prospect of privatisation, British Shipbuilders were still investing in their merchant yards. In September 1987 Scottish Industry Minister Ian Lang opened a new CADAM (Computer Augmented Design and Manufacturing) centre at Govan, with 52 workstations to allow for better planning and control procedures. Being part of a larger organisation did bring some advantages. Associated group, North East Shipbuilders Ltd, previously Sunderland Shipbuilders Ltd and Govan's rival within the Group, assisted in providing some measure of work as a stop-gap for laid-off steel trades by transferring some steel work from the ferry contracts which they were working on.

It was not until late in 1987 that the first deliveries of steel for the COSCO ships were made and the steel trades could return to work. For the planning of these sophisticated container ships, the new CADAM suite came into its own, with Planning Manager Donald McPhail describing the process:

> for each ship built, a contract build strategy must be developed, with full consideration of how the ship will be built, which sections and modules will be constructed in which production area, what manpower, facilities and equipment will be required, and, most importantly, what the timescale of each task will be, how much it will cost and where it will fit into the building programme. Preparation of this strategy begins at the tender negotiation stage, several months before actual production work begins.

And with reference to manpower costs:

> the cut-throat competition we face around the world means that these must be kept to an absolute minimum, with no avoidable delays, wastage or excessive costs. In today's world shipbuilding market, such extra costs blunt our competitive edge and make it more difficult to find the follow-on orders which will ensure that we all stay in work.

The experience of Govan Shipbuilders had been a high wire act, fraught with several break points along the way, where it could so easily have been all over. Major improvements in productivity had been made. There was unprecedented labour flexibility and cooperation and the yard had been fronted since 1980 by Eric Mackie, who was determined to keep it in business. Quite unexpectedly, in February 1988, a Norwegian engineering conglomerate approached British Shipbuilders with an enquiry about the possible purchase of a shipyard.

Since 1972 £550m of State assistance had been given to British shipyards. In its first two years of operation, British Shipbuilders made losses of £154m. In 1977/8, assistance of £98m was given, with losses of £108m making a total of £203m. In 1978/9, assistance was £85m, with losses of £46m totalling £131m.

Kvaerner Industrier AS

The COSCO container ships *Taihe* and *Puhe* would be the last commercial contracts secured at the Govan shipyard before the next major event in the remarkable story of this yard, the sale to the Norwegian Kvaerner Group in June 1988.

The decision by British Shipbuilders to divest itself of its remaining shipyards presented yet another significant threat to the existence of the Govan yard. However, out of the blue, there emerged a Norwegian engineering conglomerate intent on acquiring a UK shipyard. Oslo-based Kvaerner Industrier AS was a diversified engineering group with interests ranging from oil and gas construction, mechanical engineering, wood pulp technology and shipbuilding. Kvaerner had taken advantage of Norway's booming offshore oil and gas industry to become a world leader.[57] Moreover, it was about to begin a major programme of growth which included shipbuilding. Kvaerner saw that, while shipbuilding suffered from over-capacity and was highly competitive, particularly in the bulk carrier market in which Govan had been trading, the market for transportation of liquified gas by sea would be growing and less competitive.

In 1961 Kvaerner had acquired the shipyard at Moss on the eastern side of the Oslo fjord, which in 1973 took the name Moss Rosenberg Verft after merging with the Stavanger based Rosenberg yard. At Stavanger, Kvaerner had pioneered the design and construction of LNG tankers

Although contracted for by Govan Shipbuilders, the container ships *Taihe* and *Puhe* were completed by Kvaerner; *Taihe* seen here on No 1 berth. The very substantial arrangement of staging and the number of handrails present on the structure of the ship indicate the great number of improvements made in health and safety. (*David Miller*)

based on the spherical tank system. The first such ship was delivered in 1973. By 1986/7 however, shipbuilding at Stavanger had ceased because of the high cost of construction in Norway and the emergence of the more lucrative construction associated with North Sea production platforms. With an accumulation of expertise in the design and construction of gas ships and with potential

Stern of *Taihe* shortly before launching on 25 January 1989. Note the water brakes fitted to the underwater hull to help slow the vessel down as she enters the water.

orders in the pipeline, Kvaerner was reluctant to let go of this niche shipbuilding sector, foreseeing that the market for gas ships would grow in the 1990s.

To sustain its ambition, Kvaerner needed to acquire shipbuilding capacity, especially in the European Community (EC) where subsidies were available, unlike in Norway. The UK was of particular interest, as labour costs per hour were estimated to be 57% of those in Norway. Additionally, the Shipbuilding Intervention Fund operated by the UK Government which, in 1988, offered up to a 29% subsidy, made the UK additionally attractive.

In February 1988, Kvaerner approached British Shipbuilders about the possibility of acquiring Govan Shipbuilders, which British Shipbuilders hoped to return to private interests. What happened next would take the Govan yard, under Norwegian management, into one of the most interesting periods in its existence. It would be transformed into a modern, efficient, commercial shipyard, building high quality, sophisticated vessels equal to or better than anyone else in the world.

From March to May 1988 Kvaerner examined the performance of the Govan yard. This resulted in the following conclusions: manning levels were too high, indirect costs were too high and investment was needed to improve performance. A 5-year Plan was submitted to British Shipbuilders at the end of May, which initiated negotiations. These continued from June until 9 August, when a deal for purchase was concluded. Kvaerner immediately placed an order for two LPG carriers at Govan, with UK Government support for two further vessels. Kvaerner's aim was to transfer gas ship technology from its Norwegian yards to Govan and to raise performance of the yard to the best European levels. This meant reducing man-hours per ship by 25-28% and a reduction of build cycle times by 25%, depending on ship type. It also meant making nearly 500 employees redundant. For this, Trade Union approval was necessary. With no alternatives other than closure to hand, the unions agreed.

Kvaerner paid £6m for the yard but, under the terms of the agreement, British Shipbuilders would contribute £25m to cover the cost of redundancies and make a contribution to capital investment and restructuring. Losses that Govan Shipbuilders had incurred would stay with that company and not be carried forward into the new company. The sale and the support from the Intervention Fund offered by the British Government had to be approved by the European Commission. This amounted to payments of £1.25m investment aid and £19.6m to fund redundancies.

Kvaerner takes over. This photograph was taken in the boardroom of the Fairfield building after negotiations with Kvaerner were concluded. Eric Mackie is at left with Kvaerner executives Hans Jurgen Frank, Jan Flatseth and Jon Hatleskog.

A condition applied by the EC to granting aid was the reduction in the capacity of the yard from three building berths to two.[58] The yard would take the name Kvaerner Govan Ltd (KGL). It was agreed that LPG technology accrued over the years by Moss Rosenberg Verft should be transferred to Govan as soon as possible, to enable the latter to be self-supporting in LPG ship construction.

At this stage in the decline of merchant shipbuilding in the UK the remaining yards were Appledore Shipbuilders, Ferguson Shipbuilders and the Govan yard. North East Shipbuilders which included the modern Pallion yard in Sunderland had found no buyer and closed in 1989. Belfast shipbuilder Harland & Wolff although Government owned also remained although not under the control of British Shipbuilders.

The Kvaerner Board noted that £557,000 had been paid for the assets of Govan Shipbuilders and that British Shipbuilders had agreed to fund 500 redundancies, to pay a goodwill payment of £900 to each employee, to pay £10,195,000 for under recoveries, £300,000 towards specialist welding training and £7,450,000 for capital redevelopment. Financial and manpower data from Govan would be forwarded to Moss Rosenberg Verft monthly.[59] In September 1988, the redundancy programme began. This required 276 manual workers and 216 staff to leave by the end of 1989. Not unsurprisingly, this affected motivation levels, with absenteeism running at over 20%. It is fair to say that most of the individuals selected for redundancy considered themselves lucky, whilst probably many of the remaining employees looked forward to a similar payoff if the new owners did not succeed.[60]

While the managing board remained the same, with Eric Mackie as Managing Director, Kvaerner's Hans Jorgen Frank was appointed Chairman. Eric Mackie made it clear that he would retire after one year. Over October and November 1988, the new subsidiary company Kvaerner Govan Ltd (KGL) developed an investment programme to upgrade steel working facilities, with the aim of producing 450 tonnes of steel per week from one shift – equivalent to 20,700 tonnes per annum. This was to improve productivity in steel working by 40%. A Programme Report was prepared by Norwegian firm, Total Transportation Systems (TTS), and presented to the Board. TTS specialised in custom designed material handling and production line technology, particularly for shipbuilding, at its base in Norway. It also had a consulting company specialising in shipbuilding technology and related productivity, located at Newport News, Virginia, USA, the site of the vast Newport News Shipbuilding complex, builders of aircraft carriers and nuclear submarines for the US Navy.

Yard Redevelopment Plan 1988

The main proposals planned for Govan included:

1. Mechanised Pipe Processing System by Oxytechnik at a cost of £507,000.
2. A new steel preparation system including the merging of plate and section stockyards, a new plate and section treatment line and new panel and web lines. Based on experience from similar mechanised production line installations in other shipyards world-wide, TTS estimated that the present man-hours/tonne of steel fabrication of 40.5 would fall to 16.5 once the new equipment was installed. This would enable 450 tonnes of steel to be produced per week. Estimated cost was £3.2m.

At the building berth, several plans were considered, including widening, installing 180-ton berth cranes, and building a 240m x 40m building dock across berths 2 and 3. This would be served by a goliath crane, capable of 300-ton lifts. The scheme that was eventually chosen included a large tank assembly shop where, what was described as 'grand blocks' of up to 1200 tonnes, could be built for erection on a single, new berth system.

Steinar Draegebo[61]

The report that TTS wrote was authored by Steinar Draegebo, the Chief Executive of the Virginia, USA, office of TTS. Draegebo was a naval architect with considerable shipbuilding experience, including having been responsible for implementing pioneering CAD/CAM systems for shipbuilding and computerised production planning systems. He had taken senior roles in Canadian shipyards at Quebec and Halifax, after first having spent a few years in the Norwegian Aker Group building VLCC tankers. Draegebo had been involved in selling Norwegian shipbuilding technology to shipyards world-wide. Draegebo regarded the Govan yard's production methods as obsolete and inefficient and, with his report, he went to Kvaerner's head office in Oslo to outline the proposed improvements. This resulted not just in the award of the contract to TTS, but, in a second contract, to explore how the improvements made in steel working could be carried through to hull erection on the building berth. This report, titled 'Ship Production Strategy', was presented to KGL in May 1989.[62] Impressed with Draegebo's experience and his knowledge of the Govan yard, the Kvaerner Board offered him the Managing Director's job, given Eric Mackie's imminent retirement. At the same time, Kvaerner placed senior Norwegians from their shipyard at Stavanger and the Moss subsidiary onto the KGL board.

Steinar Draegebo, Managing Director of Kvaerner Govan Ltd (1989–1992), and responsible for sweeping changes at the yard resulting in greatly improved productivity and, for the first time in decades, profitability. (*David Miller*)

Work on what became Phase 1, Improvements in Steelworking, began on 1 July 1989 and was completed in December of that year. Phase 2, hull erection, began shortly after. It envisaged blocks of up to 1200 tonnes in size being fabricated in a new 'grand block assembly shop'. A new berth system was proposed, where the 1200-tonne blocks could be positioned and moved into place as required.[63]

In September 1989 Steinar Draegebo joined the Board as Deputy Managing Director. The order book comprised the two COSCO container ships, *Taihe* and *Puhe* and two 56,000cbm LPG carriers, while two 32,000cbm LPGs were being tendered for. The container ships were due for delivery in July (*Taihe*) and December (*Puhe*) but had been delayed because of the late delivery of units subcontracted to North East Shipbuilders Ltd. Moreover, a unit for *Taihe* had been lost overboard on the delivery voyage from Sunderland.

When he came to Govan, Draegebo was surprised at what he thought was the relaxed attitude people had and at the existing culture. One of his first actions was to abolish the four canteens used on a hierarchical basis to replace them with one for all. Subscriptions to golf clubs for directors were stopped. The five or so chauffeur driven cars for directors were disposed of. Everyone had to clock in, not just manual workers.[64]

An indication of Draegebo's style at board meetings became apparent when he gave a strong sense of how he intended to proceed. In addition to discussing progress on contracts, he outlined what he saw as the way forward. The number of hourly paid employees was to be consistent with production volume and productivity targets. He gave as an example the steel preparation and fabrication shop, where the target was 500 tonnes per week. This equated to 21 man-hours per tonne, equivalent to 270 persons working combined day and night shift. On average, there should be 1 foreman for every 20 workers and 1 manager for every 5-10 foremen. He referred to the often blatant disregard for safety rules as being unacceptable. Enforcing these rules was each foreman's responsibility.

> I took time from my schedule to meet individually with every foreman, accompanied by his immediate manager, to ensure the message was understood. I also instituted the so-called 'Goals Programme', where small teams of middle managers or foremen from Production and Administration were tasked with identifying near-term and specific targets for cost cutting and productivity improvements, and to see them implemented and reporting results. For many, this proved to be an eye-opener, showing wasteful practices that needed to be changed, including housekeeping and cleanliness that needed to be improved.[65]

Directors were to give their home telephone numbers to each other, as well as to key Kvaerner personnel. The overall objective was to turn the loss-making yard into a profitable one within three years.[66]

A review of KGLs then-existing marketing strategy resulted in a report which showed that, through brokers in the UK and Europe, self-unloading bulk carriers, LPG carriers, products carriers, container ships, cruise ships and RoRo ferries were in demand, all of which were suitable for the Govan yard. The report noted that EC 6th Directive, under which subsidies up to 28% were available, was being phased out and replaced with the 7th Directive, with subsidies scaled back. The report included a comparison of costs between European and Far East prices, which showed that, across 22 ship types listed, the Far East yards were anything between 22% (typically for a Ro-Ro ferry) and 39 % (typically for a container ship) cheaper. Only cruise ships were close to European figures at 10% cheaper)

By the end of 1989, KGL was preparing tenders for the following enquiries:

> Two 27,000cbm LPGs for Kvaerner Shipping.
> One 30,000dwt self-unloading bulk carrier for the Australian National Line.[67]
> A 66,000dwt self-discharging bulk carrier for Foster Yeoman.[68]
> A 550,000cbf reefer for Irgens Larsen.
> Four 15,000dwt newsprint carriers for Finn Lines, with 'an indicative budget to be prepared, to see if we were in the right ballpark'.

A manpower report showed that there were 2116 employees, of whom 1732 were permanent and 384 temporary.[69] By November, the redundancy programme had 319 applications, although the company had only approved 183, unwilling to let certain skilled persons go.

In December 1989 Eric Mackie retired. He had been at the Govan yard for ten years and never tired of saying that he had been instructed by British Shipbuilders to run the yard down and close it. He had chosen to do the reverse, and the continued existence of the yard was down to him. Under Mackie, seventeen ships had been built at Govan totalling 429,000 tonnes. Moreover, he was regarded with great affection and admiration throughout the yard. There was another significant retirement that month, when trade union convenor Sammy Gilmour, one of the original and highly respected shop stewards from the days of the UCS work-in, also left.

It was now down to Steinar Draegebo to implement the Kvaerner plan. On 12 February Ian Lang, Scottish Office Minister of State for Industry and Education, visited the yard to announce investment totalling £11m from the Industry Department Scotland, to fund the next phase in the reconstruction of the yard. This was principally what became known as the Tank Assembly Shop (TAS), where grand blocks would be assembled.

In an interview given to the works newspaper *Clydesider* at this time, Draegebo pulled no punches when he said that productivity at KGL was poor. 'High-cost countries such as Denmark can build ships at a third the man-hours we can. The intervention fund in the UK last year was 26% and, even then, it was difficult for KGL to

Block Build. The after block of *Havfrost* having been taken from the new Tank Assembly Shop sits at the head of the building berth ready to be manoeuvred into position in the autumn of 1990.

win orders and this year, 1990, the fund was being cut to 20%.' Asked if he therefore saw no future in KGL, he replied 'absolutely not! We have many factors in our favour, such as the skills of our employees, the know-how as well as support from Kvaerner, and the likelihood of a long period of improved market conditions later in the 1990s. However, we must dramatically improve our productivity. We must do so quickly.' He explained his threefold strategy:

> 1 To design ships that are easier to build. The introduction of new production engineering facilities will make the ships production-friendly.
> 2. New facilities and methods will enable more work to be undertaken indoors and dramatically shorten berth time.
> 3. Working as a team. There is no job security in dragging out work when a contract is coming to an end and spending more time than necessary on any one task.

He emphasised that KGL intended to be a responsible employer and a good corporate citizen and to be the kind of company one is proud to be associated with. 'To achieve these goals, we need to make a profit which is only possible if we build quality ships on time and within budget.' He added that by 1993 the yard should be able to get steel throughput up to 30,000 tonnes, equivalent to three Panamax sized ships a year, approximately twice our current output.[70]

Welding Problems

Transitioning from practices and processes at Govan Shipbuilders Ltd (GSL) to those required by the type of ships to be built by KGL would prove to be difficult and technically challenging. None would be more so than that encountered in welding, particularly welding the vast gas tanks that were the main feature of the LNG tankers on order. Under the rules of the classification society which, in this case, was DNV (Det Norske Veritas), non-destructive testing of welds, meaning ways in which welds were examined to ensure they were sound, usually by x-rays, was more stringent than had previously been the case with ships built at Govan. As a rough estimate, where previously 200 such tests were made on welds for the entire ship, the gas

A midships block weighing 1276 metric tons for *Havfrost* sitting on transporters waiting to be manoeuvred into position on the building berth.

One of the very large gas tanks under construction in the Tank Assembly Shop. Welding these structures initially resulted in a large number failures when x-rayed. Great, and ultimately successful, efforts were made to improve the welding standards required. This included greatly enlarging the quality control department.

ships required something in the region of 20,000. Inevitably, many faults were found, which necessitated the faulty welds being gouged out by caulkers and the welds redone. By March 1990, faulty welding had become such a serious issue in the construction of ships 301 (*Helice*) and 302 (*Havfrost*), that a new quality control organisation of around 40 people had to be established. In March 1990 this organisation was headed by Norwegian welding expert Leif Nordbo.

David Millar started an apprenticeship at the yard as a welder in 1975, moving four years later to the Quality Control Department. Following the takeover by Kvaerner, he was sent to Kvaerner Moss Rosenberg Verft in Stavanger to understand the ethos and non-destructive testing methods used by Kvaerner. On his return to Govan, he was made Quality Control Manager:

This was a horrendous time in my career. I practically lived in the yard for a couple of years until we got out of the mess. It was clear when we started construction of the first of the LPGs [ship 301 *Helice*] that the yard was not prepared for the change over from traditional ship types to the new ones. On a typical Govan ship, to comply with the classification society rules, the yard had to submit an inspection plan, consisting of the laid out detail of non-destructive testing required. Depending on ship size, this would include a couple of hundred radiographs and an agreed length of ultrasonic inspection and magnetic particle and dye penetrant inspection. However on the gas ships, each of the four huge, low temperature free standing gas tanks had to have 100% radiography, this equated to thousands of radiographs on the tanks alone … Well, it did not take too long on start-up to discover that the welds were not good, and the repair rates went off the scale, as radiographs were taken, repairs were called, then, when the repairs were completed and resubmitted for test, they were also failing . . . the

repairs were piling up. It was one step forward and three steps backwards.[71]

An aerial view of the first of the LPGs, *Helice*, on No 1 berth not long before she was launched on 7 September 1990. The area to the right of the hull where the joiner shop used to be has been cleared to make way for the Tank Assembly Shop. (*David Miller*)

With up to 50% faults found in units, most, but not all, were from hand welding and as a result, a programme of training welders was established to be managed by training company M I Technology, who operated a training school within the yard. The aim was to train as many welders as possible to Class A certification. At the same time, a shortage of welders led to units being sub-contracted out to structural steel company Motherwell Bridge. This was expensive, leading to a drive to attract more A Class welders to be part of the permanent workforce.

The learning curve had been steep but over time the situation began to improve and this would lead to a level of excellence that became the hallmark of KGL ships. However, in the early years it was said, not entirely in jest, that *Helice* had been welded at least three times.

The year 1990 saw a period of major change and, unsurprisingly, resulted in yet another loss. According to Draegebo, this was down to three things: productivity was worse than expected, the last two wage increases had been twice as high as was assumed when the LPG ships were contracted for, and welding problems on *Helice* caused a huge amount of repeat work. Yard Convener Davie Cooper said 'last year was a period of restructuring so losses were to be expected to an extent. After a period of nationalisation and moving to become a privately owned company, there had to be a period of change, but we are moving in the right direction and the lines of communication are more open.'[72]

Personnel

An accumulation of issues, mainly welding-related, and a walk-out by caulker burners in a dispute on which union should operate the new production lines, led to programme slippage. Management considered options to make up time by bringing in welders from UiE at Clydebank and possibly from various locations abroad, although good progress in the recruitment of Class A welders would eventually make this unnecessary. Similarly, sub-contracting units out to UiE, Babcock and Cammell Laird was considered, but ultimately, the costs were way over the budgeted figure.

In August 1990 a new wage deal was agreed, as well as a new system of shifts and the introduction of the 37-hour week. This brought about what was termed 'Bell-to-Bell' working, the elimination of ineffective time. By November 1990 the steel shop was getting closer to its target, delivering 550 tonnes per week at 34 man-hours per tonne, against the target of 29 man-hours.

Former KGL Personnel Director, Bobby Gordon, describes some of the changes that were introduced under Steinar Draegebo:

> The introduction of the double day shift was aimed at optimising the use of our facilities and increasing throughput over a 16-hour working day. It also gave us greater control of the quality of the man-hour outputs, by conducting X-ray work on the 3rd shift at night, without disturbing day production and enabling immediate feedback on quality failures. To meet the far greater quality demands, we had to recruit top class welders, capable of welding to 100% X-ray standards, and upgrade our own welders to the same standards. We were working on new products and new materials. This involved a major internal training programme.
>
> 'Bell-to-Bell' working was a response to the large number of man-hours being lost at starting and stopping times. It was estimated that the effective number of man-hours was as low as 3.5 during an 8-hour shift and this had to improve. It was also part of a culture shift, to bring home the fact that the shipyard had to be a vibrant place of work and that effective working man-hours would lead to improved competitiveness. Bell to Bell was central to our approach and was unique to KGL at that time. The elimination of all lost, unproductive, work time was essential. On the wages side, we realised we had to pay the best we could, to attract and retain the best people, so wages were raised by 14%, eliminating all the supplements which were a feature of the British Shipbuilders days. We also reduced the working week from 39 to 37 hours.[73]

Double Day shift

In May 1991 the Unions rejected the wage offered by the Company, not for financial reasons, but because of the proposed new shift working arrangements. The shift working system inherited by KGL from British Shipbuilders comprised five different shifts, with very little coordination. The proposed new system was a double day shift system (day shift and back shift) which meant a 16-hour day, with a third shift at night. The latter was essential, due to the large number of welds made during the day shifts that had to be X-rayed. The night shift was also the best time to carry out maintenance routines, leaving the day shifts unimpeded. However, not only was the offer rejected, but the workforce voted for a yard-wide strike, which began on 10 May.

Looking back at the strike events at Govan in May 1991, Steinar Draegebo made the following comments:

> It had become increasingly clear to me that the Redundancy Scheme was a fundamental problem behind the trade union's seemingly irrational reluctance to accept logical proposals to improve productivity by more flexible work patterns and also by resisting taking full advantage of the substantial investments the Company made in new technology. Inherited from the days of British Shipbuilders, the Redundancy Scheme guaranteed a generous pay-off for time-served employees who lost their jobs due to layoffs. Many employees regarded this as a 'pot of gold' that could be collected if they would be laid off, and therefore saw the prospect of job losses, if the yard would not be able to compete for new work, as a something to look forward to rather than a threat.
>
> In 1990/91, labour relations became even more strained, as elements within the trade union increasingly resisted the changes management tried to implement. The yard suffered from several, short, 'wildcat strikes'. One such event was when the trade unions objected to the Company's efforts to get tea breaks to be held at the place of work rather than in the canteen – time to walk back and forth to the canteen twice a day was a major source of unproductive time. The local press called it 'a storm in a teacup' but management saw it as a fundamental sign of lack of trade union cooperation to implement essential productivity improvements.
>
> Following the UK-wide coalminers' dispute that, some years earlier, had brought the British economy to its knees, the Thatcher Government had implemented new labour laws, that tried to restrict unauthorised strike actions. When the unions at KGL in May 1991 started yet another unauthorised

> strike, they failed to follow the correct legal procedures. By law, each striker was therefore technically in breach of his individual employment contract and the Company was entitled to terminate the contract forthwith.
>
> I saw this *de facto* illegal strike as an opportunity in one swoop, to get rid of the existing counter-productive work practices imbedded in the workers' employment contracts and replace them with new contracts that allowed for modern and flexible working practices, and, at the same time, removing the negative temptations of the Redundancy Scheme. I was convinced that the majority of the strikers, on due reflection, would understand and accept the Company's well-intended reasons and would prefer the relative job security of working for a productive company, rather than fighting for obsolete working practices that had been allowed during state ownership.[74]

Having in strict confidence, sought solid legal advice, and with the support of the new top management team he had appointed, on 17 May 1991 Draegebo wrote to every striking employee:

> Dear Employee,
> We have noted that you have elected to participate in strike action against the Company. This constitutes a fundamental breach of your employment contract, and you are therefore advised herewith that your employment has been terminated today.
> The Company intends, in the near future, to offer you re-employment, provided you are prepared to accept new contract conditions set by the Company. The new conditions will reflect fair and progressive working practices, and will eliminate outdated restrictions and costly commitments, which, if continued, threaten to put the Company out of business. As a private company, we simply cannot go on losing money, and we all know that, by eliminating restrictive working practices, we can radically improve our productivity and become a profitable yard.

Draegebo has since added context to the decision he made to carry-out the above action:

> Fearing that the Kvaerner Head Office in Oslo would object to the notion of firing striking workers – something almost unheard of in Norway with its tradition of benign labour relations – I waited until the afternoon of May 16th before sending a message to the Kvaerner corporate management in Oslo, informing them of my intended action 'unless instructed otherwise', saying that I considered the dramatic action essential to bring an end to increasingly destructive labour relations. Friday May 17th was Norway's Constitution Day and a significant public holiday, so it was unlikely that anyone at Kvaerner's Head Office would see my message until the following week, by which time termination of the striking workers had taken effect, and it was too late for anyone to interfere.[75]

This totally unexpected reaction from the Company to the strike came as a shock to most employees and created headlines nationally and internationally and substantial coverage on all national TV networks. However, following negotiations, the great majority of the workforce agreed to the new terms, and work restarted on 27 May, to the relief of all parties. This represented a major breakthrough as, not only did the new production facilities come into play, along with new steel working processes and the Tank Assembly Shop, but now the workforce was aligned, through agreement, into a pattern of shifts and simplified practices that allowed for much higher levels of productivity. At the beginning of August, output from the steel shop was raised from 500 to 650 tonnes per week. Less encouraging, however, was the estimated loss for the past year of £7m, with 1992 predicted to reach break-even. Two months later, the loss was estimated at £10m, although it was not until March 1992 that the final loss for 1991 was declared at £11.3m. At the same time, the Kvaerner Group made record profits in 1991, with the shipbuilding group reported to be highly profitable, largely due to the Masa-Yards in Finland.

By July 1991 Draegebo was able to say that productivity was rising. Assisted by the new Tank Assembly Shop, *Helice* had been delivered to owners who were delighted with the quality of the ship. Moreover, absenteeism was dropping from a high of over 22% to 3%.[76] To ensure that *Havfrost* was launched on programme, 430 employees had agreed to postpone their annual holiday.[77] By September, KGL had 202 Class A welders plus 128 agency welders on site, with the intention to train or hire an additional 180. By the autumn of 1991, the yard was beginning to see the benefit of the new production facilities, new working arrangements and higher productivity. Ship 301, *Havfrost*, was fitting out, the first blocks of Ship 303 *Helios* were under construction in the Tank Assembly Shop and steel for Ship 304, *Havis*, was working its way through the preparation bays.

In October, the order for a 35,000dwt chemical tanker worth £50m was announced, two having been previously ordered from the Kvaerner Floro yard in Norway. Although Draegebo said there were still improvements to be made, he commented that the cost of a Govan built ship now matched that of similar sophisticated ships being built at Kvaerner's sister yards abroad.[78] Having been appointed as a non-executive director to the Board of Kvaerner's shipyards in Finland and Norway and later in Germany, Draegebo was well positioned to make those comparisons.

Block Build

The capital investment programme would fundamentally change the way in which ships were built at Govan. Previous modernisation schemes since the changeover to welded prefabrication in the late 1950s had enabled the Govan yard to assemble ships from prefabricated units of typically 20 tons, rising up to 130 tons as successive upgrades in plant and craneage took place. These units, fabricated in the shops, were transported to the building berth and lifted into position on the berth using travelling berth cranes. The units would be held in position until welded together. The new Block Build system enabled blocks of 1200 tons and more to be assembled in the new Tank Assembly Shop from units fabricated in the existing Fabrication Shop. Importantly, the size of the blocks enabled much of the outfit work, such as installing piping, ducting, electrical cabling, outfit steel, etc, to be done much more productively indoors at the block stage, rather than later under much more adverse weather conditions and with much less accessibility at the berth. The pre-outfitted 'super' or 'grand' blocks would then be taken from the Tank Assembly Shop on Scheuerle self-propelled modular transporters to the new No 1 berth, where they were jacked into position and 'skidded' either up or down the berth as required, by a hydraulic skidding system. Moreover, in the case of the gas ships, the system was capable of fitting a gas tank, pre-assembled in the Tank Assembly Shop and weighing up to 1000 tonnes, into a hull block already on the berth. The new block build system dramatically reduced berth time for a hull that would have taken 45 weeks under the old unit system down to 20 weeks. This meant that, given the predicted volume of work, only one building berth was needed henceforth.

The most spectacular aspect of the new facilities was the construction of the vast Tank Assembly Shop which was located close to the Fabrication Shop, such that units could be quickly transferred from the latter to the former. To make way for this new shop, the old Joiner Shop, then serving as offices, had to be demolished. In its heyday, the Joiner Shop was a reminder of Fairfield's time as a builder of the finest passenger and passenger cargo vessels with magnificently crafted interiors. For the design of the new shop, Kvaerner approached a number of consulting engineers, finally selecting Gronstad & Tveito A/S of Bergen, Norway, to prepare the design. The construction contract for the Tank Assembly Shop was awarded to Balfour Beatty, with a completion date of March 1991. Upon completion, the building won a prize as the best new industrial building in Scotland.

The combination of the Tank Assembly Shop with pre-outfitting and ground transportation by wheeled, self-propelled modular transporters able to skid Grand Blocks into place, enabled Govan to adapt construction methods similar to what had been implemented by large modern shipyards in Asia through the use of Goliath cranes and building docks. However, Govan could do so with much lower investments and could make use of its existing berth.

Being part of an international shipbuilding group enabled Govan to benefit not just from the transfer of experience and information concerning technical matters, but to cultural exchanges such as hosting football teams from sister companies. In June 1991 Kvaerner Govan Select hosted Kvaerner Egersund Select, drawn from Kvaerner's Egersund's offshore fabrication yard. The Norwegian team was given a yard tour, a meal at Ibrox Stadium and a shopping spree in Glasgow. The event, organised by shop steward Jamie Webster, was a great success.[79]

Getting There

In December 1991 Kvaerner announced an order worth £300m for six chemical carriers of which four were to be built at Govan and two in sister yard Kvaerner Kleven in Norway. This order was a major boost for both yards which had beaten off competition from South Korean, German, Polish, Belgian (Boelwerf) and Spanish (Astilleros Espanoles) yards. Two of the Govan ships were for Norwegian Storli owners and two for National Chemical Carriers of Saudi Arabia. KGL now had an order book of six, sophisticated vessels with full employment out to 1995.

In the same month, the American trade journal *Maritime Reporter* selected *Helice* as an outstanding ocean-going ship. Draegebo commented that, once again, it reflects the quality of workmanship in this yard and was 'a credit to us all'. This was the second Govan ship to have been chosen by this journal on the basis of quality, the first being *Taihe* a few years earlier.

The launch of ship 303, *Helios*, marked the first ship to take full advantage of the new Tank Assembly Shop and the new No 1 berth's skidding system, although the after part of the ship had been built on the old unit system.

The contract for the chemical tankers introduced yet

A good general view of the yard and surrounding area showing the newly completed Tank Assembly Shop. The third of the LNGs to be built, *Helios*, is on No 1 berth, the only berth now in use because of the quick build cycle. (*David Miller*)

another new challenge for the Govan workforce – welding stainless steel. This again was something that the Norwegians had considerable experience of, having built the first stainless steel tanker, *MT Lind*, in 1960 at the Stord shipyard. With this experience in welding gas ships behind them, preparation for the new contracts included sending welders from Govan to Kvaerner's yards at Kleven, Floro and Forde, as well as classroom and practical training in Govan by welding specialists.

In May 1992 the Govan yard was reminded of past glories when a group of men who had served on the Fairfield built battleship HMS *Howe* and who were now members of the HMS Howe Association, visited the yard to commemorate the 50th anniversary of the battleship's completion. Seven men, who had worked on construction of this ship and who were still employed in the shipyard, were introduced to the former crew men. A further naval association was initiated that spring, when discussions with Vickers Shipbuilding & Engineering Ltd at Barrow began with the prospect of Govan being involved in the construction of an LPH (Landing Platform Helicopter) which would become HMS *Ocean*.

At the end of 1992 Kvaerner's Head Office appointed Steinar Draegebo to upgrade and modernise the huge Kvaerner Warnow shipyard that the Group had just acquired as former East German state-owned companies were being privatised. He was replaced as Managing Director of KGL by Kjell Mikalsen. Under Draegebo, the Govan yard had experienced the most dramatic turnaround in its fortunes. From a position in the second league, it had been transformed into one of the most productive shipyards in Europe, rising from a chronic loss making position to profitability, albeit only just. On his departure, Draegebo noted that, over his four years at the yard, the Kvaerner Group had doubled in size, with 23,000 employees worldwide and from having one shipyard, Govan, back in 1988, it now had nine shipyards in five countries. Perhaps most significant was his statement that 'I am particularly pleased with the improvements I have seen in our industrial relations. We have moved away from a period of confrontation to one of co-operation and increasing trust and openness.'[80]

Under Kvaerner and Draegebo, the Govan yard had been brought to a point where the majority of people associated with it could see that it was possible to be an efficient and competitive shipyard. The growing sense that progress was being made was reflected in absenteeism, which was at a record low, the quality of the product was beyond doubt and Draegebo's plea for individual responsibility had resulted in a highly motivated workforce. Draegebo's successor, Kjell Mikalsen, was a naval architect and, prior to his appointment at Govan, the Deputy Managing Director of Kvaerner's Floro shipyard, where he had overseen the construction of chemical tankers.

The 1993 wage agreement included a slight increase on basic wage rates, the introduction of a bonus scheme and a minor adjustment to the three shift system, with all remaining at 37 hours overall. On 25 June 1993 the first of the chemical tankers, built for Jo Tankers,[81] Ship 305 *Jo Selje*, was launched. As was the Kvaerner custom, there was no naming ceremony at the time of launch. The ship, built entirely from blocks fabricated in the Tank Assembly Shop, was the first to have her underwater paint system in place, which meant that she would not need to be dry-docked on completion. Berth time was 23 weeks.

On 7 October HM Queen Elizabeth arrived at the Govan yard to christen and name the ship *Jo Selje*. It was the first time the Queen had been in the yard since she had launched the Canadian Pacific liner *Empress of Britain* 38 years earlier on 22 June 1955. In attendance at the launch was Diderik Schnitler, Kvaerner's shipbuilding President, who announced that the Govan yard, with an order book valued at £250m, would achieve profitability by the end of 1993. He said that, in the five years since Kvaerner had acquired the yard, its financial performance had been transformed from annual losses of £45m under British Shipbuilders to showing a small profit in 1993; £31m had been invested in new plant, equipment and techniques and Govan was now comparable to the best in the world. Working practices, industrial relations, absenteeism and accidents had been greatly improved and skills upgraded.[82]

Managing Director Kjell Mikalsen (1982–1996) shows HM Queen Elizabeth II over the chemical carrier *Jo Selje*, which she christened on 7 October 1993. The ship had been launched without a name on 25 June 1993. This was Queen Elizabeth's first visit to the Govan yard since she launched the *Empress of Britain* 38 years earlier.

The 37,252dwt chemical carrier *NCC Riyad* launched on 9 September 1994. Tug *Chieftain* in attendance.

While these headline statements grabbed attention, the weekly management meetings often threw up interesting details such as, on 16 May 1994, when it was recorded that the average age for a KGL worker had gone down from 43 years to 41 years. A cryptic one-liner from the meeting on 9 May simply stated, 'Visitors from Samsung and Floro were amazed at KGL quality – it is too high'.[83]

The visit by HM the Queen was not the only visit from a reigning monarch. On 6 July 1994 King Harald V of Norway visited the yard to officially open the Tank Assembly Shop. Later that year, with three ships in production, the company employed an additional 150 steelworkers, taking the total workforce to 1750. Productivity, although greatly improved, was still not good enough, given the increasingly competitive nature of winning contracts. To raise productivity, a new campaign was launched under the catch phrase 'Govan Productivity 2000', with the aim of making a 30% improvement. This was seen as vital to win orders in the conditions then prevailing in the market. In August 1994, employment stood at:

	permanent	temporary	total
Hourly Paid	1005	209	1214 (includes 65 apprentices)
Staff	300	54	354
Total	1305	263	1568

On 12 December 1994 the Saudi Arabian ambassador to the UK, Ghazi Abdul Rahman Al Gosaibi, came to the yard to name the third of the chemical tankers, *NCC Riyad.* At the ceremony, KGL was able to report a second profitable year, albeit a slim one. The Managing Director's Christmas message to the workforce was that the increase in productivity was essential, the shipbuilding industry will, he thought, recover in the second half of the 1990s and that he hoped KGL would lead the world in building LPGs. In fact, despite the great efforts made by the company to build these sophisticated vessels to a very high standard, KGL

would not build any more vessels of that type. Nevertheless, Kvaerner's Vice President of Finance, Jan Heggelund, was able to report 'All of the Kvaerner Group yards were in profit, including Govan. Profit at Govan was not large, but compared with European yards, Govan had performed well.'[84] In 1994 Kvaerner's shipbuilding division recorded pre-tax profits of £114.5m. Govan made a profit of £2m, although its order book was the shortest in the Kvaerner Group.[85]

As KGL worked its way through its contracts, by early 1995 no new orders had been booked since the LPH in May 1993. This meant reduced steel throughput with, correspondingly, higher production costs. In April the second chemical carrier, later named *Bow Clipper*, was launched and after her departure in July, only the LPH[86] *Ocean* was left in the yard, although contracts were placed for a small products carrier, *Emily PG*, and a North Sea supply vessel, *Stirling Forth*. Looking back at 1995, Mikalsen said that, despite 1995 being a difficult year and one of low output, the yard returned a profit of £2m, commenting that 'We had redundancies, extensive training, outworking arrangements to keep the wheels turning . . . we had quite a few people working in Rosyth and the message back was very positive: the workers from Govan did an excellent job ... many of our competitors in the UK and Europe went bankrupt or struggled with big losses.'[87]

By 1996 the continued expansion of the Kvaerner Group had resulted in a company that comprised shipbuilding, ship repairing, ship owning and ship equipment manufacturing. The shipbuilding yards of which Govan was one included Masa yards in Helsinki and Turku in Finland, Warnemünde in Germany and Kleven, Ulsteinvik, Fjellstrand and Floro in Norway, as well as Philadelphia in the US. The Group also had ship equipment production sites in Norway, Sweden, Germany, the UK and Singapore and repair yards at Gibraltar and Kimek in Norway and Vyborg in Russia.

Naval Work and HMS *Ocean*

Under the terms of its formation, KGL was not eligible to tender directly for warship work. Nevertheless, in the spring of 1992 KGL minutes record a number of exploratory meetings between Steinar Draegebo and Vickers Shipbuilding & Engineering Ltd (VSEL) at Barrow for a vessel that would later take form as HMS *Ocean*. The invitation to tender for what would be a controversial contract had been issued by the MoD back in February 1992. Two yards responded, VSEL at Barrow-in-Furness and Swan Hunter on Tyneside. Swan Hunter, which had been returned by British Shipbuilders to the private sector in 1987, one year before Govan Shipbuilders, was highly dependent on the order for survival. VSEL, on the other hand, was not, being, together with its subsidiary Cammell Laird, the sole provider of nuclear submarines for the Royal Navy. However, VSEL who had not built surface ships in ten years, aspired to building surface ships again for the MoD, knowing that other major surface vessels were being

The chemical tanker *Bow Clipper* running trials on the Firth of Clyde in the summer of 1995. (*David Miller*)

The bow block of HMS *Ocean* emerging from the Tank Assembly Shop. Collectively, the KAMAG transporters were capable of taking weights up to 1500 tons.

planned. With VSEL operations at Barrow largely configured for submarine construction, the exploratory meetings resulted in KGL being asked to tender to VSEL for the hull, knowing that the Govan yard would build the hull to specification but much less expensively than VSEL could. On 9 July 1992 KGL put forward an indicative price of £94m, at which, according to the minutes, VSEL 'expressed disappointment'.[88] By August the KGL price had been reduced to £81m.

The MoD received provisional tenders from both VSEL and Swan Hunter in October 1992, which revealed a gap of £9m in favour of VSEL. Both bids were noncompliant however, and following meetings with MoD, both builders were asked to resubmit 'Best and Final Offers' on 26 March 1993. These offers, received on 22 April, showed that the price difference had grown significantly to £71m in favour of VSEL, the figures being £139.5m compared to £210m.

Part of the reason why such a big difference in price occurred was due to differing interpretations of the specification and, in particular, what should be built to naval engineering standards and what should not. The other reason lay in VSEL's decision to subcontract the hull to KGL, viewed as a highly efficient commercial yard. Building to commercial standards was generally cheaper than building to naval standards, so contracting the hull out to KGL made sense, the intention being that, once completed, the hull would sail under its own power to Barrow where the military outfit would begin.

The subsequent award of the contract to VSEL dealt a serious blow to shipbuilding on the North East Coast. Swan Hunter, a shipyard that meant as much to the Tyne as the Govan yard did to the Clyde, needed the contract to stay in business. The discrepancy between the price submitted by both builders was so large in favour of VSEL that the Barrow yard was accused of trying to eliminate Swan Hunter by 'buying' the contract. Such was the outcry that the National Audit Office was asked to investigate this claim. A thorough investigation into the procurement process ensued. This was published at the end of July 1993, finding that both builders were treated in a fair and even-handed manner. With regard to the perception of KGL, the report noted that:

> The vessel will be built to the highest commercial standards, with which we are fully cognisant and experienced, and to specified Naval Engineering Standards. Kvaerner Govan see 'commercial' standards as relating to their work on technically demanding vessels such as Liquid Petroleum Gas Carriers. In their opinion, in many cases, these vessels were being built to higher standards (for example fire, cabling and safety) than the Naval Engineering Standards they will be required to meet for the LPH.[89]

Production of the LPH was scheduled to start in the summer of 1994 with a July 1995 launch, and delivery to VSEL in February 1996, and a final in-service date of March 1998. This contract meant recruiting more employees at Govan, an initial 80 additional CADAM draughtsmen, with more manual workers to follow. The contract was signed on 31 August 1993. In the event, work in the steel shops started on 30 May 1994, with the launch scheduled for 11 October 1995. A KGL production report, reproduced here as an example, dated 1 September 1995 stated:

> Structural steelwork on programme to be completely surveyed by launch. Tanks are two weeks behind programme, but still targeted for a Week 41 launch: installation of port rudder and shaft lines are also critical for launch. One percent (5757 metres) of cable had been pulled to 25 August. Resources are being steadily increased to achieve the maximum weekly rate of 2.7% (17,655 metres) of cable pulled. Total spend to 15 August is 988,000 man-hours and predicted completion is 1,709,000, which would represent a 10% saving.[90]

Launched on programme, the forward port poppet collapsed during the launching run causing the hull to make contact with the concrete berth as it entered the water, distorting 30 metres of steel work near the keel. Over the Christmas period, the vessel was drydocked at Scotstoun and the damage repaired. During fit-out, electrical work which had been contracted out to another Kvaerner company, KVI, was seriously behind target, resulting in Siemens being brought in to speed things up. This, together with the inevitable 'Variations to Contract' and delays between Barrow and Govan, led to a contract extension and revised delivery date. *Ocean* finally left Govan on 18 November to run contractor's trials on the Firth of Clyde lasting ten days. The ship arrived at Barrow to begin the military fit-out on 27 November 1996.

The final cost for Kvaerner's construction of *Ocean* was:

Budget £88m with £2.5m for VTCs (variations to contract)

Labour	£15.98m
Overheads	£24.76m
Material	£30.79m
Subcontracts	£10.87m
Other	£0.75m
Margin	£7.35m
Total	£90.5m
Man-hours	2,106,000

Working with VSEL on HMS *Ocean* raised the possibility of further naval work. At Barrow, much of the old shipyard had been demolished to allow concentration on submarine construction, with the consequent loss of accumulated surface ship experience. There was also recognition at Barrow that the Govan yard had taken great strides to become efficient, especially in steelworking, typically not something possible building to Naval Rules or with naval contracts. With two LPDs (Landing Platform Dock) subsequently HMS *Albion* and HMS *Bulwark* and two RFA (Royal Fleet Auxiliary) oil tankers, subsequently *Wave Knight* and *Wave Ruler*, on the horizon, the possibility of further collaborative work with VSEL was discussed by both parties.[91] It was understood, however, that the LPDs had more naval content than the LPH and were consequently less likely to involve KGL.

In May 1994 VSEL announced that they had been successful in submitting a tender (VSEL was the only yard to tender) to the MoD for the construction of the two LPDs. Although no construction work came to Govan, VSEL did however ask KGL to advise on ship design, build strategy and steelwork sub-contracts.[92]

The tender for the oil tankers however, was open to more shipyards although, to be eligible, KGL had to collaborate with an existing builder already on the MoD list. At the same time, KGL declined involvement in the feasibility study for the oilers, which were still some way off, preferring to concentrate their limited new project resources on what they termed 'real' tender prospects. Despite that, KGL stated its intention to 'continue to press both MoD and VSEL to maximise commercial build techniques to provide value for money.'[93]

Throughout 1995 the yard received numerous enquiries for ships, most of which required a significant effort to price, in the knowledge that few of them would be secured.

Ongoing enquiries, November 1995:

Type	Origin of enquiry	Size	Status
E833 Chemical tankers (4)	KKF Stolt	23,000dwt	Indicative price prepared
E829 LPG carrier (2)	Havtor	35,000cm	Outline spec. being updated
E805 Auxiliary oiler (2)	BAeSEMA	16,000dwt	BAeSEMA to act as prime
E798 LPG Carrier (2 or 3)	Lauritzen Kosan	5760cm	Urgent tech response required
E791 Products tanker*	Pritchard-Gordon	6150dwt	Contract signed but subsequently cancelled.[94]
E790 Chemical tanker* (1+1)	Odfjell-Kvaerner	36,800dwt	Contract signed
E778 PSV (1+1)	Gulf Offshore	82m	Repeat of 310, awaiting feedback
E725 RoRo (4)	Khi Da	122m	KGL withdrew from enquiry
E722 Command ship*	Kvaerner A/S	200m	Contract still to be resolved
E711 Chemical tanker	Sea Chem	41,000dwt	Indicative price for 1 and 4 prepared
E661 Ro-Ro (2+2)	Exxtor (Shipcraft)	159m	Awaiting documents
E637 Mining vessel	De Beers Marine	140m	Awaiting documents[95]

* converted into contracts.

Sea Launch Commander

Despite the constant drive to improve productivity in the face of ever-increasing overseas competition, especially from the Far East, Kvaerner recognised the great skill base which the Govan yard had built up and the quality of the work it produced. Govan could turn its hand to most ship types, which made it Kvaerner's yard for 'one offs'. While this was a credit to the yard, it meant that it would not benefit from repeat orders and associated reduced costs, where succeeding vessels of the same type were cheaper to build as had been experienced recently with the LPGs.

The towering side of *Sea Launch Commander* being manoeuvred into the fitting-out basin after being launched on 12 December 1996.

HMS *Ocean* was a good example of a one-off. Another contract of this type began as Enquiry E722 for a highly unusual ship, eventually named *Sea Launch Commander*, but known in the yard as Ship No 312. In 1995, a company called Sea Launch was established, drawn from interests in the USA, Russia, Norway and Ukraine. As the majority shareholder, Boeing, the US element, managed the consortia. The object was to launch satellites from a platform at sea near the equator, an arrangement which conferred certain advantages over land-based launches. The contract to build a command ship originated with the Kvaerner Masa-Turku yard in Finland, which wanted to partner with Govan for all of the quality reasons given above. However, by August 1995, while Kvaerner favoured Govan, Boeing favoured the Kvaerner Warnow yard in Germany on cost. This prompted Mikalsen to visit the Department of Trade and Industry to make the case for additional funding. This was eventually successful, as the decision to nominate Govan to go forward with a bid was made on 23 August. Govan was successful, with a final signing of the contract taking place on 15 December 1995 and a delivery date for the ship of April 1997. Design was undertaken by the Turku yard in what would prove to be a technically challenging build, with numerous design changes.

Sea Launch Build Timeline

In September 1995, Production Director, John Dunlop, put forward the following build timeline in weeks:

1995	Production start	week 48
1996	Unit build start	week 03
	Block build start	week 11
	First block to berth	week 21
	Launch	week 47
1997	Completion	week 13
	Delivery	week 17[96]

While *Sea Launch Commander* was being programmed, on 10 November 1995 the terms of contract for a repeat chemical carrier were signed in Bergen. This was for *Jo Sypress*, based on the design of *Jo Selje*.

By the end of 1995, three vessels were now in production; HMS *Ocean* was fitting out, steelwork on the offshore supply vessel *Stirling Forth* was underway and work on the small products carrier *Emily PG* about to start. Despite the relatively small amount of steel going through the shops, a modest profit of £0.5m for 1995 was made, with £5m projected for 1996. Looking back at 1995, Kjell Mikalsen noted that it had been a year of low activity, which had meant some redundancies and training programmes to keep the wheels turning, while *Stirling Forth* and *Emily PG* were readied for production. He referred to a number of European competitors who had gone out of business.

On 1 July 1996 Mikalsen stepped down as Managing Director to be replaced by Sigbjørn Ellingsen, who had joined the yard from Kvaerner's Rosenberg yard as Technical Director the previous year. By the time that Ellingsen had taken up his position, the order book had picked up significantly with, in addition to HMS *Ocean*, *Sea Launch Commander*, the chemical carrier *Jo Sypress*, *Stirling Forth* and *Emily PG* in various stages of construction.

Although greatly improved, further productivity gains were still sought, largely as a result of the growth of low-cost shipbuilding in South Korea. The new Managing Director said that 'The quality is good here, but we have a big challenge to get a greater profit out of the yard – this is the real protection for everyone's future. I believe we can do a lot more by improving the production and planning systems within the yard, but it is not going to be easy.'[97] To motivate the workforce and advise on changes within the company, PA Consultants were brought in to look at ways in which productivity across the yard could be raised. A new drive under the heading 'Govan 2000' was initiated, which lasted throughout 1995, to be replaced with a new campaign called 'On Time'. The main problem remained raising productivity to match overseas competition. By the following March however, the situation had deteriorated to the point where Ellingsen felt it was necessary to write to every employee:

> It is well documented that Kvaerner Govan lost a lot of money last year. It is also evident that we will make further losses this year. From a business viewpoint, of course, this cannot continue and we are left with two choices: either we manage to improve and survive or we will close.

He went on to say that, 'while the Russian contract for *Sea Launch Commander* was welcome, it was not enough to keep the workforce together. KGL was spending at least 25% more man-hours in Technical and Production compared to competitors . . . our competitors are employing drastic cost-cutting and downsizing measures to ensure their competitiveness and we must do likewise.'[98]

The year 1997 was to bring mixed fortunes ,which began in March with the depressing news that the £200m contract for the two RFA oilers submitted jointly with BAe SEMA, had been won by VSEL. This was a significant blow to the Govan yard, which now had no alternative but to scale back the Govan workforce of 1200 by 500. However, in the following month, Ellingsen was able to announce that Govan had won a £27m contract to build two platform supply vessels for Toisa Ltd. While these vessels, eventually

Toisa Intrepid and *Toisa Invincible,* were small compensation for the loss of the oilers, every contract counted. This order provided work until the summer of 1998 and enabled the redundancies to be scaled back. This left RoRo ferries and survey ships[99] as distant prospects.

On 22 September *Sea Launch Commander* was named in the shipyard by Allen Ashby, President of Sea Launch who said 'Govan has done a very wonderful job for us with this boat. It was not an easy project, because it is different from anything that anyone has ever built before.'[100]

Events in November 1997 came thick and fast, the first of which was the announcement of new contracts for two North Sea supply vessels for Farstad Supply A/S of Norway (*Far Strider* and *Far Supplier*), while a third order was announced, this time for an oil well test vessel for Brøvig Offshore.[101]

In the same month, Kvaerner's shipbuilding group announced pre-tax profits of £75m for the first 9 months of 1997. Govan was forecast to make a loss for the year, while redundancies made necessary by the dwindling order book reduced the workforce from 1200 to 800.[102] Despite moving back into the red, also partly blamed on the strong pound, and the reduction in the workforce, it seemed likely that the yard was on the way back up when Kvaerner announced a £7m investment package for Govan. Ellingsen said this was a significant endorsement of their belief that Govan had a strong future. In an article for the company newsletter *Clydesider* he said: 'No later than just a few short weeks ago, we faced a very real threat of closure. Our order book was very quickly running out, with little or no prospect of us being able to secure profitable contracts.'

The investment package of improvements included a new outfit paint shop and the Foran 3D CAD system, to gain efficiencies both in productivity and lead time. Also proposed was a new main office block. Until now, KGL's main office was the magnificent Honeyman & Keppie designed building erected in 1890 for the Fairfield Shipbuilding & Engineering Co Ltd, a building which fully reflected the leading position which that Company then held. However, KGL wanted to 'present a more modern, professional image to our clients'. The proposed location of the new office building was by the fitting out basin, which required several old buildings to be demolished.

On 27 March 1998 Princess Margriet of the Netherlands visited the yard to name ship 314 *Jo Sypress.* By 1998, however, the Kvaerner group of companies was lurching into major debt. Although not its worst performer, the vast expansion of Kvaerner's shipbuilding division over the course of the 1990s had resulted in the acquisition of thirteen shipyards, making it the largest shipbuilder in Europe. Despite specialising in niche markets, competition, especially from an emergent South Korea where a dramatic increase in shipbuilding capacity had taken place, coupled to exchange rate issues, had resulted in dwindling order books, particularly at Govan. Across the Group, losses for 1998 were estimated to be in the region of £110m.

Back at Govan, as part of Kvaerner's policy of rotating managers, Sigbjørn Ellingsen had left Govan at the beginning of August to be replaced by Gunnar Skjelbred, previously the Managing Director of Kvaerner's operations in Gibraltar. It was an unfortunate time for Skjelbred as, within a few months, major upheavals at Kvaerner headquarters in Oslo were about to make fundamental changes.

Exit Kvaerner

On 14 October 1998 Kvaerner's Chief Executive, Erik Tonseth, the prime mover behind the acquisitions policy, was forced to resign following serious losses and shareholder pressure. His departure signalled a strategic review of Kvaerner's operations, with the disposal of loss-making divisions a distinct prospect. This was correctly read as a threat to the future of the Govan yard, where the order book was running down. Seven months later, on 13 April, Kvaerner's new Chief Executive, Kjell Almskog, announced that its entire shipbuilding operation would be put up for sale, commenting that over the last year the position of shipbuilding had been hit hard by the falling value of the Korean currency and overcapacity in the industry.

With what seemed like depressing regularity, the Govan yard was, yet again, faced with a fight for survival. It was not just bad news at Govan, as Kvaerner's industrial gas turbine plant at Clydebank, formerly John Brown Engineering, which employed 750, was also put up for sale, as was the UK's last marine engine builder Kvaerner Kincaid at Greenock.

Kvaerner: A Postscript

Looking back, Paul Friedberg, KGL's Production Director, said that, despite the dire situation that the Kvaerner shipbuilding group was in by the late 1990s, they tried hard to win orders, particularly in the floating production part of the oil and gas sector. The contract for *Crystal Ocean*, a very complex DP3 floating production and storage vessel for Brøvig Offshore, was one positive outcome and had shown that the yard could be competitive both within commercial and offshore shipbuilding segments.[103] *Crystal Ocean* was built on time in 23 months and, moreover, made a small profit, which motivated the yard to show senior management at Kvaerner Oslo and London what it was capable of. The yard had a willingness to perform and the unions' understanding of what it takes to be competitive helped the yard to become attractive to an outside take-over.[104]

David Millar, former Quality Manager at the Govan yard, had this to say:

> The influence that Kvaerner brought to Govan was extremely positive, if not liked by everyone. When Kvaerner took over the yard, it was still under British Shipbuilders, it was a tired yard, with no real recent investment, some of the production equipment was in use during the war, the management structure was antiquated with too many levels, directors, managers, assistant managers, foremen, chargehands etc. It had separate canteens for directors, managers, foremen and the workforce. Kvaerner came in and changed all that, reduced the management tiers and made people more accountable, invested a lot of money on the new Tank Assembly Shop and equipment, changed working practices to be more flexible. It was exciting times working there during this period and the yard was becoming more competitive and able to look at taking contracts that other yards would struggle with.[105]

The Govan yard had come a long way under Kvaerner, with greatly improved productivity, good industrial relations, a highly skilled workforce and a modernised shipyard. As yard convener Jamie Webster put it, 'Life in the ultra-competitive, commercial, shipbuilding market under Kvaerner had made us a flexible, adaptable and progressive workforce.'[106] Despite this, the yard now faced what one newspaper described as almost certain closure. In July 1999, the first round of redundancy notices was issued, giving the statutory 90 days' notice. In the wake of the closure announcement, the Labour Government moved quickly to set up a task force to find a buyer for the yard. Headed by Sir Gavin Laird, formerly head of the AEEU Union, the task force included Scottish Office Industry Minister Lord MacDonald, Bill Scott and Malcolm Clark and was supported by senior civil servants. The task force was greatly assisted by the support given by Yard Convener, Jamie Webster, backed by the entire work force which at that time comprised 800 permanent employees and 400 subcontractors. Offers of interest were received from Jaap Kroese who had bought the Swan Hunter yard on Tyneside, and Ferguson Shipbuilders Ltd. However, the real hope lay with GEC Marconi, owners of the Yarrow and VSEL shipyards.

To place what happened next in its proper context, a little back-tracking is necessary. Kvaerner's change of direction and withdrawal from shipbuilding played out against the restructuring and consolidation of the defence interests of GEC (General Electric Company), a vast British based engineering conglomerate. Restructuring resulted in a bewildering series of acquisitions and name changes which, in essence, for its shipbuilding assets, had begun years earlier, in 1985, when GEC's defence division had bought Yarrow (Shipbuilders) Ltd in Scotstoun, builders of frigate sized warships, from the nationalised British Shipbuilders Corporation.

In 1987 GEC renamed its very successful defence interests GEC-Marconi, and under its aegis, a series of further acquisitions were made, including, in 1995, VSEL (Vickers Shipbuilding & Engineering Ltd), the Barrow-based builders of the Royal Navy's nuclear submarines. Further repositioning within the GEC-Marconi empire had resulted in the formation of Marconi Electronic Systems in 1998 which, in January 1999, spawned another offshoot called Marconi Marine to bring together the Barrow and Scotstoun shipyards under one management.

However, grander moves were afoot, namely the creation of a British defence company of the scale and magnitude to compete against European and American competitors in world markets. To achieve this, it was necessary to merge Marconi Electronic Systems with British Aerospace (BAe).

Meanwhile, at this time another crisis was in the making at Govan as already described. In November 1997 Managing Director Sigbjørn Ellingsen had announced the order for two further supply vessels, *Far Strider* and *Far Supplier*, for Farstad Supply A/S of Norway. With work running low, this order secured employment until the summer of 1998. *Sea Launch Commander* was in the final stages of fitting out and due for delivery that month while the chemical carrier *Jo Sypress* was due for delivery in March that year.

In the early months of 1999, with the two supply vessels in the final stages of completion, the yard's order book had thinned out, with only *Crystal Ocean* and *Balder Viking* in progress.

While several tenders were in preparation, including that for the MoD Ro-Ro ferries, with no new contracts in place, once again it looked like the end of what had been a very long road for the yard.

It was at this point, April 1999, that Kvaerner announced its decision to sell off all its shipbuilding interests, much at the same time that the merger between British Aerospace PLC and the Marconi Electronic Systems Business of the General Electric Company PLC, took place. The combined group was subsequently named BAE Systems. The events between April and July of that year, when an agreement was sought for the sale of Kvaerner Govan to GEC, were fraught and on a razor's edge.

However, GEC's decision to re-enter surface shipbuilding at Barrow had caused difficulties, resulting in significant delay in the construction of the two fleet oilers, the second of which, *Wave Ruler*, had not been laid down. Ordered in March 1997, and due for delivery in 2002, the £200m contract was in serious difficulty.[107] By transferring *Wave Ruler* to Govan, GEC's oiler contract would maintain some momentum and also keep the Govan yard open, at least until completion of the ship.

On 7 July GEC made an offer of £1m for the Govan yard on the basis that it might close the yard after completion of *Wave Ruler*. GEC also made it clear that Kvaerner would have to cover redundancy costs of about £9 million should the yard close. Kvaerner rejected this offer, as they thought they could get more for the yard by selling off its assets. At the same time, Kvaerner issued the 240 redundancy notices which were due to take place on 16 July.

As a highly effective campaign, led by Yard Convener, Jamie Webster, gained momentum, talks continued, this time with the intervention of John Reid, Secretary of State for Scotland. On 15 July 1999 a Heads of Agreement was signed by Kvaerner and Marconi Electronic Systems for the Govan yard. The yard was sold for £2.25m, with Kvaerner agreeing to pay for redundancies, should there be any, for up to 42 months after the sale. Clydeport, the River Clyde's harbour authority, agreed to purchase the land and buildings at the Govan yard for £2.25m, lease the site back to Marconi over a 20-year period, and also carry out environmental remediation work at the site should shipbuilding cease.[108] The deal was not finalised until 14 December 1999, with the newly devolved Scottish Executive impressing upon the MoD the importance of the Ro-Ro sealift ships contract for the Govan yard.[109]

ERA FOUR 2000–2024

BAE Systems

On 30 November 1999 BAE Systems was registered, following BAe's acquisition of Marconi Electronic Systems for £7.7 billion. From an operational perspective, however, the most immediate task was the integration into a single facility of its two Clyde shipyards at Govan and Scotstoun that best utilised their respective complementary capabilities and capacities. Each had a long history, different facilities, different expertise and operational cultures. Similarly, integration with Barrow-in-Furness, an altogether much larger works, had to be forged. To assist in this process, BAE Systems formed a new entity, BAE Systems Marine in 2000, which included the two Clyde yards and the submarine yard at Barrow. Altogether, BAE Systems Marine now accounted for approximately 80 per cent of the UK's warship design, build and integration skills base, with the balance at that time made up by Swan Hunter on Tyneside and Vosper Thornycroft (VT) at Southampton.

At the time of the official handover to BAE Systems in December 1999, the only contract in the yard was the small support vessel *Balder Viking*, with the oiler *Wave Ruler* in the early stages of being transferred from Barrow. Much more significantly for the Govan shipyard, however, was the environment in which it would now be operating. Throughout its existence, with the exception of two world wars, the yard had traded commercially on the open market. The Govan yard was now part of one of the world's largest defence contractors, responsible in the UK for many

Aerial view of part of the yard taken on 10 March 2005. The LSD (Landing Ship Dock) *Cardigan Bay*, which was launched on 9 April that year, is on the building berth and sister vessel *Mounts Bay* in the basin. The SBOH (Ship Block and Outfit Hall) is being further extended. Note the poor condition of the 250-ton cantilever crane which would be demolished two years later.

The stern block of *Defender* emerging from the SBOH on self-propelled modular transporters each of which had a maximum capacity of 400 tonnes.

aspects of national defence. Moreover, the Strategic Defence Review of 1998[1] had laid out the case for major additions to the Royal Navy in the form of two aircraft carriers and a programme of future escorts to replace Type 22 and Type 23 frigates, as well as Type 42 destroyers.[2] In July 2000, the Defence Secretary announced that the first batch of up to twelve Type 45 destroyers would be procured, to be built by BAE Systems, with the first expected to enter service in November 2007.[3]

Since the collapse of the Fairfield Shipbuilding & Engineering Co in October 1965, and its resurrection under various guises since, the shipyard had failed to trade profitably, with the exception of a few years under Kvaerner, when minor profits were made. Dependent on Government money to keep going, and with the constant threat of closure, the workforce had become habituated to the fight for survival.

Now, the beginning of a stable future outside the commercial market beckoned. However, this too would be subject to threats of various kinds, financial as well as closure, but at least the next few years seemed to be secure. Moreover, the whittling away of most of the British commercial shipbuilding industry down to statistical irrelevance had left only two surface yards capable of building larger sized ships in their entirety, Govan and Swan Hunter's Wallsend yard on Tyneside.[4]

At Govan, the last ship ordered under Kvaerner, *Crystal Ocean*, the sophisticated well-test vessel, was delivered in November 1999, leaving the yard with no work except *Balder Viking*, although by this time the relocation of the fleet oiler *Wave Ruler* was in progress, with steel first cut at Govan on 10 February 2000. The subsequent progress of this contract highlighted the difference in surface ship capabilities between Govan and Barrow, with Govan's steelwork and commercial shipbuilding expertise enabling the contract to be brought much closer to specified completion dates.

	Steel cut	Launched	Completed
Wave Knight	22 October 1998	29 September 2000	8 April 2003
Wave Ruler	10 February 2000	9 February 2001	27 April 2003

The years 2000 and 2001 presented BAE Systems with the problem of maintaining a credible shipbuilding capability at Govan, principally its skilled workforce, through lack of work, although the Type 45 destroyer and future aircraft carrier programmes were not far off. In December 2000, to help keep the yard going, the contract to build another support ship, *Stirling Islay*, a sister to Ferguson Shipbuilders' *Stirling Iona* and *Stirling Jura*, was won from Stirling Ship management. Further work followed the unfortunate closure of the small Ayrshire shipyard Ailsa-Troon Ltd. In July 1998 BAeSEMA[5] had contracted ten, water jet landing craft from the Ayrshire yard. After only two had been launched, and with no new work booked, the Troon yard was closed in December 2000, resulting in the contract for the remaining eight landing craft being switched to Govan. This helped to maintain employment levels.

Royal Fleet Auxiliary 'Bay' Class Landing Ship Dock

In April 2000 the MoD issued a tender for two auxiliary dock landing ships, subsequently known as the 'Bay' class. Although BAE Systems tendered for the work, the contract was placed with Swan Hunter on Tyneside in December of that year. Early in 2001, with much of the Govan yard idle, BAE Systems endeavoured to bring new work to Govan by offering to build two further 'Wave' class tankers for the MoD at a reduced cost. The MoD declined. However, with Govan still short of work, in November 2001 the MoD placed the order for two further 'Bay' class ships, partly in recognition that the yard 'would not be able to sustain the necessary skills capacity to meet the needs of the forward naval warship programme.'[6]

Now a class of four ships, the first two, *Largs Bay* and *Lyme Bay*, were contracted from Swan Hunter in December 2000 and laid down in October 2001. *Mounts Bay* was laid down at Govan in August 2002, followed by *Cardigan Bay* in October 2003. The Govan ships were completed in December 2005 and August 2006 respectively. However, events at Swan Hunter were proving to be problematic, with the shipbuilder struggling to control costs and adhere to construction milestones. Significant delays and major cost over-runs resulted in their first ship, *Largs Bay*, being delivered over one year late, in April 2006. The second ship, *Lyme Bay*, while structurally complete, was similarly late and over budget. Moreover, as lead yard, Swan Hunter was

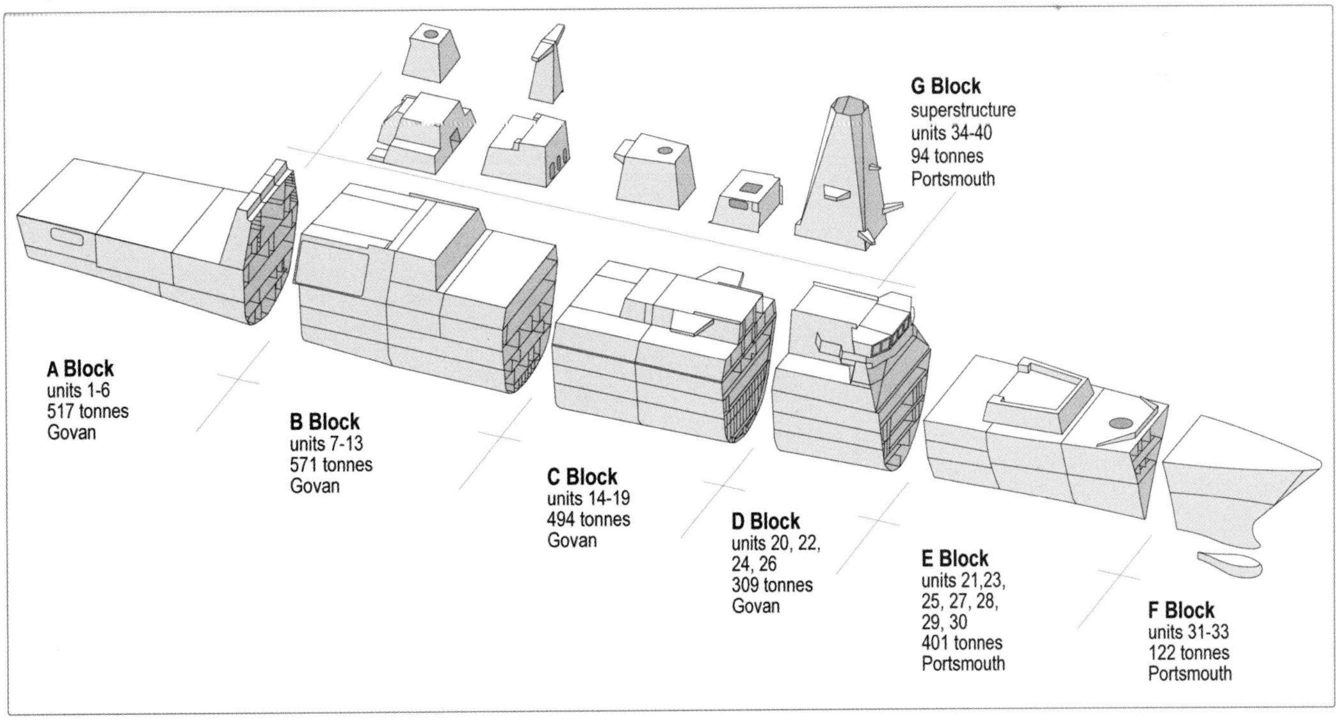

Block diagram of the Type 45 destroyers showing units per block, block tonnage and where fabricated.

Blocks for HMS *Duncan* on the building berth on 15 January 2010. D block (with the ship's bridge) is being welded to forward blocks E and F while C block has been positioned on the berth and will be hydraulically skidded into position using a block transition area for welding to C block.

responsible for design services to Govan, the supply of which was greatly disrupted and adversely affected production at Govan. In his statement to the House of Commons, the Secretary of State for Defence, Des Browne, said that the 'major programme issue has been the poor performance and cost growth at Swan Hunter, which has also impacted on the cost of the ships being built by BAE Systems.' As a result, Swan Hunter was stripped of the contract for *Lyme Bay* with lead yard services and design authority transferred to BAE Systems.[7] The incomplete *Lyme Bay* was towed from the Tyne to Govan, arriving there on 22 July 2006 for completion.

Type 45 Destroyers

With a gestation period stretching back to 1993, it was not until December 2000, that BAE Systems Marine was awarded the 'Demonstration for Manufacture' contract. This contract was worth £1.2 billion for the first batch of three Type 45 destroyers out of an anticipated total class of twelve.

A block build strategy was adopted, which enabled a high level of fitting-out to take place without reference to adjoining blocks. It also permitted the simultaneous construction of several blocks, speeding overall build times, as well as permitting the geographically dispersed construction, which served political considerations in terms of maintaining key skills and levels of employment in the few shipbuilding sites remaining across the UK. In this respect, the MoD echoed earlier Admiralty strategies from the twentieth century, aimed at maintaining warship building capacity in shipbuilding centres across the country during periods of cyclical downturn in commercial ship demand.

The four sites to be included in the Type 45 build were Vosper Thornycroft's Southampton yard and BAE Systems' Barrow and two Clyde yards. Because of its recent accumulated experience in building 12 out of 16 Type 23 frigates, as well as preceding types going back decades, it was decided that the first of class, *Daring*, would be assembled and fitted-out at Scotstoun from blocks built at Govan and Southampton. The following ships would be assembled and fitted out at Barrow from blocks fabricated at Govan, Scotstoun and Southampton. The Type 45 programme was subsequently reduced from 12 to eight units in 2004 and again, from eight to six, in 2008. The in-service date originally envisaged for the first of class was November 2007, with all six in service by 2011.

To transfer blocks weighing up to 1000 tons built at each of the four sites around the west and south coasts of Britain, a 1350-tonne, 90m long barge, *Woolston*, was built at VT's new £50 million shipbuilding hall at Portsmouth Naval Base.[8] The original build plan for the Type 45 destroyer comprised six blocks plus superstructure, which were to be built at the following locations:

Blocks	**where fabricated**	**Assembly, fit-out and completion**
First of Class (*Daring*)		
A & D	Glasgow	Scotstoun
B & C	Barrow	Scotstoun
E & F	Portsmouth	Scotstoun
Ships 2-6		
A & D	Glasgow	Barrow
B & C	Glasgow (Govan)	Barrow
E & F	Portsmouth	Barrow

(See diagram on page 203)

Sixth and last of the *Daring* class destroyers, HMS *Duncan* is launched on 11 October 2010. *Duncan* would be the last ship to be launched conventionally at the yard.

However, by the end of 2002 it became apparent that assembly and completion at Barrow would add significant risk to the programme, mainly because of issues arising there concerning the *Astute* class nuclear submarines, the first of which had been laid down in January 2001. In January 2003 it was agreed with MoD that Barrow would no longer participate in the Type 45 programme and that assembly of all six would be at the two Clyde yards, with Scotstoun assembling the first of class as originally intended and Govan all five thereafter. All blocks would be fabricated at Govan with the exception of Block E/F, the bow forward of the bridge, which would be at Vosper Thornycroft's new shipbuilding facility at Portsmouth replacing its shipyard in Southampton.

Block build precluded the option of keel laying, which was traditionally considered to be the start of a ship's life on the building berth. Henceforth, the cutting of the first steel plate assumed the start and 'first steel' was cut at Govan for *Daring*'s Forward Machinery Room on 28 March 2003, while work on the main build began at Scotstoun on 11 August. At Portsmouth, on completion of the new shipbuilding hall in late 2003, work on Block E/F and masts began. Thereafter, Govan fabricated most of the steelwork for the five remaining ships, with Portsmouth providing the forward block and masts. Fresh investment at Govan focussed on the plate preparation areas, including a new plasma cutter, robotic section cutter, laser cutter and a seam welder at the panel line. The build programme for the Type 45 destroyers came to an end in August 2012 when *Duncan* left Scotstoun to begin sea trials, eventually being commissioned into the Royal Navy in September 2013.[9]

All six Type 45s were launched from a building berth and, in the case of the five built at Govan, from its No 1 berth, served by four 80-ton capacity berth cranes. These cranes, as much as anything, had contributed to the visual appearance of a shipyard recognisable for miles around. However, with block build moving fabricated blocks of 1000 tons and more, berth cranes and building berths had

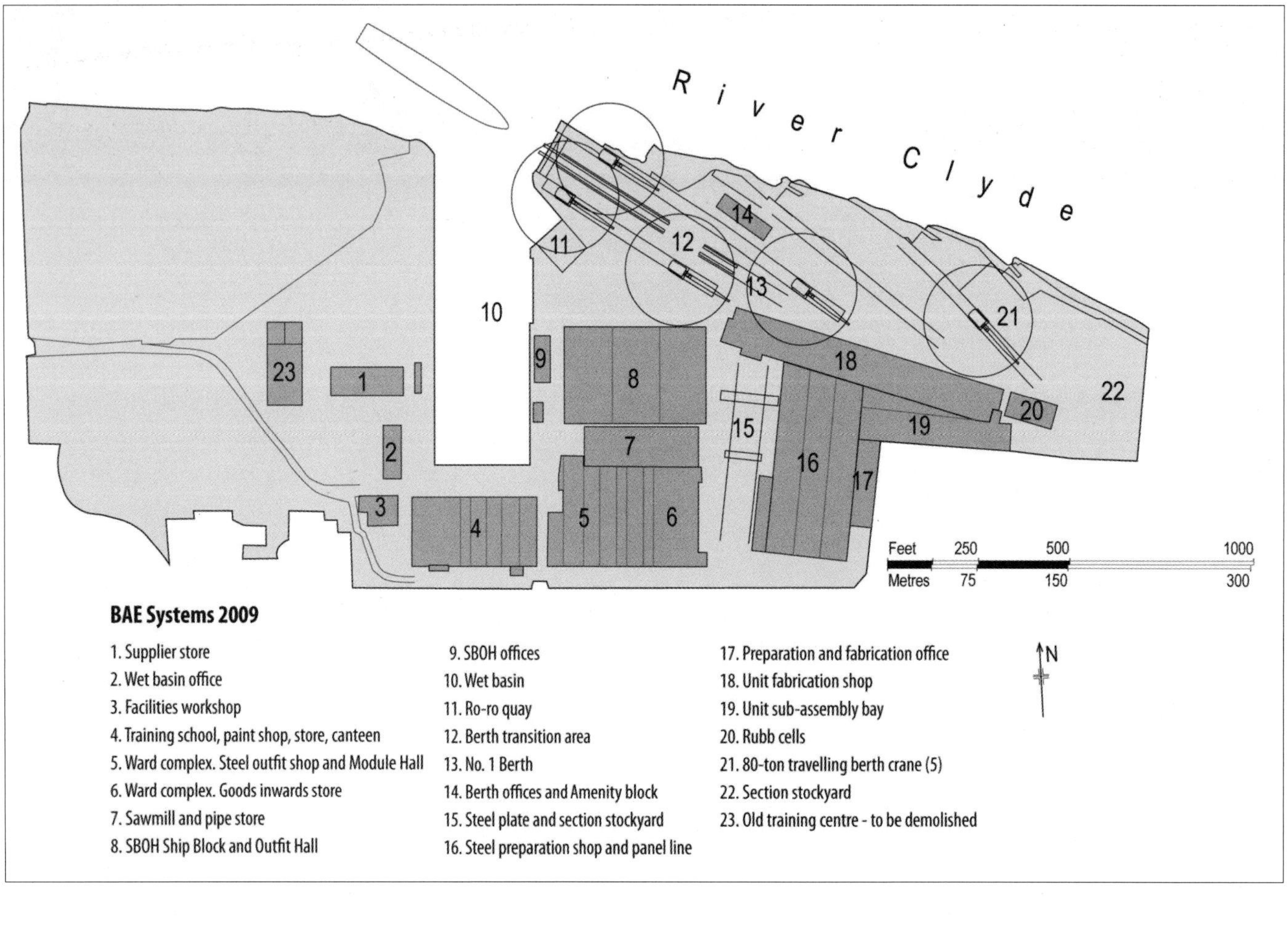

The Govan yard as it was during the T45 programme – and not significantly changed from Kvaerner days. Units were fabricated in the fabrication shop then built up into blocks in the Ship Block and Outfit Hall (SBOH), formerly known as the Tank Assembly Shop. Blocks were assembled in sequence on No 1 berth and launched, then taken to Scotstoun to complete fitting-out. Berth cranes, increasingly irrelevant, have been reduced from seven to five.

In April 2012, the forward part of Lower Block 04 of HMS *Queen Elizabeth* was moved out of the SBOH in preparation for joining up with the after section which is still in the SBOH.

become a thing of the past. The launch of the last Type 45 destroyer, *Duncan*, on 11 October 2010, brought to an end over 800 dynamic launches at the Govan yard, going back to 1864. In future, blocks would be assembled on the hard standing and loaded out onto a barge as a completed hull or as a block for assembly elsewhere. The need to provide a load out area at Govan contributed to the demise of the 250-ton giant cantilever crane, designed and erected at the Fairfield yard by Sir William Arrol & Co Ltd in 1911. Despite its A-listing, plans for its removal were first considered by Kvaerner, as block build once again circumvented its use by preinstalling heavy items such as main machinery. In 2007 permission was sought for its removal to make way for a load out area and ro-ro quay for the delivery of the Type 45 blocks from Portsmouth and this was duly granted by Historic Scotland. Over the course of 2013 and 2014, the five remaining 80-ton berth cranes were dismantled, leaving Govan shipyard shorn of its visual signposts, but better adapted to the requirements of modern shipbuilding.

The Aircraft Carrier Alliance

In July 2008, after Vosper Thornycroft's shipbuilding assets had been merged with BAE Systems to form BVT Surface Fleet, the Government announced the signing of contracts with the Aircraft Carrier Alliance (ACA) for the construction of two aircraft carriers. The ACA was a partnership between BAE Systems, Babcock International, Thales Group and the MoD, with BAE Systems in the lead. The need to disperse construction across shipyards in the UK was met by placing contracts with BAE Systems at their Clyde and Portsmouth shipyards, Appledore Shipbuilders in Devon, Cammell Laird at Birkenhead and A&P Tyne on Tyneside. The former Royal Naval Dockyard at Rosyth, privatised in 1997 when Babcock International acquired the site, was chosen for integration, where the aircraft carriers would be assembled and completed. This work was carried out in the largest of the three dry docks available, assisted by a 1000-ton capacity goliath crane acquired for the purpose.

The construction of the two largest blocks for *Queen Elizabeth*, LB03 and LB04, was allocated to the Govan yard. Work on the 8500-ton LB03 began in the Ship Block Outfit Hall (SBOH), formerly the Tank Assembly Shop, in June 2009. Units were fabricated in the adjacent fabrication shed and taken to the SBOH for assembly. On completion in May 2011, LB03 was rolled by remote controlled Scheuerle transporters onto the barge *AMT Trader* for transportation to Rosyth. The contract to take the block the 600-mile journey around the north of Scotland was managed by Malin Group Ltd. Work on LB04, the largest and most complex of all the blocks, weighing over 11,000 tons and comprising both engine rooms, began in January 2010 and was completed in November 2012. Unlike the previous block, which had been taken around the north of Scotland to Rosyth, poor weather forecasts forced the decision to take the longer journey in calmer waters round the south coast of England. The 750-ton aft island was constructed at Scotstoun using steel units supplied by Govan. Construction of *Prince of Wales* followed a similar pattern, with first steel cut at Govan in April 2011 and completion in January 2016.

When shipyard Convener Jamie Webster was interviewed in 2012 and asked about the Govan yard's survival since facing closure in 1999, he delivered this upbeat message:

> First of all we build really good ships, we have managed change in an exemplary fashion, we have demonstrated commitment and passion and sometimes we have landed in the right place at the right time. It comes down to what sacrifices and compromises you make in order to remain [in shipbuilding]. Thirteen years on, we are building the two largest ships that the Royal Navy has ever had and from then [1999] until now we have started about 1000 apprentices. That is a future for young people and a major achievement.
>
> The 13 years with BAE Systems have been the happiest of my life in the yard because they are a good employer, they have invested a lot in the people and the workforce, they value the workforce and they are not remotely anti trade union. We have our disagreements but we always resolve them.[10]

However, the yard was still to face further challenges.

Rationalisation

As work on the Type 45 destroyers and blocks for the *Queen Elizabeth* class aircraft carriers passed its peak, BAE Systems announced a much anticipated restructuring and rationalisation of its shipbuilding operations in November

2013. With future building programmes and, in particular, the Type 26 Global Combat Ship then at the beginning of the design process likely to require significantly less manufacturing capacity, the decision was taken, in collaboration with the Ministry of Defence, to cease shipbuilding at Portsmouth. Much to the relief of shipyard workers on the Clyde, BAE Systems considered that the most effective location for the manufacture of the future Type 26 programme was to consolidate all of its surface shipbuilding operations in Glasgow, a decision based on skills, operational delivery and cost. Moreover, BAE Systems planned fresh investment to create a world-class capability on Clydeside which would deliver an affordable Type 26 programme.

Opposite: The completed Lower Block 04 of *Queen Elizabeth* on the barge *AMT Trader* in November 2012 waiting to depart to Rosyth. LB04 was the largest of the blocks weighing over 11,000 tons and housed both engine rooms.

Despite the promise of this new investment on Clydeside, the consequences of restructuring were the inevitable reduction in jobs, with 1775 redundancies of which 940 were in Portsmouth[11] to be made during 2014. A further 885 jobs were to go at Filton, Glasgow and Rosyth by 2016. The cost of the restructuring was borne by the Ministry of Defence.

Meanwhile, to assist in maintaining employment on the Clyde until the Type 26 Programme was ready for manufacture, blocks LB05 and UB07 and UB14 for *Prince of Wales* were transferred from Portsmouth. Additionally, a

HMS *Forth* on the hard standing from where she will be moved onto a barge and taken to the Scotstoun yard for completion. Photograph taken on 14 August 2016.

The naming ceremony for the Batch 2 'River' class offshore patrol vessel HMS *Trent* on 13 March 2018 with cadets in attendance.

contract for three offshore patrol vessels was announced, to be built on the Clyde. At this time, both shipbuilding sites in Glasgow employed 3200 people on the *Queen Elizabeth* class carriers and on the design of the Type 26 Global Combat Ship.

On 7 January 2014 the first public indications that the Govan yard might not have a future came when BAE Systems issued a press release, describing two investment options aimed at creating a '21st century complex warship facility' at its Clyde shipbuilding sites for future ship programmes. The Company said its proposals were the result of a three year-long study, where processes were compared to 34 shipyards worldwide.[12]

Option One, costing £200m, was a single-site solution, intended for Scotstoun, which included a new 330m covered dock hall, steelwork preparation and fabrication facilities, and a paint cell with offices and welfare facilities. The proposal also included a new quay, where ship blocks could be transported by sea. Work would begin in January 2015 with completion in 2017. Under this proposal, the Govan yard would close.

Option Two, costing £100m, was a two-site option, where, at Govan, extensions to the existing fabrication and main ship build hall were proposed, while the building slip would be levelled for a new transfer quay. At Scotstoun, upgrades to the dry docks were proposed, along with strengthening at the deep water berth to support mobile cranes. If this option was selected, work at Govan would begin in February 2015, for completion in 2017, and at Scotstoun, from 2017 to 2018. Public consultation began soon after, with planning applications for both proposals submitted in March and a final decision expected by the end of the year.

With an existing major office complex and dry docks at the Scotstoun site, the rationale behind the single-site option was clear, and it was no secret that BAE Systems favoured it. If selected, it would finally mean the end for the Govan yard, which would be phased out by 2018 once the offshore patrol vessels were completed. This view was reinforced when demolition of the Scotstoun shipyard began in late 2014, in preparation for the proposed frigate factory. In 2015, however, plans for the single site at Scotstoun were dropped. In the end, funding for this project, set against the reduction in the number of Type 26 frigates from thirteen to eight, proved to be more than the MoD were prepared to support.

Offshore Patrol Vessels

In 2009 the MoD signed a Terms of Business Agreement with BAE Systems (then BVT Surface Fleet) to provide continuity of employment as may be necessary for major projects as then envisaged, primarily the aircraft carriers and the Type 26 Global Combat Ship. It was foreseen that the termination of these projects would result in a reduction in capacity and manpower. In 2014, again to maintain the capability of its surface yards on the Clyde through lack of work, the MoD announced its intention to build three 'River' class offshore patrol vessels (OPVs), *Forth*, *Medway* and *Trent*. This order was increased in 2015 from three to five ships, the additional vessels being named *Tamar* and *Spey*. Work began at Govan on the first OPV, *Forth*, in October 2014 with the last, *Spey*, completed in January 2021 at £44m under its original cost of £680m.[13]

In April 2021, two blocks of HMS *Glasgow* on the hard standing at Govan being prepared for joining.

An atmospheric drone shot of HMS *Glasgow* taken on 25 November 2022 with a large swathe of Glasgow in the background. The ship is being prepared for transfer from the hard standing to the Malin Group's semi-submersible lifting barge CD01. (*Copyright © 2023 BAE Systems. All rights reserved*)

Type 26 Global Combat Ship

The assessment phase for this programme began in March 2010, when BAE Systems was awarded the contract to design what was referred to as the Global Combat Ship. Approximately 550 design engineers worked in Bristol, Glasgow and Portsmouth to develop the specification for the vessels, with manufacturing expected to begin by the end of 2014. The intention was to build thirteen ships of the class, but the Strategic Defence and Security Review of 2015 reduced the number to eight, with the sums originally allocated for the five remaining vessels reallocated to develop the simpler and less expensive Type 31e frigates.[14]

HMS *Glasgow*

On 2 July 2017 BAE Systems were awarded a £3.7 billion contract to build the first three Type 26 frigates, with the first steel being cut soon after at Govan for the first of class, HMS *Glasgow*. This was the ninth ship to be given the name *Glasgow* by the Royal Navy and the second to be built at the Govan yard (in 1909 the 'Town' class cruiser *Glasgow* had been launched at the Fairfield yard). The forward block of the latest *Glasgow* was moved from the SBOH on 19 April 2021 and later joined on 1 May 2021 by the aft block.

On 25 November 2022 the structurally complete ship was manoeuvred using Sarens self-propelled modular trailers onto a barge, managed by Malin Group, and taken down river to Glen Douglas, where the ship was floated off and then towed back up river to Scotstoun, where fitting-out commenced in dry dock.

Fit for the Future

On 15 November 2022 BAE Systems was awarded the £4.2 billion contract to build the remaining five Type 26 frigates, all of which will be constructed at the Govan yard, with fit-out at Scotstoun. In anticipation of this contract, in May of that year BAE Systems submitted an outline planning proposal for a new £100m ship assembly hall at Govan. Having been unable to provide the new frigate factory at Scotstoun in 2015, this meant utilising the build strategy for the Type 26 frigates similar to that adopted earlier for the Type 45 destroyers – units fabricated in the fabrication shop, blocks assembled in the SBOH, with final assembly on the hard standing and fit-out at Scotstoun. This utilised the production facilities installed at the Govan yard by Kvaerner almost thirty years earlier. The scale of the Type 26 contract, however, fully justified a significant investment in new shipbuilding facilities at the Govan yard. Several site layouts were considered before the decision was made to infill the fitting-out basin (wet basin) and build the new assembly hall there.

When completed in 2024, the new facility, measuring 170 metres long by 81 metres wide and 49 metres high, will allow two Type 26 sized hulls to be built side by side, simultaneously.

In July 2022 a spokesperson for BAE Systems said:

> The opportunity to provide a new modern ship building hall of this nature would allow BAE Systems to adopt improved shipbuilding techniques together with improved construction access and state of the art, dedicated, on-site office and amenities accommodation. It would also significantly improve ship building capability in Govan, which in turn will safeguard employment for the ship building and affiliated trades in the Glasgow city region and secure continued investment in training and education, in innovative technologies and in BAE Systems' supply chain. [15]

In 2023 Sven Hanssen, BAE Systems Manufacturing and Operations Director, added detail to the investment the Company was making at its Scotstoun and Govan yards under four headings: An Applied Skills Academy, Wet Basin Hall, New Panel Line, and Digital Shipbuilder.

> The key thrust is that BAE Systems has a massive investment programme underway on the Clyde that will make our business fit for the future. All of the above four initiatives are intertwined as we seek to generate a digital thread through all of our activities from concept design to fitting the part on the ship. Equally, we will enable our workforce from shop floor to board room through upskilling in new technologies and ways of working. Our new and upgraded facilities will leverage technology and processes to significantly improve the working environment for our employees and finally, to make our yards safer, more efficient, and able to deliver improved quality.[16]

The BAE Systems Govan yard showing the new shipbuilding hall under construction on 19 July 2024. (*Photo by courtesy of Bob Shaw*)

Going forward, links to past achievements, albeit tentative, are there too, with recent visits to the Govan yard by representatives of the Australian and Canadian navies to assess the Type 26 frigate design, which they subsequently adopted.[17] *Parramatta*, the first destroyer built for the Royal Australian Navy, was constructed at Fairfield in 1909. Similarly, a host of vessels for Canadian owners, most notably the Canadian Pacific Railroad, have been built at Fairfield.

Despite the odds, which have seen the virtual elimination of the merchant shipbuilding industry in the UK and set against the enormous changes that have taken place over the last 160 years, the shipyard built on the grounds at Fairfield Estate in 1864 is still working, building highly advanced ships and with new investment that will make it a state of the art shipbuilding facility. Interestingly, the vast new shipbuilding assembly hall under construction as these words are written, will be a close neighbour to the 1871 A-listed John Elder Engine Works building, where pioneering work in the development of marine steam propulsion took place, changing transportation at sea. Perhaps more than anything else these two remarkable structures describe the arc of history that has kept, and is keeping, the Govan yard at the forefront.

APPENDICES

Appendix A Business results 1834 to 1884

The results of the business first started by Charles Randolph in 1834 up to 1884 prior to the incorporation of the Fairfield Shipbuilding & Engineering Co Ltd in 1885. This covers the period from general engineers to acclaimed shipbuilders and marine engineers five decades later.

Randolph & Co.

From 30 September 1834 until 30 June 1837

Wages	Value of work	Profit
£4,483	£13,750	£2,977

Randolph Elliott & Co.

From 1 July 1837 until 1 September 1852

Wages	Value of Work	Profit
£47,570	£182,544	£29,970

Randolph Elder & Co.

From September 1852 to June 1868

Wages	Tonnage	IHP	Value of Work	Profit
£607,911	46,002	65,035	£2,229,092	£188,201

John Elder & Co.

From July 1869 until 17 September 1869

Wages	Tonnage	IHP	Value of Work	Profit
£201,136	24,133	14,592	£639,525	£73,803

John Elder & Co. (under executors of John Elder's estate)

From 18 September 1869 until 30 June 1870

Wages	Tonnage	IHP	Value of Work	Profit
£143,716	10,423	13,549	£344,660	£10,412

John Elder & Co. (under Ure, Jamieson and Pearce)

From 1 July 1870 until 30 June 1878

Wages	Tonnage	IHP	Value of Work	Profit
£1,625,541	170,556	134,587	£5,237,451	£328,485

John Elder & Co. (under Wm. Pearce)

From 1 July 1878 until 31 December 1884

Wages	Tonnage	IHP	Value of Work	Profit
£1,980,076	191,774	250,214	£6,720,769	£575,064

Marine steam engines constructed by Randolph Elder & Co. 1854–1859

No.	Ship name	Owner	Builder	Date
1	Wm. McCormick	Londonderry S S Co	A Stephen & Sons.	1854
2	Brandon	London & Limerick S S Co	Barclay Curle	1854
3	Beaver	G&J Burns	Robert Steel & Co.	1854
4	Otter	G&J Burns	Robert Steel & Co.	1855
5	Pantaloon ?	James Paterson		January 1855
6	Petrel ?	"		January 1855
7	Rapid ?	"		January 1855
8	Brisbane	Australasian S N Co	Robert Steel & Co.	1854
9	Columbian ?	Wm. Simmons & Co		July 1855
10	Mangerton	Robert Steel & Co		1855
11	Prince Arthur	Denny & Rankin		1855
12	Khersonese	James Johnson, Liverpool	Robert Hickson & Co.	1856
13	Valparaiso	Pacific S N Co	John Reid & Co.	1856
14	Inca	Pacific S N Co	Robert Steel & Co.	1856
15	Circassian ?	James Johnson, Liverpool		1856
16	Panther	Burns & McIver	Robert Steel & Co.	1857
17	Tempest	Handyside & Henderson	Sandeman & McLaurin, Whiteinch.	1856
18	Don Alfonso	Glasgow & Lisbon S P Co		March 1859
19	Don Pedro	Glasgow & Lisbon S P Co		March 1859
20	John Bell	John Bell, Glasgow	A Stephen & Sons.	1857
21	Royal Bride ?	J M Hyde & Co, Bristol		March 1858
22	Admiral	Rigaer Dampfschiffahrt Gesellschaft	R Napier & Sons.	1858
23	Callao	Pacific S N Co	John Reid & Co.	1858
24	Lima	Pacific S N Co	R Napier & Sons.	Re-engined 1859
25	Bogota	Pacific S N Co	R Napier & Sons.	Re-engined 1859

Appendix B North Atlantic Record Breakers 1879-1893

Fairfield and Competitors

	Fairfield	J&G Thomson	Barrow	Harland & Wolff
1879	*Arizona*/Guion			
1880				
1881	*Alaska*/Guion	*Servia*/Cunard	*City of Rome*/Guion	
1882				
1883	*Oregon*/Guion	*Aurania*/Cunard		
1884	*Umbria*/Cunard			
1885	*Etruria*/Cunard			
1886				
1887				
1888		*City of New York*/Inman		
1889		*City of Paris*/Inman		*Teutonic*/White Star
1890				*Majestic*/White Star
1891				
1892				
1893	*Campania*/Cunard *Lucania*/Cunard			

Appendix C **Progress of Construction**

The four ships shown here give a good indication of the sequence of construction for riveted ships. Note that construction of *Transylvania* was subject to delay because the onset of the Depression reduced Anchor Line's ability to make stage payments.

Ship No.	600		616		634		640	
Building berth	No. 5		No. 2		No. 1		No. 3	
Name	**Transylvania**		**HMS Norfolk**		**Empress of Japan**		**Worcestershire**	
Owners	Anchor Cunard		HM Government		CPR		Bibby Bros	
Description	Passenger		Cruiser		Passenger		Passenger	
Dimensions	572.5'x70'x51.25'		590'x68'3"x43'6"		640'x83.5'x485"		482'x64'x36.5"	
Gross tonnage	16293		-		-		-	
Load displacement	23200		10,000		-		-	
Contract date	1918	3 May	1924	13 May	1928	26 June	1929	11 April
Keel laid, commenced	1919	1 Dec		15 Sept		9 Aug		2 Sept
Keel laid, completed		15 Dec		24 Oct		10 Aug		2 Sept
Frames commenced in shed		25 Dec		24 Nov to Jan 1925		17 Sept		31 July
Frames erected, commenced	1920	14 May		24 Dec		17 Dec		25 Nov
Frames erected, completed	1924	3 Dec	1925	28 Aug	1929	28 Feb	1930	26 Feb
Double bottom, commenced	1920	13May	1924	27 Nov	1928	5 March	1929	10 Oct
Double bottom, completed		12 July	1925	30 Jan		13 March	1930	17 Oct
Stem erected		18 Oct	1924	25 Nov	1929	2 March		10 Feb
Stern frame, erected		14 May	1924	24 Feb	1928	11 Dec		10 Feb
Shaft brackets, erected		18 Oct	1925	8 Aug to 10 Sept	1929	15 Jan		18 Feb
Shell plating, commenced	1921	19 March		12 Feb		31 Jan	1929	17 Sept
Shell plating, completed	1924	27 Aug		16 Dec		21 June	1930	29 May
Shell riveting, commenced	1921	6 May		10 March	1928	19 Oct		8 March
Shell riveting, completed	1924	6 Nov	1926	13 Jan	1929	3 Nov		2 July
Water testing, commenced		27 Aug	1925	10 May		3 June		13 May
Water testing, completed	1925	30 Jan		2 July	1930	30 April		7 Oct
Joiner work in shop, commenced		7 Jul	1926	2 Feb	1929	13 March		16 May
Joiner work at ship, commenced	1924	15 Dec		15 March		4 July		4 Aug
Boring stern, commenced	1925	26 Jan		6 Feb		4 Oct		12 Aug
Boring stern, completed		24 Feb		27 Feb		2 Nov		21 Aug
Propellers shipped		26 Feb		21 Aug		4 Dec		12 Sept
Rudders shipped	1924	22 Dec		22 Jan		19 Sept		10 July
Launched	1925	11 March		30 March		17 Dec		8 Oct
Boilers on board, commenced		1 April		8 April		20 Dec		1 July
Boilers on board, completed		9 April		19 April	1930	10 Jan		1 July
Ready for machinery		22 May		11 June	1929	19 Dec		13 Oct
Shipping main machinery, commenced		25 May		18 June	1930	31 Jan		15 Oct
Shipping main machinery, completed		16 June		13 Aug		5 Feb		25 Oct
Closing in, commenced		17 June		12 July		26 Feb		25 Oct
Steam on boilers		13 Aug		-		25 April	1931	20 Jan
Engines turned		21 Aug	1927	25 March		28 April		27 Jan
Left builder's dock		25 Aug		23 April		4 May		4 Feb
Vessel dry docked		25 Aug		23 April at Govan		6 May at Belfast		18 Feb*
Vessel undocked		28 Aug		26 April		8 May		4 Feb
Proceeded down river		1 Sept		2 May		5 May		4-6 Feb
Trial trip		1-3 Sept		20 Jan		8-12 May		-
Due for delivery		-		18 June		-		9 Feb
Delivered		3 Sept		28 June		9 June		4 Feb

Appendix D World War One Output

No	Name	Type	Built	Tons Gr	Tons Disp	HP	HP1A
497	VALIANT	battleship	1916		27500	75000	
499	AVENGER	AMC	1916	14744			
500	MELAMPUS	destroyer	1915		1037	25000	
501	MELPOMENE	destroyer	1915		1040	25000	
502	WATERWITCH	despatch	1915		400	500	
503	RENOWN	battlecruiser	1916		26500	112000	
504	MANDATE	destroyer	1915		1037	25000	
505	MANNERS	destroyer	1915		1040	25000	
507	E.37	submarine	1916		662	1600	
508	E.38	submarine	1916		662	1600	
509	MINDFUL	destroyer	1915		1034	25000	
510	MISCHIEF	destroyer	1915		1025	25000	
511	ONSLAUGHT	destroyer	1916		1025	25000	
512	ONSLOW	destroyer	1916		1025	25000	
513	E.47	submarine	1916		662	1600	
514	E.48	submarine	1917		662	1600	
516	OBSERVER	destroyer	1916		1025	25000	
517	OFFA	destroyer	1916		1025	25000	
518	ORCADIA	destroyer	1916		1025	25000	
519	ORIANA	destroyer	1916		1027	25000	
520	PHEASANT	destroyer	1916		994	25000	
521	PHOEBE	destroyer	1916		1044	25000	
522	K.13	submarine	1917		1980	10000	
523	K.14	submarine	1917		1980	10000	
524	TEMPEST	destroyer m'layer	1917		1065	27000	
525	UNDINE	destroyer	1917		1085	27500	
526	CARDIFF	cruiser	1917		4190	40000	
527	RODNEY	battlecruiser	1919				
530	VENDETTA	destroyer	1917		1295	27000	
531	VENETIA	destroyer	1917		1296	27000	
532	L.15	submarine	1918		895	2400	
533	L.16	submarine	1918		895	2400	
534	WALRUS	destroyer m'layer	1918		1300	27000	
535	WOLFHOUND	destroyer	1918		1300	27000	
536	WAR PATRIOT	tanker	1918	5197		2500	
537	WAR PIONEER	cargo	1919	5256		2500	
539	L.55	submarine	1918		960	2400	
540	TRAIN FERRY No.3	trainferry	1917	2672			410 nhp
544 to 561	Ac.1151 to Ac1168	lighter x 18	1917	2340			
571	SIKH	destroyer	1918		1075	27000	
572	SIRDAR	destroyer	1918		1075	27000	
573	SOMME	destroyer	1918		1075	27000	
574	IRVINE	minesweeper	1918		800	2200	
575	KENDAL	minesweeper	1918		800	2200	
576	KINROSS	minesweeper	1918		710	2200	
577	LYDD	minesweeper	1919		800	2200	
580	CARLISLE	cruiser	1918		4290	40000	
581	COLOMBO	cruiser	1919		4290	40000	
582	SPEAR	destroyer	1918		1075	27000	
583	SPINDRIFT	destroyer	1919		1075	27000	
586	WANDERER	destroyer	1919		1325	27000	
	WW1 total			**30,209**	**108,082**	**1,029,400**	**410**

Refit and Repair 1914/1918

Warrior	Armoured cruiser
Grafton	Cruiser
Active	"
Askold	Russian cruiser
Caroline	Light cruiser
Cambrian	"
Boadicea	"
Kildonan Castle x2	Auxiliary cruiser
Gloucestershire	"
Jason	Torpedo Gun Boat
Gossamer	"
Skipjack	"
Spanker	"
Leda	"
D4	Submarine
D8	"
K22	"
L8	"
E38	"
L15	"
L16	"
R11	"
Myrtle	Minesweeper

World War Two Output

No	Name	Type	Built	Tons disp.	HP
666	PHOEBE	cruiser	1940	5450	62000
668	KELVIN	destroyer	1939	1760	40000
669	HOWE	battleship	1942	35000	110000
672	IMPLACABLE	aircraft carrier	1944	25690	148000
673	NAPIER	destroyer leader	1940	1773	40000
674	NESTOR	destroyer	1941	1773	40000
675	MUSKETEER	destroyer	1942	1920	48000
676	ORKAN	destroyer	1942	1920	48000
679	BELLONA	cruiser	1943	5770	62000
680	ORIBI	destroyer	1941	1540	40000
681	OFFA	destroyer	1941	1540	40000
682	PANTHER	destroyer	1941	1540	40000
683	PARTRIDGE	destroyer	1942	1540	40000
684	ONSLAUGHT	destroyer	1942	1610	40000
690-693	LCT.7 to LCT.10	LCT1	1941	904	2800
694	GOATHLAND	destroyer escort	1942	1087	19000
695	LA COMBATTANTE	destroyer escort	1942	1087	19000
701	STARLING	sloop	1943	1430	4300
702	WOODCOCK	sloop	1943	1470	4300
703	VENUS	destroyer	1943	1710	40000
704	VERULAM	destroyer	1943	1710	40000
705	WAKEFUL	destroyer	1944	1710	40000
706	WESSEX	destroyer	1944	1710	40000
707	CAMPERDOWN	destroyer	1945	2315	50000
708	FINISTERRE	destroyer	1945	2315	50000
709	THESEUS	aircraft carrier	1946	13350	40000
723	LST.3037	LST3	1945	2256	5500
724	LST.3038	LST3	1945	2256	5500
	WW2 Total			**124,136**	**1,158,400**

Appendix E Govan Shipyard Timeline

1834
Randolph Elliot & Co

1852
Randolph Elder & Co

1868
John Elder & Co

1885
The Fairfield Shipbuilding & Engineering Co Ltd

1905 - 1919
Coventry Ordnance Works
(with John Brown & Co Ltd and Cammell Laird Ltd)

1966
Fairfield (Glasgow) Ltd

1968
Upper Clyde Shipbuilders Ltd Govan Division

1972
Govan Shipbuilders Ltd

1972 - 1984
Scotstoun Marine Ltd
(a subsidiary of Govan Shipbuilders Ltd)

1988
Kvaerner Govan Ltd

1999
Marconi Electronic Systems Ltd
2000
BAE Systems Marine Ltd
2008
BVT Surface Fleet Solutions Ltd
2009
BAE Systems Surface Ships Ltd
2011
BAE Systems Maritime
2012
BAE Systems Maritime - Naval Ships

Appendix F **Board of Directors** A selection from the various Boards that directed the course of the company from the 1880s onwards.

Fairfield Shipbuilding & Engineering Co Ltd

1889
Sir William G Pearce (Chairman)
Richard Barnwell (Managing Director)
Richard Saxton White (Shipyard Manager)
Sir William McOnie
Sir James Roberton

1893
Sir William G Pearce (Chairman)
Richard Barnwell (Managing Director)
Sir William Arrol
Dr Francis Elgar
Andrew Laing (Engine Works Manager)
Sir William McOnie
Richard Saxton White (Shipyard Manager)

1905
Sir William G Pearce (Chairman)
Alex Gracie (Managing Director)
Colonel Mosely
Alexander Sampson (Shipyard Manager)
Sir Digby Morant
W M Rhodes
George Strachan (Secretary)

1918
Sir Alex Gracie (Chairman and MD)
Restal Ratsey Bevis
Sir Douglas Brownrigg
Alex Cleghorn (Engine Works Manager)
James Hamilton (Shipyard Manager)
Admiral Sir Digby Morant
George Strachan

1921
Robert Alfred Workman (Chairman)
Alexander McAusland Kennedy (MD)
Edward McKay Edgar, (Senior partner in Sperling & Co Ltd)
Sir John Esplen
William Orr Workman
George Strachan

1933
Sir Alexander Kennedy (Chairman and MD)
George William Barr (Deputy MD)
Sir Douglas Brownrigg
Robert Strachan

1945
Sir William Lithgow (Chairman)
George William Barr
Lord Elgin
Henry Lithgow
Robert Strachan
G Strachan

1959
Sr John M Erskine (Chairman)
James Lenaghan (MD)
Dr A W Davis
Vice-Admiral E W Longley-Cook
A S Nicholas
D D McPhie (Secretary)

1965
J E Boyd (Chairman)
James Lenaghan (MD)
J H B Chapman
A W Davis
Derek B Kimber (Deputy MD)
G R Rickman
A H White
W S Wilson
R Atkinson (Secretary)

Fairfield (Glasgow) Ltd

1968
Iain M Stewart (Chairman)
Professor K J W Alexander
E O T Blanford (MD)
Lord Carron
A Cunningham
James D Houston
R C Ingram (Secretary)
Derek J Palmar
Sir Jack Stamp
H C Stenhouse

Upper Clyde Shipbuilders Ltd

1971
Anthony E Hepper (Chairman)
Sir Charles Connell and Sir Eric Yarrow (Deputy Chairmen)
Ken Douglas (MD)
K A Alexander
T H Burleigh
J B Corfe
J R Duff
H Farrimond
M O Hughs
A I Mackenzie
J F Starks
J F Stephen
R A Williamson, (secretary)

Govan Shipbuilders Ltd

1985
Dr P A Milne (Chairman)
Eric D Mackie (MD)
William Gallagher (Deputy MD)
J S Shand
J Wilson
R A McCann
J J Shannon
David M Smith
J F Milne
J G Kelly (Secretary)

Kvaerner Govan Ltd

MANAGEMENT STRUCTURE 1991
Steinar Dragebo (Managing Director)
David M Smith (Commercial Director)
Norman MacPherson (Quality manager)
Heikki Kytola (Project Director)
Kjel Inge Vedholm (Finance Director)
Tom Batemen (Management Services Director)
Sten Solberg (Steelwork Operations Director)
Dirk Dijkstra (Engineering Operations Director)
Robert Gordon (Personnel & Employee Relations Director)

Appendix G
Select Biographies

The Fairfield Shipbuilding & Engineering Co Ltd from 1880 to1950

Richard Barnwell 1849–1898
Born in Canterbury in 1849. After training at the Union Bank of London Ltd, Barnwell was employed at John Elder & Co Ltd in 1880 which was then under the control of William Pearce. In 1885, he became a partner and in the following year when the business was converted into a limited company under the title Fairfield Shipbuilding & Engineering Co, he was made Managing Director. He was also chairman of the Herne Bay Pier Company, managing partner of the Northern Pacific Steamship Company, a director of the Isle of Man Steam Packet company and the Liverpool and North Wales Steamship Co, the Scottish Oriental Steamship Company and a board member of the Royal Insurance Company. He died after a long illness in 1898 at the age of 48. His wife, a step-sister of Lady Pearce, died in 1893.

Sir George William Barr CBE 1881–1956
Born in Barrow-in-Furness, he began an apprenticeship at Barrow with the Naval Construction & Armaments Co which was soon to become Vickers Sons & Maxim Ltd. He rose to become Assistant Shipbuilding Manager there but left in 1912 to take the position of Shipyard Manager at the Government owned dockyard at Cockatoo Island, Sydney, NSW. Soon after the First World War broke out, he returned to the UK to become Shipyard Manager at William Beardmore's Naval Construction Works at Dalmuir. Following the end of the war, Barr returned to Barrow as Shipyard Manager, becoming a director of the company in 1923. Ten years later, he was appointed Deputy Managing Director at Fairfield, becoming Managing Director in 1946. He was awarded the CBE in 1945 and knighted in 1951. In November 1947, he was made Chairman of the Liverpool & North Wales Steamship Co Ltd in which Fairfield had an interest dating from 1891.

Archibald Douglas Bryce-Douglas 1840–1891
Bryce-Douglas was born in the seaside town of Saltcoats on the Ayrshire coast. Expected to follow his father into the ministry, he preferred a practical, hands-on pathway which resulted in partial apprenticeships as a joiner and a millwright with local firms before moving to Glasgow and Randolph Elder's Centre Street engine works. In the early 1860s, however, he set off on what would be an interesting and eventful period, travelling to New Zealand and then, after working his passage to the Pacific Coast of South America, a period of one year with the Peruvian Navy. Following that he became a sea-going engineer with the Pacific Steam Navigation Co (PSNC) for which Randolph Elder & Co, had manufactured many compound engines. After returning home and gaining his chief engineer's ticket, he became assistant engineer of the PSNC and, in 1869, their Superintendent Engineer at Callao, where the steamship company was headquartered. About 1876, Bryce returned to Glasgow to become head of Fairfield's engine works. During this time, he was responsible for the machinery for the record breaking ships *Arizona*, *Alaska* and *Oregon* as well as *Livadia* and *Petr Veliky*. In 1888 Bryce was offered the post of Managing Director of the Naval Construction and Armaments Co at Barrow, where he remained until his early death in 1891 at the age of 50.

Isabella Elder 1827–1905
Born in Tradeston, Glasgow in 1827, she married John Elder in 1857 and, for a brief period after her husband's death in 1869, managed the shipyard before handing over to her brother John Ure, John Jamieson and William Pearce. Isabella, who, on the death of her husband had been left very wealthy, became increasingly concerned about the well-being of Govanites which she demonstrated through various generous acts including the purchase of 36 acres of ground opposite the shipyard which she opened in 1885 as Elder Park. She also provided a school for domestic economy, a training home for nurses, the Elder Cottage Hospital in Govan and the Elder Park Library. Isabella Elder donated generously to the University of Glasgow, which included the founding of the John Elder Chair of Naval Architecture and Mechanical Engineering in 1883. She also pioneered medical education for women creating Queen Margaret College, subsequently part of Glasgow University. In 1906 a bronze statue of her was unveiled in Elder Park where it remains today. In 2015 the Isabella Elder Building at the University of Glasgow was opened.

Dr Francis Elgar 1845–1909
Born in 1845, Elgar became a naval architect after an apprenticeship at Portsmouth Dockyard, followed by a three-year programme of study in ship construction which he passed with distinction. In 1871 he became assistant to Sir Edward Reed, the former Chief Constructor for the Admiralty, and in 1874 manager of Earle's shipyard in Hull until 1876. He then spent two years in Japan advising the Japanese Government on naval construction, after which he returned to London where he set up as a consultant naval architect. His first association with Fairfield came in 1882, when he investigated the cause of the sinking of the Fairfield built Orient liner *Austral* at Sydney. In 1883 he was involved in the investigation into the *Daphne* disaster at Alexander Stephen's yard when the steamer turned over in the river at the end of the launching run. In the same year, he became the first professor of naval architecture in Britain, occupying the John Elder Chair of Naval Architecture at Glasgow University. While there, he was awarded an honorary LLD. In 1886 he was appointed Director of Dockyards by the Admiralty. In 1892 he resigned to take up an appointment as a director and consultant naval architecture at the Fairfield yard. In October 1904 Elgar tendered his resignation. This was deeply regretted by the Fairfield Board although he continued as the Company's consulting naval architect. In 1907 Elgar became chairman of Cammell Laird at Birkenhead and, following the death of Sir William G Pearce in November 1907, Elgar returned to Fairfield as Chairman. He died on 17 January 1909 aged 63.

Sir Alexander Gracie 1860–1930
Born at Dunvegan on the Isle of Skye, Gracie served an engineering apprenticeship with James & George Thomson at their works in Glasgow after which he worked at William Denny's Dumbarton works for two years before returning to J&G Thomson, which by this time had relocated to

Clydebank, as Assistant Manager of their engine works under Thomas Bell. He left Clydebank to become became manager of the engineering department at Fairfield in 1895, a Director in 1901, Managing Director in November 1904 and finally Chairman in 1909. He was a member of the distinguished Committee on Designs which sat in 1904 to give advice and approval to Admiral Sir Jackie Fisher's new concept battleships and battlecruisers, the Dreadnoughts. Gracie was awarded an MVO (Member of the Royal Victorian Order) by King Edward VII in 1908 following the Atlantic speed record set by Fairfield built battlecruiser HMS *Indomitable*. He was knighted in 1918. In 1919 he resigned as Chairman and as Managing Director in June 1920 following the change in ownership of the Fairfield Company. He held a number of directorships, including Cammell Laird and English Electric.

Percy Archibald Hillhouse 1869–1942
Born in Derby in 1869, he studied naval architecture at Glasgow University, graduating in 1888, after which he worked at A&J Inglis shipyard in Glasgow. In 1893, he joined the drawing office at J&G Thomson's yard in Clydebank. Between the years 1898 and 1902, Hillhouse was the first Professor of Naval Architecture at Tokyo Imperial University. In 1902 he returned to Glasgow and the Fairfield shipyard where he became assistant to chief naval architect Dr Francis Elgar, succeeding him as chief naval architect in 1909. Hillhouse remained in this post until 1921 when he was appointed to the John Elder Chair of Naval Architecture at Glasgow University, previously held by Sir John Harvard Biles. Hillhouse remained as Professor of Naval Architecture at Glasgow University until 1942.
In January 1915 he was onboard the steam submarine *K13* which sank on trials in the Gareloch, an ordeal which lasted for 56 hours before he, with 46 others, was rescued.

Sir Alexander McAusland Kennedy 1860–1939
Born in Dumbarton in February 1860, Alexander Kennedy began what would be a rapid and distinguished career by starting an apprenticeship at the shipyard of Archibald McMillan & Sons, in his home town. Taking classes in naval architecture, he obtained a first class diploma with honours from the South Kensington Institute of Science, receiving a gold medal. At age 25, in 1885, he became Chief Draughtsman at McMillan's, Assistant Manager by age 29 and in 1896, General Manager at 36. In 1904 he left McMillan's to be Managing Director of the William Hamilton shipyard at Port Glasgow, subsequently absorbed into Lithgows. In February 1917,he was appointed Managing Director of the Northumberland Shipbuilding Co at Howdon, North Tyneside. Kennedy most likely managed the acquisition spree of merchant bankers Sperling & Co, which led to Fairfield being part of the resulting Sperling Combine. Kennedy became Managing Director of Fairfield and, later, Chairman. In 1921 he was knighted. He remained at Fairfield until March 1937, when, at the age of 76 and at his own request, he resigned the Chairmanship and from the Board at Fairfield. He died at his home at Culzean Park, Gosforth, Newcastle-upon-Tyne in January 1939. Kennedy sat on various Government bodies during the First World War and had several directorships. Between 1917 and 1919, Kennedy was President of the Clyde Shipbuilders' Association He became President of the Shipbuilding Employers' Federation in 1919. He was a committee member of Lloyds Register of Shipping, and Chairman of the Shipbuilders Investment Co Ltd.

Sir James Lithgow 1883–1952
Sir James Lithgow was born in 1883 in Port Glasgow, son of William Lithgow, a shipbuilder there. He attended Glasgow Academy, after which he elected to join the family business in 1901 by becoming an apprentice at Russell & Co's shipyard in which his father was a partner. On his father's death in 1908, James and his brother Henry assumed ownership of the business. James served in the army during the First World War but was recalled from France in 1917 to become Director of Merchant Shipbuilding to ensure that output targets for merchant ships were met by shipyards. Further yard acquisitions followed including the Glasgow based marine engineers David Rowan & Co. In 1918 Russell & Co was renamed Lithgows Ltd, and further expansion of the business took place. In 1925 he was made a baronet. In the late 1920s and early 1930s, he was the architect of National Shipbuilders Security Ltd, formed to mitigate the effects of the depression in the shipbuilding industry by buying redundant or ailing shipyards and closing them. James also undertook reorganisation of the steel industry in Scotland. In 1935 James and his brother added the Fairfield shipyard to their already vast shipbuilding empire after it had run into serious financial difficulties. During the Second World War James was made Controller of Merchant Shipbuilding and Repairs as well as being a Lord Commissioner of the Admiralty. On his death in 1952, Sir James Lithgow was at the head of the largest privately owned shipbuilding company in the world.

Sir William Pearce 1833–1888
Born in Brompton, Kent, William Pearce trained as a naval architect at Chatham Royal Dockyard. In 1863 he took up employment as a ship surveyor in the River Clyde area for Lloyds Register but in the following year he became Shipyard Manager at Robert Napier's Govan shipyard. During this period there, he became involved in the design and construction of a number of Atlantic liners. In 1869 following the death of John Elder, his association with Elder's yard began when he became Shipyard Manager. In 1878 on the retirement of his two partners, Pearce became the sole proprietor of the business. Pearce was an energetic and enterprising businessman and under him the shipyard became the most prestigious in the UK through the design and construction of a series of express, record breaking Atlantic liners as well as the construction of the novel yacht design, *Livadia*, for the Czar of Russia. In 1885 he was elected Conservative Member of Parliament for the newly created Govan constituency. In the same year he changed the company name from John Elder & Co, to the Fairfield Shipbuilding & Engineering Co, converting it to a limited liability company at the same time. He was created a baronet in 1887.

Appendix H

Govan Workspace and Fairfield Heritage

Govan

From the late 1950s a general industrial decline started in the UK which had a profound effect on Govan which had been home to a great many industries, some, but not all, related to shipbuilding. This decline was paralleled by the poor condition of the built environment where it was estimated that by the 1950s 59% of housing in Govan was in the lowest category for sanitation and structural condition. Under the terms of the Town and Country Planning (Scotland) Acts for Comprehensive Development Areas, much of Govan, its tenements and factories were scheduled for demolition and by the mid-1980s much of it had disappeared.

Govan Workspace

It was about this point in time, 1981, that a local social enterprise called Govan Workspace was formed with the broad aim of creating jobs, revitalising Govan and beginning the process of making it once again an attractive place for people and businesses to be located. Govan Workspace did this by acquiring properties which were significant landmarks closely related to the development of Govan but had become vacant and derelict. These properties were then repurposed, modernised and equipped as rentable spaces to stimulate the local economy.

The Fairfield 1891 Building

Soon after BAE Systems took over the shipyard from Kvaerner, they no longer required the shipyard office buildings which comprised the A-listed 1890 building, the 1903 drawing office extension on North Elder Street and the 1956 Fairfield Rowan building on Govan Road. The reason for this decision lay in the office buildings the Company had at their Scotstoun site, the former Yarrow yard, which were significantly larger and much better adapted to their needs. Clydeport, owners of the land and buildings at the Govan site, were notified and, in 2001, the buildings were duly vacated, boarded up and left to their fate. Over the following years the deteriorating nature of the 1890 building in particular resulted in it being placed on the Buildings at Risk Register.

Fairfield: Business and Heritage Centre

In 2008 Govan Workspace began investigating a new role for the building, acquiring the property in March 2009 with plans to convert the engine and ship drawing offices into commercial offices with a heritage space in the beautifully panelled ground floor former management offices. Its rescue from near terminal decline led to significant public sector backing and ultimately to a £5.8m restoration programme. Completed in June 2014, the project ensured that this magnificent, iconic building remains a central part of Govan's built environment, contributing once more to the local economy and providing, through Fairfield Heritage museum, a focal point for the celebration of the community's outstanding shipbuilding history.

Top: Govan in the early 1980s. Dwarfed by its shipyard cranes, the Fairfield building stands in a scene of near devastation with entire streets of tenements demolished. (*Stephen Farmer*)

Middle: Fairfield Heritage museum open to the public, telling the story of the Fairfield shipyard from its origins in the 1860s to the present day with BAE Systems.

Bottom: Organised by Fairfield Heritage, public talks are offered on subjects concerning Govan's great industrial history. Here, Brian Newman is delivering a talk on shipyard cantilever cranes designed and manufactured in Glasgow.

Appendix J Queen Elizabeth Class

Block Breakdown: by whom and in date sequence

(Indicative dates)

Queen Elizabeth

		Steel Cut	Complete
LB01	Babcock (Appledore)	December 2008	February 2011
Starboard sponsons	Babcock	April 2009	November 2012
LB03	BAE Govan	June 2009	June 2012
CB05/06	Babcock	September 2009	December 2010
LB04	BAE Govan	January 2010	November 2012
LB02	BAE Portsmouth	February 2010	May 2012
CB03	A&P Tyne	March 2010	August 2011
CB02	Cammell Laird	July 2010	May 2012
Port sponsons	Babcock	July 2010	May 2013
LB05	BAE Portsmouth	January 2011	May 2012
CB04	Cammell Laird	April 2011	November 2012
UB07	BAE Portsmouth	June 2011	November 2012
UB14	BAE Portsmouth	October 2011	June 2012

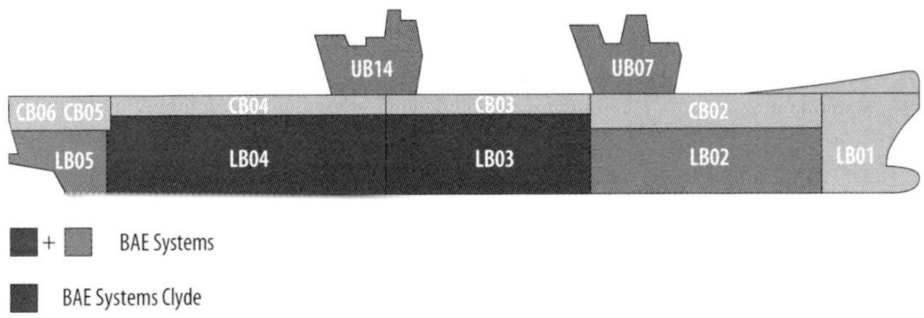

Prince of Wales

		Steel Cut	Complete
LB03	BAE Govan	April 2011	February 2014
Starboard sponsons	Babcock	January 2012	April 2015
Port sponsons	Babcock	June 2012	November 2015
LB04	BAE Govan	August 2012	August 2015
LB02	BAE Portsmouth	September 2012	December 2014
CB05/06	Babcock	September 2012	August 2014
LB01	Babcock (Appledore)	December 2012	March 2015
CB03	A&P Tyne	January 2013	August 2014
CB02	Cammell Laird	June 2013	April 2015
UB07	BAE Scotstoun	November 2013	March 2015
CB04	Cammell Laird	January 2014	September 2015

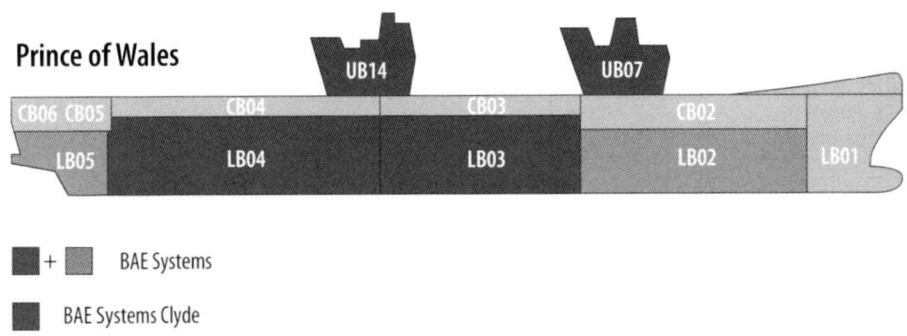

Appendix K **Ships in scale** A representative selection of ships built by the Fairfield yard 1860-2024

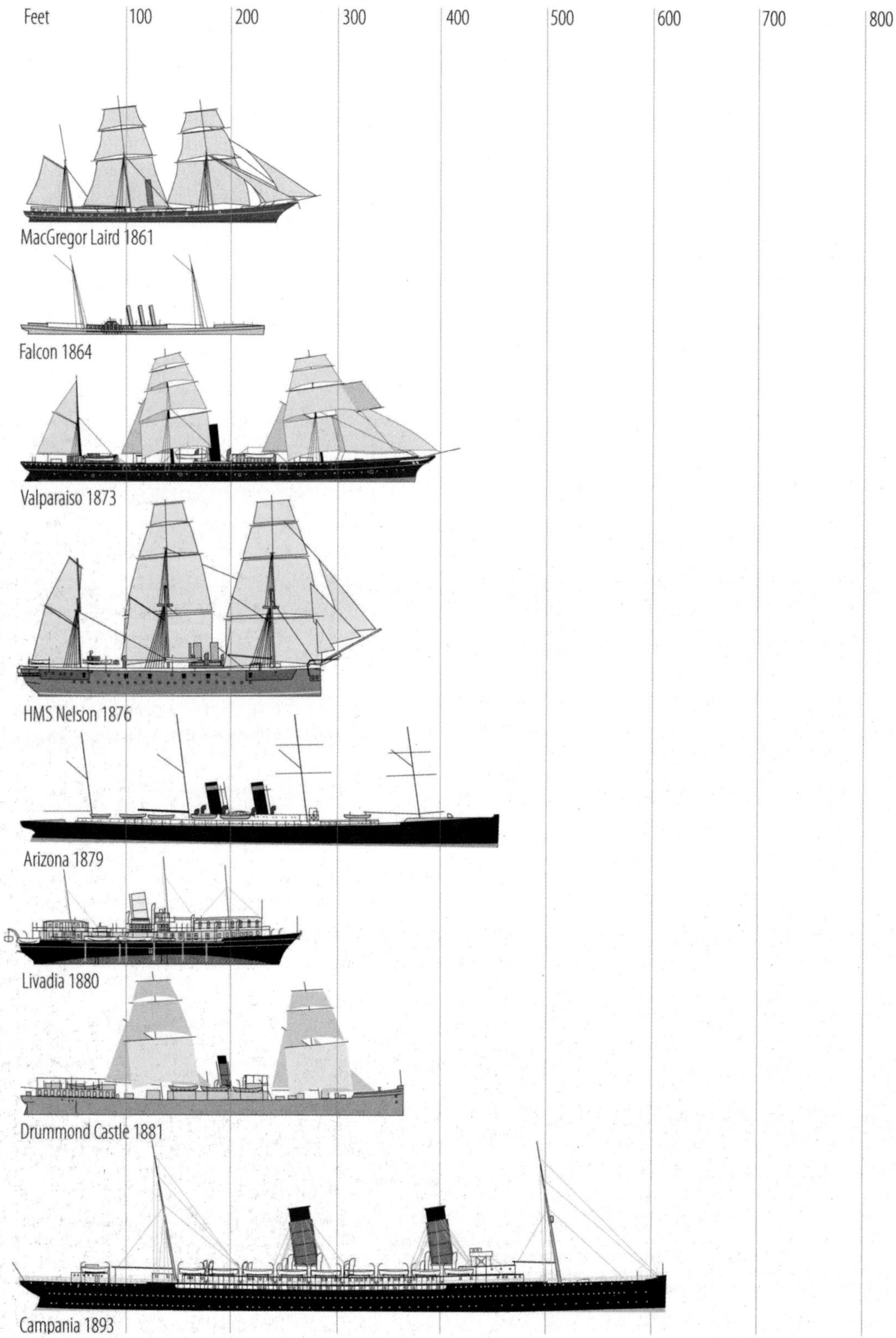

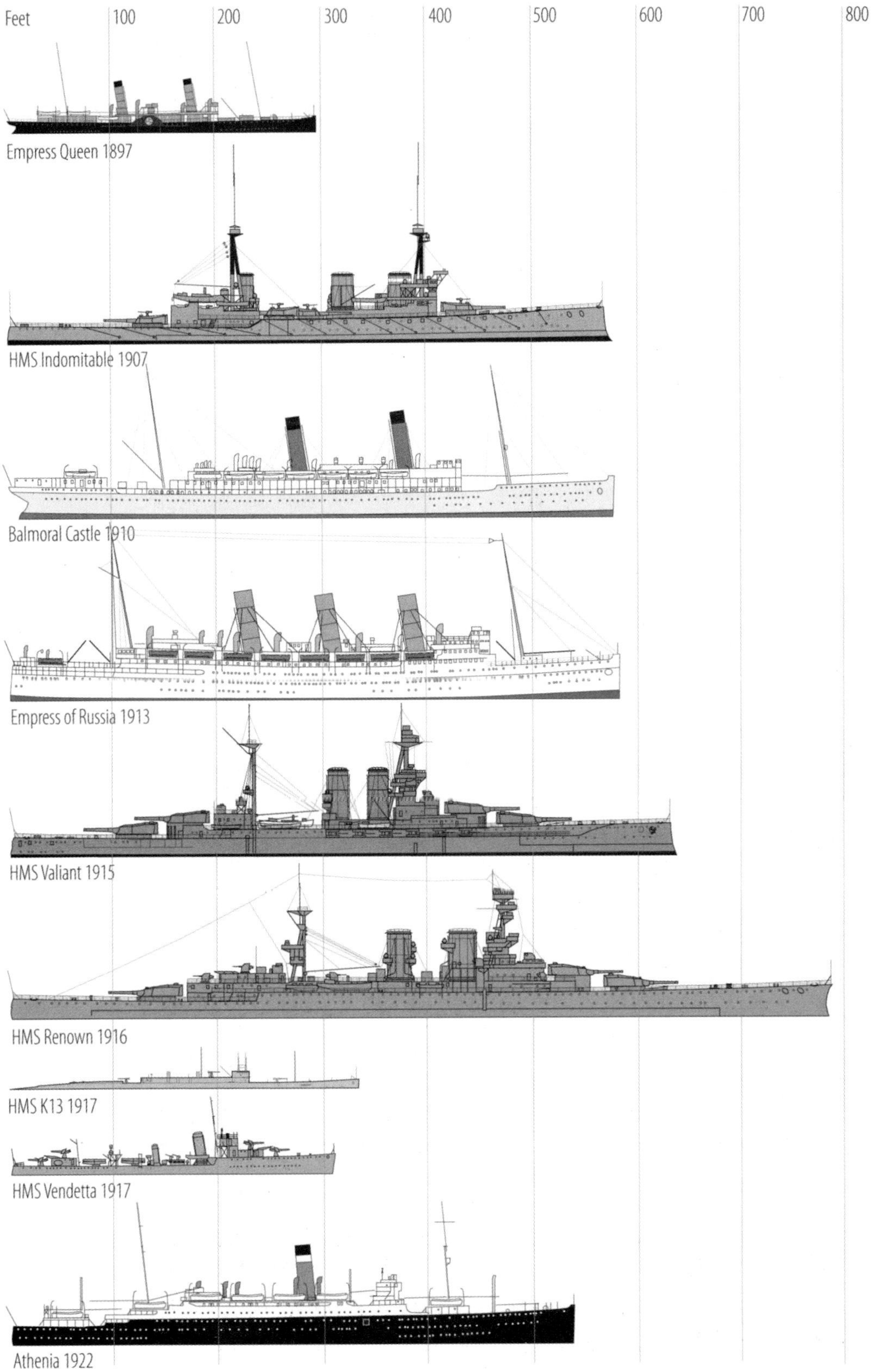
Feet
100
200
300
400
500
600
700
800
Empress Queen 1897
HMS Indomitable 1907
Balmoral Castle 1910
Empress of Russia 1913
HMS Valiant 1915
HMS Renown 1916
HMS K13 1917
HMS Vendetta 1917
Athenia 1922

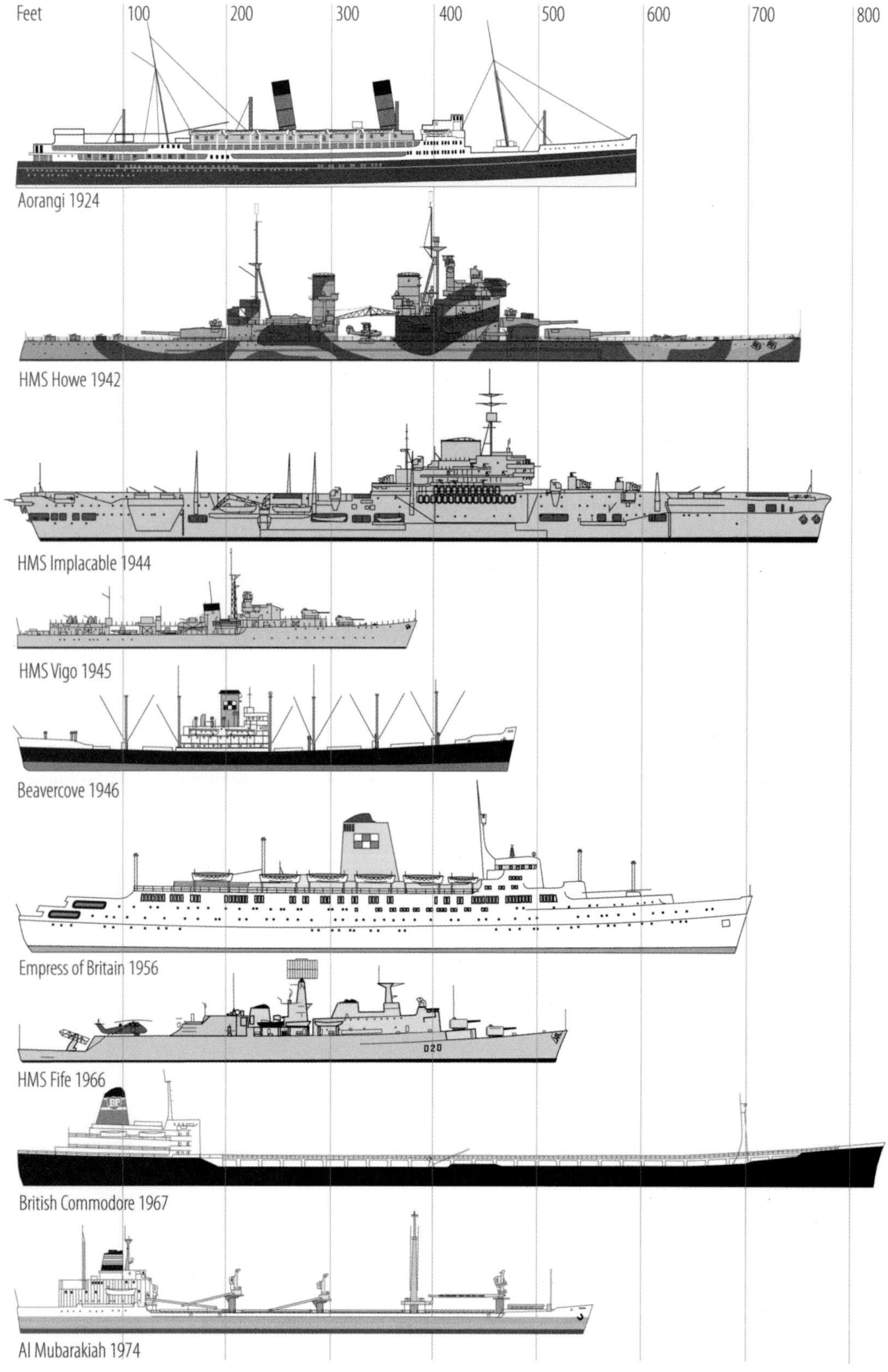
Feet
100
200
300
400
500
600
700
800
Aorangi 1924
HMS Howe 1942
HMS Implacable 1944
HMS Vigo 1945
Beavercove 1946
Empress of Britain 1956
HMS Fife 1966
D20
British Commodore 1967
BP
Al Mubarakiah 1974

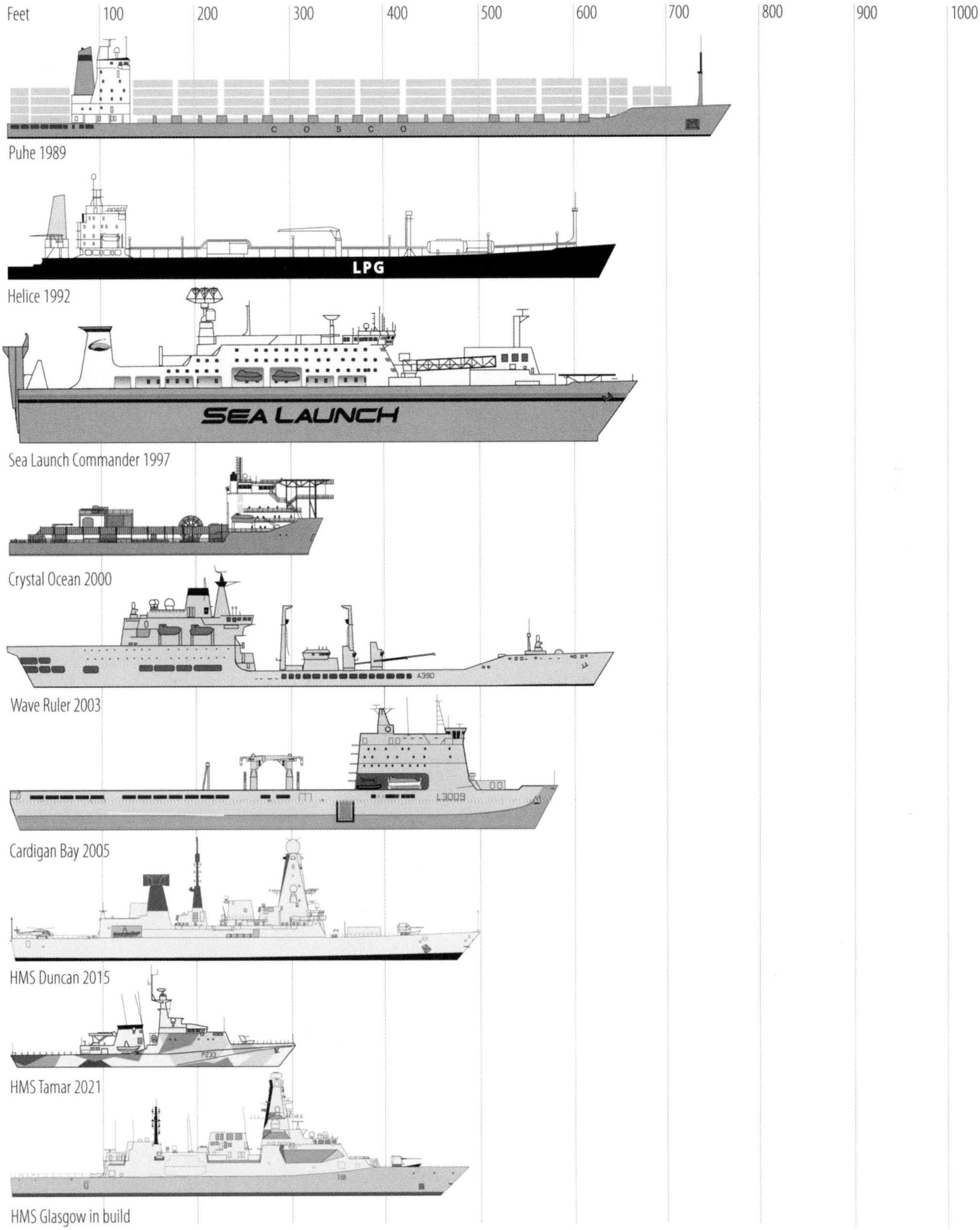
Feet
100
200
300
400
500
600
700
800
900
1000
COSCO
Puhe 1989
LPG
Helice 1992
SEA LAUNCH
Sea Launch Commander 1997
Crystal Ocean 2000
A390
Wave Ruler 2003
L3009
Cardigan Bay 2005
HMS Duncan 2015
HMS Tamar 2021
HMS Glasgow in build

Appendix L Ship list

Dates: (where known) laid down, launched and completed.

Particulars: length and beam (expressed in feet unless otherwise indicated), tons (usually gross or displacement for warships), Engine codes: d- diesel; de - diesel electric; gt - gas turbine; p - paddle; pe - petrol; sr - steam reciprocating; st - steam turbine; te - turbo electric; w - wind. Power output (hp, ihp or shp) and speed (knots).

List compiled from several sources including, with thanks: the Scottish Built Ships Database (https://www.clydeships.co.uk/) the British Shipbuilding Database, Marine Technology Special Collection, Newcastle University. original shipyard lists held by Fairfield Heritage.

Randolph Elder & Company at Govan Old Shipyard

No.	Name(s)	Type	Owner	Dates	Particulars		Fate
-	floating dock	for Surabaya	?	2000t			
	floating dock	French Gov't	?-Jan1863-?	1000t	For erection in Cochin, China		
14	MACGREGOR LAIRD	passcargo	African S S Co	?-3Dec1861-20Feb1862	245x30, 966t	sr, 200hp, 9.5k	Lost 1872
15	TALCA	"	Pacific S N Co	?-Sep1861-Nov1861	190x30, 708t	p, 260ihp	Scuttled 1880
16	COLLEEN BAWN	"	Drogheda Stm Pkt Co	?-2 Apr1862-May1862	220x28, 697t	p, 1300ihp	Scrapped 1901
17	EMERALD ISLE	schooner	P Ternan	?-16Jan1862-Jan1862	81.2x20.1, 95t	w	
18	MURILLO	cargo	MacAndrew & Co	?-16May1862-14Jun1862	192x28, 614t	sr, 100hp	Foundered 1865
19	VELAZQUEZ	"	M Saenz & Co	?-7Jan1863-?	192x28, 549t	sr, 100hp	Stranded 1883
20	ALESSANDRO VOLTA, *Aldo, Kairon, Aghios Georgios*	passcargo	Navigazione Generale Italiana	?-26Aug1862-Sept1862	203.4x25.4, 610t	sr, 471ihp	Deleted 1935
21	CHILE	"	Pacific S N Co	?- Aug1863-5Oct1863	265x36, 1672t	p, 400hp, 10k	Sold Chile 1880
22	QUITO	"	"	?-29Oct1863-13Jan1864	271x32.8, 1388t	p, 320hp, 13k	Sold Peru1864
23	CALABAR	"	"	?-9Feb1864-11May1864	255x31, 1122t	sr, 860 ihp	Sold China1877
24	PAYTA	"	"	?-19Jul1864-28Sep1864	263.7x38.4, 1344t	p, 400hp	Wrecked 1881
25	Lighter	Pacific S N Co					
	ST GEORGE	floating dock	Callao Floating Dock Co	?-24Apr1866-?	300x76x21, lift 5000t		
	floating dock	French Gov't	1865	300x70, lift 5000t	For service at Saigon		

at Fairfield Shipyard

No.	Name(s)	Type	Owner	Dates	Particulars		Fate
26	FALCON	cargo/blockade	Donald Macgregor	?-12May1864-2Jun1864	220x28, 446t	p, 1400ihp, 17.75k	Sold 1865
27	FLAMINGO, *Galgo*	"	"	?-26May1864-23Jun1864	220x28, 446t	p, 1400ihp	Sold 1865
28	PTARMIGAN	"	"	?-11Jun1864-13Jul1864	220x28, 446t	p, 1400ihp	
29	CONDOR	"	"	?-29Jun1864-10Aug1864	220x28, 446t	p, 1400ihp	Ran aground 1864
30	EVELYN	"	"	?-1Sep1864-25Oct1864	220x28, 446t	p, 1400ihp, 17k	Sold 1866
31	KONING WILLEM III	passcargo	Netherland Indies S N Co	?-11Feb1865-10Jun1865	225x51x12, 778t	sr, 610ihp	Wrecked 1857
32	KONINGEN SOPHIA	"	"	?-Apr1865-7Sep1865	225x51x12, 774t	sr, 610ihp	Scrapped 1896
33	PACIFIC	"	Pacific S N Co	?-28Jan1865-10Apr1865	267.4x40.2, 1631t	p, 450hp, 13k	Dismantled 1881
34	LIMENA	"	"	?-9Jun1865-9June1865	267.2x40.2, 1623t	p, 400hp, 12k	Hulked 1880
35	SANTIAGO	"	"	?-27May1865-3Aug1865	267x40.2, 1619t	p, 450hp, 13k	Wrecked 1869
37	OTENTOSAMA (Launched as *Tycoon*)	yacht	Ross, Barbour & Co	?-14Jan1865-22Feb1865	122.8x15.25, 95t	sr, 50hp	Sold 1867
38	BARON SLOET VAN DE BEELE	passcargo	Netherland Indies S N Co	?-24Jul1865-3Oct1865	751t	sr, 480ihp	
39	CITY OF LIMERICK, *Warrior, Vandyck*	"	London, Belgium, Brazil & R Plate S N Co	?-27Jul1865-15Aug1867	238x32, 1109t	sr, 1060ihp	Hulked 1892
40	SATARA	"	British India S N Co	?-27Oct1866-12Dec1866	245x32, 1237t	sr, 1060ihp	Scrapped 1891
41	CITY OF BRUSSELS, *Desirade*	"	London, Belgium, Brazil & R Plate S N Co	?-10Feb1868-10Mar1868	268x32.5, 1226t	sr, 225hp	Scrapped 1906
42	NELLY	yacht	Thomas Vaughan	?-Apr1865-30Jun1865	114.2x15.1, 91t	p, 26hp	
43	BARON BENTINCK	passcargo	Netherland Indies S N Co	?-13Oct1865-13Oct1865	202x28, 643t	sr, 390ihp	
44	MANDINGO, *Alps*	"	African S S Co	?-20Oct1865-9Jan1866	262.3x32.4, 1153t	sr, 275hp, 12k	Wrecked 1907
45	LAGOS, *Andes, Gortz*	"	"	?-4Dec1865-26Feb1866	262.2x32.4, 1151t	sr, 275hp	Scrapped 1908
46	RAKAIA, *Ebro, Baldomero Iglesias*	"	Panama, New Zealand & Australia R.M. Co Ltd	?-31Jan1866-12Apr1866	273x34.1, 1456t	sr, 350hp	Scrapped 1898
47	PANAMA	"	Pacific S N Co	?-17Mar1866-24May1866	267x40.1, 1642t	p, 450hp, 13k	Hulked 1881
48	EARL OF BELFAST, *Newry*	"	R Henderson	?-31Mar1866-21Apr1866	176.6x24, 395t	sr, 70hp	
49	MURILLO, *Huelva, Roelas*	cargo	J Wilson	?-14Apr1866-23May1866	200x30, 738t	sr, 100hp	Collision 1887
50	ZURBARAN	"	"	?-28Apr1866-16Jun1866	200x30, 739t	sr, 560ihp	Foundered 1899
51	CAMPEADOR	"	G MacAndrew	?-17May1866-10Jul1866	190x28, 642t	sr, 350ihp	Collision 1882
52	COLON	"	"	?-4Jul1866-8Aug1866	190x28, 642t	sr, 350ihp	Wrecked 1871

53	BALBOA	cargo	Serapio Acebal	?-28Jul1866-25Aug1866	190x28, 643t	sr, 380ihp	Foundered 1879
54	CORTES	"	"	?-18Aug1866-5Sep1866	190x28, 642t	sr, 350ihp	Abandoned 5 April 1872
55	PIZARRO	"	John Wilson	?-17Sep1866-6Oct1866	190x28, 643t	sr, 350ihp	Wrecked 1888
56	ALVARADO	"	"	?-18Oct1866-3Nov1866	190x28, 643t	sr, 350ihp	Wrecked 30 Jan 1880
58	CORMORAN	fishing	Soc des Pecheries de l'Ocean	?-1Sep1866-22Sep1866	90.5x16.1, 77t	sr, 30hp	
59	HERON	"	"	?-1Sep1866-22Sep1866	90.5x16.1, 74t	sr, 30hp	
61	ARICA	passcargo	Pacific S N Co	?-12Mar1867-26Apr1867	204x30, 742t	p, 1430ihp, 12.5k	Wrecked 13 Jan1869
62	QUITO	"	"	?-18Apr1867-13Jun1867	200x30, 743t	p, 1430ihp 12.5k	Hulked 1882
63	SUPE	"	"	?-5Mar1867-1Apr1867	145x25, 298t	sr, 220ihp	Sold 1882
70	Tug for River Godavari	rivertug	Indian Govt.	?-27Apr1867-?	130x25x1, 382t	sternwheel, 175ihp	
71	Tug for River Godavari	"	"	?-27Apr1867-?	130x25x1, 382t	sternwheel, 175ihp	
72	Tug for River Godavari	"	"	?-27Apr1867-?	130x25x1, 382t	sternwheel, 175ihp	
73	NORMANDY, *Touareg*	cargo	G Gibson & D R MacGregor	?-2May1867-7Jun1867	181.2x26, 387t	sr, 80hp	Wrecked 1894
74	PELICAN	fishing	Soc des Pecheries de l'Ocean		?-19Jun1867-6Jul1867		90.8x16.1, 78t sr, 30hp
75	VINE, *Kong Sigurd*	passcargo	D R Macgregor & Co	?-31Aug1867-30Sep1867	190x26.75, 417t	sr, 410ihp	Foundered 26 Nov 1902
76	MIDGE	gunvessel	Admiralty	?-21May1868-?	603t	sr, 580ihp	Sold 15 Mar 1907
79	SAINT CLAIR, *St Colm*	passcargo	Aberdeen, Leith, Clyde & Tay Shipping. Co	?-18Feb1868-21Mar1868	199x26.6, 588t	sr, 750ihp	Scrapped 193
80	GALATEA	sailing ship	S R Graves & Son	?-19Sep1868-17Oct1868	230x38.4, 1520t	w	Foundered 1880
81	BIAFRA, *Energique, Inayet*	passcargo	African S S Co	?-5Jun1868-3Aug1868	265x32, 1080t	sr, 1010ihp	Wrecked 1902
82	DON	"	D R MacGregor & Co	?-7May1868-30May1868	241.6x30.3, 1152t	sr, 120hp	Lost 13 Feb 1871
83	GUINEVERE, *Luna*	sailing ship	J McCunn	?-20Aug1868-7Sep1868	197x34, 879t	w	
86	DUNA	passcargo	D R MacGregor & Co	?-Jun1868-18Jun1868	181.2x26.3, 594t	sr, 70hp	Sold Russia 1868
87	CITY OF RIO DE JANEIRO, *Teniers*	"	Tait & Co	?-Jul1868-3Sep1868	273x33.2, 1597t.	sr, 300hp	Scrapped 1892

John Elder & Company

No.	Name(s)	Type	Owner	Dates	Particulars		Fate
88	HEREFORD	sailing ship	Merchant Shipping Co	?-24Jul1869-19Aug1869	230x37.2, 1510t	w	Wrecked 8 April 1907
89	COLUMBUS	"	J Lidgett & Sons	?-2Dec1868-24Dec1868	182.4x31. 744t	w	Scrapped 1905
90	CITY OF YORK	"	G Smith & Sons	?-22Sep1869-26Oct1869	212x35.8, 1254t	w	Stranded 12 July 1899
91	MAGELLAN	passcargo	Pacific S N Co	?-30Dec1868-8Mar1869	359.6x41. 2856t	sr, 500hp, 11.5k	Scrapped 1893
93	CARRICK CASTLE, *Nautilus*	sailing ship	Thos Skinner & Co	?-3Oct1868-30Oct1868	197.5x34. 879t	w	Scrapped 1932
92	PATAGONIA	passcargo	Pacific S N Co	?-1Mar1869-23Apr1869	353x41, 2867t	sr, 500hp, 11.5k	Wrecked 1 Oct 1895
94	DOUNE CASTLE, *Florinda*	sailing ship	T Skinner & Co	?-18Mar1869-15Apr1869	197.1x34, 911t	w	Burned 1904
95	ARAUCANIA	passcargo	Pacific S N Co	?-29Apr1869-2Jul1869	354.8x41, 2877t	sr, 500hp, 12k	
96	CORDILLERA	"	"	?-26Jun1869-3Sep1869	343.5x40.5, 2860t	sr, 2130ihp, 13k	Wrecked 20 Sept 1884
97	BONNY	cargo	British & African S N Co	?-?-18Jan1869	260x301263.27t	sr, 222nhp	Scrapped 1910
98	ROQUELLE	"	"	?-?-26Feb1869	260x30, 1267.43t		
99	WILLIAM MILLER, *Sappho*	"	D R MacGregor	?-10Apr1869-5May1869	215.8x27.7, 889t	sr, 100hp	Stranded 1883
100	DANUBE, *Salvor, Nervion*	passcargo	"	?-9Jun1869-1869	215.6x27.7, 887t	sr, 100hp	Scrapped 1936
101	CONGO, *Ville de Marseille*	cargo	British & African S N Co	?-11Mar1869-30Mar1869	261.2x30.2, 1267t	sr, 200hp	Scrapped 1907
102	BENIN	passcargo	African S S Co	?-10May1869-10Jul1869	275x33, 1530t	sr, 1080ihp, 11k	Collision 12 March 1881
104	ELBE	"	Royal Mail S P Co	?-20Nov1869-1869	350x40.2, 3063t	sr, 600hp, 12k	Scrapped 1902
106	AFRICAN, *Pico*	coaster	Peter Brash	?-8Jul1869-27Jul1869	156.2x22.1, 234t	sr, 40hp	
107	SIN NANZING	passcargo	J F H Troutmann	?-21Aug1869-2Nov1869	225.4x32.5, 1142t	p, 250hp	
108	AREQUIPA	"	Pacific S N Co	?-20Dec1869-10Mar1870	225x35, 1065t	p, 1720ihp, 11k	Hulked 1883
109	ITALY	"	National SS Co	?-2Apr1870-22Jun1870	389x42.3, 4169t	sr, 500rhp	Scrapped 1894
110	JOHN ELDER	"	Pacific S N Co	?-29 Aug1870-3Dec1870	370x41, 3832t	sr, 3000ihp, 12k	Stranded 17 Jan 1892
111	LOANDA	"	British & African S N Co	?-19May1870-23Jul1870	1474t	sr, 900ihp, 10.5k	Scrapped May 1897
112	BAN RIGH, *Libertador, Marroquin, Bolivar*	"	Aberdeen S N Co	?-3Mar1870-21Apr1870	240x30, 958t	sr, 1385ihp	
113	ATACAMA	"	Pacific S N Co	?-17Aug1870-1870	290x38.3, 1822t	sr, 300hp, 11k	Wrecked 1877
114	LIBERIA	"	British & African S Co	?-4Jul1870-19Aug1870	278x31, 1471t	sr, 975ihp, 10.5k	Missing April 1874
115	VOLTA	"	"	?-7Sep1870-1870	278x31, 1475t	sr, 250hp, 10k	Scrapped 1894
124	TAGUS	"	Royal Mail S P Co	?-21Jan1871-3Jun1871	358x41.1, 3298t	sr, 600hp	Scrapped 1897
125	MOSELLE	"	"	?-20May1871-12Sep1871	344x40.5, 3252t	sr, 3300ihp	Wrecked 29 Oct 1891
126	COQUIMBO	"	Pacific S N Co	?-12Dec1870-3Feb1871	280x38, 1821t	sr, 1400ihp, 11k	Hulked 1901
127	VERITE	cargo	Cie de Nav Paquet	?-26Nov1870-1870	220x29.5, 1070t	sr, 170hp	Burned 22 July 1887
128	CHIMBORAZO, *Cleopatra*	passcargo	Pacific S N Co	?-21Jun1871-26Sep1871	370x41, 3847t	sr, 3000ihp, 12k	Scrapped 1897

129	CUZCO	passcargo	"	?-17Oct1871-25Dec1871	370x41, 3845t	sr, 3000ihp, 12k	Scrapped 1905
130	ACONCAGUA, *Egypte*	"	"	?-6Jun1872-?	404.8x41.4, 4106t	sr, 600hp, 13k	Scrapped 1896
131	PRINS VAN ORANJE, *Maria*	"	SMN	?-4May1871-28Jun1871	320x39, 2736t	sr, 1625ihp	Scrapped 1903
132	WILLEM III, *Quang Se, Glenorchy, Pina*	"	"	?-8Mar1871-4May1871	320x39, 2735t	sr, 1625ihp	"
133	PRINS HENDRIK	"	"	?-17Aug1871-24Oct 1871	350x39, 2736t	sr, 400rhp	Wrecked 1873
134	CONRAD	"	"	?-14Nov1871-6Feb1872	352x39.5, 3065t	sr, 400hp	Scrapped 1897
135	CITY OF LONDON, *Lochbroom*	"	Aberdeen S N Co	?-8Jun1871-10Aug1871	241.9x30.5, 979t	sr, 275hp	Scrapped 1937
136	HYDRA [engines only]	coastdefnc	Admiralty	?-28Dec1871-31May1876	225x45, 3480t	sr, 1470ihp	Scrapped 1903
138	SANTIAGO	passcargo	Pacific S N Co	?-14Dec1871-13Mar1872	251.6x35.6, 1451t	p, 300hp	
139	TRUXILLO	"	"	?-12Feb1872-7Jun1872	251.5x35.7, 1449t	p, 300hp	Scrapped 1885
140	Steamer for China service	riverferry	E D Barbour		220x35		Assembled in China
141	Barge for China service		"		80x20		"
142	Barge for China service		"		80x20		"
143	ILO	passcargo	Pacific S N Co	?-11Mar1872-1May1872	280x38, 1794t	sr, 300hp	Hulked 1882
144	SORATA	"	"	?-20Oct1872-19Dec1872	390x42, 4014t	sr, 3250ihp, 12.5k	Scrapped 1895
145	ILLIMANI	"	"	?-16Dec1872-4Mar1873	390x42, 4022t	sr, 3250ihp	Stranded 1879
146	COTOPAXI	"	"	?-15Mar1873-26May1873	390x42, 4028t	sr, 3250ihp, 12.5k	Wrecked 1889
147	TACORA	"	"	?-23May1872-3Aug1872	363x41, 3524t	sr, 3165ihp	Wrecked 1872
148	HUDDERSFIELD	"	Manchester, Sheffield & Lincoln. Rly Co	?-23Sep1872-16Nov1872	231x30.2, 1077t	sr, 120hp	Collision 1903
149	HALIFAX	"	"	?-17Dec1872-21Feb1873	230.9x30.2, 1079t	sr, 120hp	Stranded 1876
150	KONING DER NEDERLANDEN	"	SMN	?-5Sep1872-23Oct1872	350x39, 3063t	sr, 1775ihp	Foundered 1881
154	ALBATROS	fishing	Soc des Pecheries de L'Ocean	?-23Jul1872-10Aug1872	90.7x61.1, 80t	sr, 30hp	Wrecked 1891
155	ISLAY	passcargo	Pacific S N Co	?-?-14Aug1873	1588t	p, 300hp	Hulked 1883
156	OROYA	"	"	?-27Jan1873-30Sept1873	1596t	p, 400hp	
157	POTOSI	"	"	?-14May1873-9Jul1873	421.6, 4219t	sr, 600rhp, 13k	Scrapped 1897
158	FAIRFIELD	barge	John Elder Co	1873	68x19, 108t	sr	
159	ELYSIA	passcargo	Henderson Bros	?-28Jun1873-10Oct1873	350x35, 2733t	sr, 678hp, 13k	Scrapped 1898
160	VALPARAISO	"	Pacific S N Co	?-30Jul1873-26Sep1873	379.2x41.8, 3574t	sr, 550hp	Wrecked 1887
161	IBERIA	"	"	?-6Dec1873-16Feb1874	420x44.5, 4671t	sr, 4000ihp	Scrapped 1903
162	LIGURIA	"	"	?-18Feb1874-12May1874	420x44.25, 4666t	sr, 4075ihp, 14k	"
164	PAOUTING, *Hae-Ting, Li-Ting, Hae-Ting*	cargo	Shanghai S N Co	?-20Dec1873-18Feb1874	210x33, 1061t	sr, 800ihp	Hulked 1908
166	PRINSES AMALIA	passcargo	SMN	?-19Mar1874-6June1874	383x40, 3429t	sr, 400rhp	
167	COMO	lakepass	Società Italiana	?-?-29Dec1873	174.5x35, 336t	p, 450ihp	Scrapped 1953
168	LECCO	"	"	?	174x36.5, 336t	p, 450ihp, 12k	Capsized 1927
169	LOBO	tender	Pacific S N Co	?-10Mar1874-21Apr1874	89.7x18.4, 106t	sr, 30hp	Wrecked 1875
170	CALEDONIAN, *Guasca*	passcargo	Ardrossan Shipping Co	?-30Apr1874-5June1874	201x26.1, 521t	sr, 150hp	Collision 1907
171	BENGUELA	"	British & African S N Co	?-15Jun1874-17Jul1874	300x33.75, 1860t	sr, 270hp	
172	CAMEROON	"	"	?-16May1874-22Jun1874	300x33.75, 1862t	sr, 280hp	Wrecked 1904
173	BLAIR ATHOLE	sailing ship	Thomson & Gray	?-29Jun1874-13Jul1874	259.1x40, 1777t		Missing 1893
174	ZEALANDIA	passcargo	J F Ure, W Pearce & Jamieson	?-Oct1875-26Nov1875	377x37.1, 2730t	sr, 500hp	Wrecked 1917
175	AUSTRALIA	"	"	?-15Nov1875-28Dec1875	376.9x37.5, 2737t	sr, 2160ihp	Burned 1891
176	VOORWAARTS	"	SMN	?-14Jul1874-?	368x37.2, 2716t	sr, 1800ihp	Stranded 1899
177	PRINS HENDRIK, *Sultan*	"	"	?-30Sep1874-?	367.2x37.1, 2735t	sr, 400rhp	Foundered 1897
178	CHUSAN	"	Shanghai S N Co	?-17Sep1874-28Sep1874	300x50, 1381t	p, 300rhp, 12k	Wrecked 1874
179	SZECHUEN, *Kiang Chang*	riverpscgo	"	?-29Mar1875-?	300x50, 1400t	p	Wrecked 1878
180	MEURTHE	cargo	French owner(s) ?	?	1000t	sr	
181	FULMINANTE	cablelayer	Argentine Govt	?-9Nov1874-?	181.1x30, 619t	sr, 120hp	
182	CASTLE ROY, *Ravens Hall*	sailing ship	George Gilroy	?-24Dec1874-21Jan1875	259.4x40, 1754t	w	Missing 1890
183	RAJANATTIANUHAR	passcargo	D R Macgregor	?-Feb1875-16Mar1875	245.7x31.9, 1304t	sr, 175hp	Wrecked 1884
184	GILROY, *Rhone, Dybvaag*	sailing ship	George Gilroy	?-5Apr1875-23Apr1875	259.2x39.9, 1767t	w	Wrecked 1906
185	EGERON, *Hok Canton*	passcargo	Stoomv.Reederij Banda	?-Feb1875-?	176.6x26, 507t	sr, 75hp	Deleted 1924
187	CLAUD HAMILTON	passferry	Great Eastern Rly	?-3Jun1875-?	251.6x30.2, 962t	p, 350hp	Scrapped 1914
186		Admiralty	Machinery for HMS *Inflexible*				
188	PARIS, *Flamingo, La Belgique, Glendale*	passferry	London, Brighton & S Coast Rly	?-15Jun1875-18Sep1875	220x25.2, 483t	p, 775ihp, 13k	Wrecked 1905
189	NELSON	arm'd cruiser	Admiralty	2Nov1874-4Nov76-26Jul81	280x60, 7630t	sr, 6624ihp	Scrapped 1910

190	BAY OF BENGAL	sailingship	J & G Bulloch & Co	?-3Aug1875-17Sep1875	250x39, 1595t	w	Missing 1905
191	BAY OF BISCAY	"	"	?-16Sep1875-19Oct1875	250x39, 1611t	w	Missing 1880
193	GAMBIA	passcargo	British & African S N Co	?-7Jul1875-20Aug1875	300x34.25, 1880t	sr, 1185ihp	Stranded 1877
194	SIN NANZING, *Vladimer, Riusei Maru*	cargo	Wm Keswick	?-17Nov1875-?	220.9x30.3, 1166t	sr, 175hp	Collision 1907
196	MUDLARK	bargehoper	Bombay Port Trust	?-22Mar1876-?	145.3x25.1, 304t	sr, 60hp	Deleted 1925
197	SNIPE	"	"	?-24Mar1876-?	145.2x25.1, 305t	sr, 60hp	"
198	CURLEW	"	"	?-22Mar1876-25Mar1876	145.3x25.1, 305t	sr, 60hp	"
199	COOLEEN	barque	W M Barkley	?-8Mar1876-24Mar1876	173x29.83, 657t	w	Wrecked 1894
200	BARNSLEY, *Gomes VI*	passcargo	Man'r, Sheff'd & Li'n. Rly Co	?-20May1876-10Jun1876	185x27, 603t	sr, 545ihp	Wrecked 1909
203	CORISCO	"	British & African S N Co	?-26Jul1876-25Aug1876	300.9x34.5, 1856t	sr, 280hp, 11k	Stranded 1885
204	ISOLA, *Else, Jessie Craig*	barque	J Kelso	1876	180x30.3, 680t	w	Scuttled 1953
205	THALIA	"	Workman Bros	?-18Sep1876-?	197.6x31.8, 792t	w	bandoned 1878
206	COMUS	corvette	Admiralty	?-3Apr1878-23Oct1879	225x44.5, 2380t	sr, 2000ihp	Scrapped 1904
207	CHAMPION	"	"	?-1Jul1878-?	225x44.5, 2380t	sr, 2340ihp	Scrapped 1919
208	CARYSFORT	"	"	?-26Sep1878-13Mar1880	225x44.5, 2380t	sr, 2300ihp	Scrapped 1899
209	CLEOPATRA, *Defiance III*	"	"	?-1Aug1878-?	225x44.5, 2380t	sr, 2300ihp	Scrapped 1931
210	CURACOA	"	"	?-18Apr1878-?	225x44.5, 2380t	sr, 2300ihp	Scrapped 1904
211	CONQUEST	"	"	?-28Oct1878-?	225x44.5, 2380t	sr, 2000ihp	Scrapped 1899
212	SHEFFIELD, *Selda, Sheffield, Huseyniye, Seyyar*	passcargo	Man'r, Sheff'd & Li'n. Rly Co	?-Jan1877-?	201x27.1, 644t	sr, 90hp	Scrapped 1961
213	MEDWAY	"	Royal Mail Stm Packet Co	?-?-11Jul1877	3687t	sr, 600hp	Scrapped 1899
214	PRINSES MARIE, *Prinzess Marie, Germania*	passferry	SMN	?-24Oct1877-21Dec1877	278x34.75, 1618t	pp, 3025ihp, 16k	Scrapped 1902
215	PRINSES ELISABETH, *Prinzess Elisabeth*	"	"	?-10Dec187-5Mar1878	278x34.75, 1606t	p, 3025ihp, 16k	Scrapped 1910
217	BRIGHTON	"	London, Brighton & S Coast Rly	?-2Fe1878-15Mar1878	220x27.5, 531t	p, 1660ihp, 15.5k	Scrapped 1927
218	VICTORIA	"	"	?-7Mar1878-26Apr1878	221.3x27.7, 534t	p, 1791ihp, 16k	Wrecked 1887
219	GABOON, *Galaxidion*	cargo	British & African S N Co	?-5Mar1878-6Apr1878	298.5x34.25, 1863t	sr, 280hp, 10k	Scrapped 1912
220	ZEEMEUW	gunboat	Netherlands Navy	?-19Jan1878-25Jun1878	126x20, 350t	sr, 300ihp	
221	HAVIK	"	"	?-18Jun1878-20Jun1878	126x20, 350t	sr, 300ihp	
222	ARIZONA, *Hancock*	passenger	Liverpool & Great Western SS	?-10Mar1879-23May1879	450x45.5, 5147t	sr, 6000ihp, 15k	Scrapped 1926
223	PRINSES MARIE	passcargo	SMN	?-24Feb1879-?	328.4x38.1, 2760t	sr, 400rhp	Stranded 1898
224	ORIENT, Oric	passenger	Orient S N Co	?-5Jun1879-?	445x46, 5386t	sr, 5500ihp, 15k	Scrapped 1909
225	DAPHNE	cutter	Wm Pearce	?-8May1879-?	70x14.21, 42t	w	
226	KINFAUNS CASTLE, *Moskva*	passcargo	Castle Mail S P Co	?-3Nov1879-17Dec1879	3507t	sr, 3000ihp, 12k	Scuttled 1914
227	RAMBLER	gunvessel	Admiralty	?-?-26Jan1880	157x29.6, 774t	sr, 690ihp, 10k	Sold 1907
228	RANGER	"	"	?-12Feb1880-30Apr1880	157x28.5, 835t	sr, 760ihp	Scrapped 1954
229	NIGER, *Cockatrice, Moorhen*	gunbtriver	"	?-10Mar1880-10May1881	160x24.67, 600t	p, 620ihp	Sold 1899
230	VARA	experiment			23x15.25, 10t		
231	LIVADIA, *Opit*	royalyacht	H I M Czar Alexander II	?-7Jul1880-6Oct1880	246.83x153, 7262t	sr, 10500ihp, 14k	
232	PRINS HENDRIK	passferry	SMN	-25Mar1880-11May1880	278x34.75, 1592t	p, 4000ihp, 16k	Wrecked 1897
233	COANZA	passcargo	British & African S N Co	?-26Feb1880-22Mar1880	290x36.2, 1518t	sr, 1152ihp	Wrecked 1893
234	MALEMBA, *Guarany, Baron Lambermont*	"	B"	?-8May1880-31May1980	290x36.2, 1521t	sr, 1178ihp	Wrecked 1904
235	MOISE	"	Cie Generale Transatlantique	?-12Apr1880-26May1880	313.5x33.5, 1862t	sr, 2400ihp, 14k	Scrapped 1923
236	SAINT AUGUSTIN	"	"	?-28Apr1880-4Jun1880	313.5x33.5, 1812t	sr, 2400ihp, 14k	Foundered 1913
237	ISAAC PEREIRE	"	"	?-12Jun1880-12Jul1880	313.5x33.5, 1798t	sr, 2400ihp, 14k	Wrecked 1906
238	ABD-EL-KADER	"	"	?-26May1880-26Jun1880	313.5x33.5, 1811t	sr, 2400ihp, 14k	Scrapped 1922
239	EL MENSAJERO	riverpass	East Argentine Rlys	?-29May1880-?	130x25, 382t	p, 300ihp	
240	MAIPU	training	Argentine Navy	?-18Oct1880-?	253.6x30.87, 723t	p, 1700ihp, 14k	Scrapped 1918
242	TORFRIDA	steamyacht	R G Duff		?-24Oct1881-?	119.5x17.6, 120t	sr, 50rhp
243	LEON 13# *Albano*	passcargo	Marques de Campo	?-7Oct1880-?	315.7x36.4, 2202t	sr, 1838ihp	Wrecked 1886
245	GARTH CASTLE, *Ismailia, Brunette*	"	Castle Mail S P Co	?-16Dec1880-15Feb1881	365x43.25, 3705t	sr, 2500ihp, 12.5k	Scrapped 1923
246	DRUMMOND CASTLE	"	D Currie & Co	?-17Feb1881-6Apr1881	365x43.25, 3705t	sr, 3200ihp, 12.5k	Wrecked 1896
247	ALASKA, *Magallanes, Alaska*	passenger	Guion S S Co	?-15Jul1881-Oct 1881	500x50, 6932t	sr, 10000ihp, 16.5k	Scrapped 1902
248	ELBE	passcargo	Norddeutscher Lloyd	?-2Apr1881-?	416.5x45, 4510t	sr, 5400ihp, 15k	Collision 1895
249	AUSTRAL	"	Orient S N Co	?-24Dec1881-29Apr1882	455x48, 5588t	sr, 5800ihp, 15k	Scrapped 1903
251	PRINS ALEXANDER, *Aurelia*	"	SMN	?-24Aug1881-?	345.5x38.2, 3041t	sr, 400rhp	Scrapped 1904
253	CHILOE	"	Pacific S N Co	?-?-19Dec1881	321x37.3, 2309t	sr, 400hp, 11k	Wrecked 1892

255	LOCH NESS, *Dag*	cargo	Dundee Loch Line S S Co	?-?-5Jun1882	225.8x31.7, 1047t	sr, 130hp	Torpedoed 1916
256	SHERBRO	passcargo	British & African S N Co	?-18Feb1882-19Apr1882	299.9x36.2, 1650tt	sr, 200hp, 10k	Scrapped 1909
257	STIRLING CASTLE, *Nord America*	cargoliner	Stirling Castle SS Owners CL	?-21Jan1882-23Mar1882	420x49.75, 4423t	sr, 7000ihp, 16k	Stranded 1910
259	SANTA CECILIA, *Selma, Chasina*	steamyacht	Lord Alfred Paget	?-?-3Nov1881	140x22, 203t	sr, 49nhp	
260	PRINS FREDERIK	passcargo	SMN	?-?-22Feb1882	350, 2997t	sr, 2500ihp	Collision 1890
261	INSULINDE, *Burgemeester Den Tex, Ile de France, Urquijo*	"	"	?-?-1May1882	350x39.4, 2997t	sr, 2500ihp	Scrapped 1915
262	PRINSES WILHELMINA, *Lombok, Orleannais*	"	"	?-?-1Aug1882	325.5x37.1, 2578t	sr, 350rhp	Collision 1909
263	WERRA	"	Norddeutscher Lloyd	?-?-4Jul1882	430.3x45.9, 5109t	sr, 6300ihp	Scrapped 1901
264	BRITTANY	passferry	London, Brighton & S Coast Rly Co	?-31May1882-13Jul1882	231x27.7, 579t	p, 270ihp, 17k	Scrapped 1904
265	NORMANDY	"	"	?-29Jun1882-23Aug1882	230x27.5, 579t	p 3164ihp, 17k	Scrapped 1909
266	BOADICEA, *Seaflower*	steamyacht	Wm Pearce	?-17May1882-31May1882	166x24, 305t	sr, 115hp	Scrapped 1946
267	FULDA	passenger	Norddeutscher Lloyd	?-15Nov1882-27Feb1883	430x45.75, 5124t	sr, 6300ihp, 16k	Sold for scrap 1899
268	SUMATRA, *Nivernais*	passcargo	SMN	?-?-30Aug1882	325x37.1, 2565t	sr, 2000ihp	Torpedoed U 1918
269	HAWARDEN CASTLE, *Cyril*	"	Castle Mail Packets Ltd	?-11Jan1883-7Apr1883	380x48.2, 4241t	sr, 3900ihp, 13k	Collision 1905
270	NORHAM CASTLE, *Martinique*	"	:	?-26Feb1883-16May1883	380.6x48.2, 4241t	sr, 3950ihp, 13k	Scrapped 1932
271	SANTA MARIA, *Chilco, Lady Pam*	steamyacht	Lord Alfred Paget	?-?-12Nov1882	154x22, 235t	sr, 49nhp	Beached 1947
272	MONGKUT *Shikoku Maru, Kotobuki Maru*	cargo	Scottish Oriental S S Co Ltd	?-?-9Feb1883	240.3x34.1, 1354t	sr, 157nhp	Deleted 1922
273	KONG BENG, *Taiwa Maru*	cargo	"	?-24Feb1883-28Mar1883	240.2x34.1, 1359t	sr, 175hp	Wrecked 1918
274	OREGON	passenger	Liverpool & Gt Western S S Co	?-23Jun1883-26Sep1883	501x54.3, 7375t	sr, 13000ihp, 7k	Collision 1886
275	TAICHIOW, *Taishu Maru*	cargo	Scottish Oriental S S Co Ltd	?-?-14Apr1883	240.2x34.1, 1359t	sr, 157nhp	Scrapped 1932
276	CALABAR	passcargo	British & African S N Co	?-19Jun1883-21Jul1883	1756t	sr, 1500ihp, 13k	Wrecked 1898
277	LADY TORFRIDA, *Tamara, Tavel, Assos*	steamyacht	Sir Wm Pearce	?-10May1883-19May1883	200x25.5, 400t	sr, 1020ihp, 13.5k	Scrapped 1934
278	WILLEM PRINS VAN ORANJE	passferry	SMN	?-?-26Apr1883	278.9x35.1, 1633t	p, 3000ihp	Scrapped 1909
280	TONGARIRO, *Zibenghla*	passcargo	New Zealand Shipping Co Ltd	?-?-23Aug1883	4163t	sr, 600hp	Scrapped 1910
281	AORANGI	"	"	?-?-2Oct1883	389x46, 4163t	sr, 600hp, 13k	Scrapped 1925
282	RUAPEHU, *Zayathla, Gwalior*	"	"	?-19Nov1883-28Dec1883	389x46, 4163t	sr, 600hp, 13k	Scrapped 1911
283	EIDER	passenger	Norddeutscher Lloyd	?-15Dec1883-13Mar1884	430x46.83, 5129t	sr, 6970ihp, 16k	Stranded 1892
284	EMS, *Lake Simcoe*	emigrant	"	?-?-27Feb1884	429.8x47, 5129t	sr, 1000nhp, 16k	Scrapped 1905
285	UMBRIA	passenger	Cunard S S Co Ltd	?-25Jun1884-8Oct1884	500x57, 7718t	sr, 14500ihp, 18k	Scrapped 1910
286	ETRURIA	"	"	?-20Sep1884-10Mar1885	500x57, 7718t	sr, 14500ihp, 18k	"
287	KAIKOURA, *Zaida*	passcgoref	New Zealand Shipping Co Ltd	?-?-8Sep1884	430x46, 4474t	sr, 4000ihp, 13k	Scrapped 1907
288	RIMUTAKA, *Zamania*	"	"	?-19Nov1884-10Jan1885	430x46, 4474t	sr, 4000ihp, 13.5k	Scrapped 1911
289	MACKAY-BENNETT	cablelayer	Commercial Cable Co	?-18Sep1884-18Nov1884	259.3x49.1, 1718t	sr, 265nhp	Scrapped 1965
290	LYONS, *Lyon, Lyons, La Nina, Hora*	cargo	London, Brighton & S Coast Rly Co	?-19Jan1885-18Feb1885	190x26, 487t	sr, 1233ihp, 15k	Scrapped 1974
291	ITALY, Italie	"	"	?-?-26Jan1885	190x26, 487t	sr, 1302ihp, 15k	Deleted 1912
292*	PARTHIA, *Victoria*	passcargo	William Pearce		3167t	sr	Scrapped 1956
	Built for Cunard by Wm Denny in 1870. Ship acquired from Cunard by Wm Pearce to cover part cost of new builds *Umbria* and *Etruria*. Given a ship number and refitted and re-engined and sold on to Guion Line in 1885.						
293*	BATAVIA	passcargo	William Pearce				
	Built for Cunard by Wm Denny in 1870. Ship acquired from Cunard by Wm Pearce to cover part cost of new builds *Umbria* and *Etruria*. Given a ship number and refitted and re-engined and sold on to Canadian Pacific in 1887.						
294	BANGKOK	passcargo	Scottish Oriental S S Co	?-?-28 May 1885	140x28, 307t	sr, 537ihp	
295	IBIS	rivergunboat	Admiralty	?-March 1885-?	100x18, 299t	p	
306	ALEXANDRA	hospital	Red Cross	1885	145t	sternwheeler	

(SMN short for Stoomvaart Maatschappij Nederland)

Fairfield Shipbuilding & Engineering Co Ltd

No.	Name(s)	Type	Owner	Dates	Particulars		Fate
307 and 308		stern wheel steamers for British Government for service on the Nile. 85x18feet. Sailed from Fairfield on 25 June 1885 and 16 July 1885.					
309		(cancelled)					
310	ALLER	passcargo	Norddeutscher Lloyd	?-22Dec1885-13April1886	438x47.8, 4964gt	sr, 7500hp, 17k	Scrapped 1903
311	TRAVE	"	"	?-18Feb1886-18May1886	455x48, 4966gt	sr, 7500hp, 17k	Scrapped 1908
312	SAALE, *J.L.Luckenbach, Princess, Madison*	"	"	?-21April1886-31July1886	438x47.8, 4965gt	sr, 7500hp, 17k	Scrapped 1924
313	DUITSCHLAND, *Zeeland*	passferry	SMN	?-23Dec1886-2March1887	286x35.3, 1677gt	sr, 4287hp	Scrapped 1922
314	NEDERLAND	"	"	?-1March1887-27April1887	286x35.3, 1677gt	sr, 4605hp	Scrapped 1910
315	ENGELAND	"	"	?-25Nov1886-5April1887	286x35.3, 1648gt	sr, 4000hp	Scrapped 1911

No.	Name	Type	Owner	Dates	Dimensions	Engine	Fate
316	VICTORIA	passferry	London Chatham & Dover Rly Co	?-20May1886-14July1886	310x34.1, 1030gt	sr, 900nhp	Scrapped 1905
317	ORMUZ, Divona	passcargo	Orient Steam Nav Co	?-29Sept1886-29Dec1886	465.5x52.1, 6031gt	sr, 1400nhp, 15.5k	Scrapped 1922
318	ANNAM	"	Soc FR d'Entreprises Coloniales	?-8July1886-?	120x23, 191gt	sr, 300hp	-
319	PRINCE ABBAS	riverpass	Thomas Cook & Sons	?-July1886-9Aug1886	160x20, 315gt	sr, 300hp, 9.5k	-
320	PRINCE TEWFIK	"	"	?-July1886-?	160x20, 315gt	sr, 300hp, 9.5k	-
321	QUEEN VICTORIA	passferry	Isle of Man, Liver'l & Manc SS C	?-29March1887-21May1887	330x39, 1657gt	sr, 1100hp, 20k	Scrapped 1920
322	PRINCE OF WALES, *Prince Edward*	"	"	14April1887-26May1887	330x39, 1568gt	p, 1100hp, 20k	Scrapped 1920
323	PAGAN	rivergunboat	Indian Govt	?-12Nov1886	150x30, 630gt	sternw, 300hp, 8.5k	
324	BHAMO	"	"	?-24Nov1886	150x30, 630gt	sternw, 300hp, 8.5k	
325	LAHN, *Russ, Dniester*	passenger	Norddeutscher Lloyd	?-6Sept1887-12Nov1887	448x49, 5661gt	sr, 9000hp, 18k	Scrapped 1927
326	EMPRESS	passferry	London Chatham & Dover Rly	?-4June1887	324.6x34.9, 1213gt	p, 765hp, 20.5k	Scrapped 1906
327	DEVAWONGSE	cargo	Scottish & Oriental S S Co	?-15Nov1887-10Jan1888	270.5x37, 1646gt	sr, 211hp	Scrapped 1931
328	MAGICIENNE	cruiser2ndcl	Admiralty	?-12May1888-18July1889	265x41, 2950t	sr, 9000hp, 20k	Sold 1905
329	MARATHON	"	"	?-23Aug1888-18July1889	265x41, 2950t	sr, 9000hp, 20k	Sold 1905
330	ROUEN, *Duchess of Buccleuch*	passferry	London Brighton & S Coast Rly Co	?-12April1888-1June1888	250.6x29.1, 761gt	p, 450hp, 19k	Scrapped 1909
332	PARIS, *Verdun, Verdun II, Paris*	"	"	?-1June1888-11July1888	250.6x29, 761gt	p, 450hp, 19k	Scrapped 1924
333	LADY TORFRIDA, *Valfreyia, Star of India, Rodos*	steamyacht	Sir William Pearce	?-2April1888-?	216.5x27, 547gt	sr, 700hp, 11.75k	Scrapped 1951
334	CHOW FA, *Kuo Hsing, Heng An, Ta Hsin*	passcargo	Scottish & Oriental SSCo	?-31July1888-24Aug1888	270.5x37, 1646gt	sr, 250hp	Wrecked 1932
335	STANLEY	"	Canadian Government	?-16Oct1888-Nov 1888	208x32, 914gt	sr, 257hp, 15k	Scrapped 1936
336	DRESDEN, *Helius, Bezm-I Alem*	"	Norddeutscher Lloyd	?-1Dec1888-?	390.5x47, 4802gt	sr, 470nhp, 13k	Sunk gunfire 1914
337	MÜNCHEN, *Gregory Morch, München*	"	"	?-23Jan1889-25Feb1889	390.7x46.8, 4801gt	sr, 470nhp, 15k	Scrapped 1910
338	COBRA, *St.Tudno, Cobra*	passferry	G&J Burns	?-2March1889-4May1889	265x33, 987gt	p, 4000hp, 18k	Scrapped 1922
339	CHINA	passcargo	China SS Co.	?-29June1889-6Sept1889	440x48, 4940gt	sr, 953nhp	Scrapped 1925
340	CALAIS-DOUVERS, *Mona*	passferry	London Chatham & Dover Rly	?-6April1889-6May1889	324x36, 1212gt	p, 6000hp	Scrapped 1909
341	KARLSRUHE	passcargo	Norddeutscher Lloyd	?-31Aug1889-?	415x48, 5057gt	sr, 3200hp, 13k	Scrapped 1908
342	STUTTGART	"	"	?-26Oct1889-?	415x48, 5048gt	sr, 3200hp, 13k	"
343	NORMANNIA, *Patriota, L'Aquitaine*	"	Hamburg-Amerika PAG	?-9Feb1890-?	500x57.5, 8716gt	sr, 16244hp	Scrapped 1906
344	RAMESES THE GREAT	riverpass	Thomas Cook & Co	1889	221x30, 400gt	p, 500ihp	Burned 1916
345	LADY TORFRIDA, *General Fernandez*	yachtsteam	Sir William Pearce	?-14Dec1890-23Jan1890	166x27.1, 364gt	sr, 74nhp	Scrapped 1934
348	DUNOTTAR CASTLE, *Caribbean*	passcargo	D Currie & Co	?-22May1890-28Aug1890	420x49.8, 5465gt	sr, 6700hp, 16k	Foundered 1915
349	ADDER, *Rio De La Plata*	passferry	G & J Burns	?-18April1890-3June1890	280x31, 951gt	p, 658nhp, 19k	Wrecked 1918
350	SOMDETCH PHRA NANG	passcargo	Scottish Oriental SS Co	?-18June1890-12July1890	270x37.1, 1652gt	sr, 1750hp	Missing 1891
351	PHRA NANG	cargo	Scottish Oriental S S Co	?-16July1890-4Sep1890	270x36.75, 1603gt	sr, 1750hp	Scrapped 1931
352	DARMSTADT, *Karadeniz*	passcargo	Norddeutscher Lloyd	?-27Sep1890-26Nov1890	415x48, 5316gt	sr, 3300hp, 13k	Scrapped 1923
353	GERA, *Valparaiso*	"	"	?-8Nov1890-13Dec1890	415x48, 5319gt	sr, 3300hp, 13k	Torpedoed 1917
354	OLDENBURG, *Akdeniz, Ak-Deniz*	"	"	?-13Dec1890-27Jan1891	5318gt	sr, 3300hp, 13k	Scrapped 1930
355	WEIMAR, *Santiago, Armonia*	"	"	?-9Feb1891-?	415x48, 4996gt	sr, 593nhp, 12k	Torpedoed 1918
356	TYNWALD, *Western Isles, Eastern Isles, Western Isles*	passferry	Isle of Man Stm Pkt Co	?-11May1891-30June1891	937gt	sr, 585nhp	Scrapped1952
359	LOO SOK		Scottish Oriental SS Co	?-9March1891-?	270.1x37,1604	sr,1350hp	Wrecked 1912
360	ST TUDNO, *Tudno*	passferry	Liverpool & North Wales SS Co	?-9April1891-11May1891	265x32.5, 794t	p, 585nhp	Scrapped 1921
361	PASS OF MELFORT	barque	Gibson & Clark	?-3Nov1891-?	298x44, 2346t	w	Wrecked 1905
364	CAMPANIA	passenger	Cunard S S Co	?-8Sep1892-17April1893	600x65.25, 12950t	sr, 30000ihp, 21k	Collision 1918
365	LUCANIA	"	"	?-2Feb1893-8Aug1893	600x65.25, 12952t	sr, 30000ihp, 21k	Burned 1909
366	KOH-I-NOOR	excursion	Victoria Steam Boat Association	?-12April1892-23May1892	300.4x32.1, 884t	p, 433nhp	Scrapped 1918
367	POLE STAR, *Ophir, Sophia S, Darien II*	lightship	Northern Lighthouses Commis	?-12Oct1892-?	459t	sr, 1000ihp	Scrapped 1951
368	CHILTONFORD, *Chile, Asailia*	barque	Chiltonford Ship Co	?-19Nov1892-?	298x44, 2348t	w	War loss 1917
370	HOUND, *Mary M, Lesbos, Korytza*	passcargo	G & J Burns	?-1March1893-?	250.3x32.1, 1061t	sr, 2200ihp	Scrapped 1949
371	ROYAL SOVEREIGN	excursion	London & East Coast Express SS Ltd	?-17April1893-?	300.2x33.1, 891t	p, 444nhp	
372	CHELMSFORD, *Inverlogie*	barque	Briggs, Harvie & Co	?-7Nov1893-?	299x44, 2347t	w	War loss 1917
373	TANTALLON CASTLE	passcargo	Castle Mail Stm Pkt Co	?-23Jan1894-?	440.3x50.5, 5636t	sr, 8000ihp, k16	Stranded 1901
374	MACHEW	"	Scottish Oriental SS Co	?-2June1894-5July1894	270.1x37, 1600t	sr, 211nhp	Scrapped 1921
375	LA MARGUERITE	excursion	Palace Steamers Ltd	?-23April1894-16June1894	330x40, 2205t	p, 857nhp, 20k	Scrapped 1925
376	GIRALDA	steamyacht	Col. Harry McCalmont	?-28Aug1894-8Nov1894	275x35, 1240t	sr, 6500ihp, 20k	Scrapped 1921
377	ARUNDEL CASTLE, *Birma, Mitau, Josef Pilsudski, Franck Hellmers, Wilbo*	passcargo	Castle Mail S P Co	?-2Oct1894-?	415x45.5, 4588t	sr, 3300ihp, 13k	Scrapped 1924
378	HANDY	destroyer	Admiralty	?-9March1895-?	194x19.37, 275t	sr, 4000ihp	Scrapped 1916
379	HART	"	"	?-27March1895-?	194x19.37, 275t	sr, 4000ihp	Scrapped 1912
380	HUNTER	"	"	?-28Dec1895-?	194x19.37, 275t	sr, 4000ihp	"
381	VENUS	cruiser2ndcl	"	?-5Sep1895-15June1896	350x54, 5600t	sr, 8000ihp	Scrapped 1921

382	DIANA	cruiser2ndcl	"	?-5Dec1895-15June1897	350x54, 5600t	sr, 9600ihp, 19.5k	Scrapped 1920
383	KEONG WAI, *Hwah Kun, Hua An, Hua An No.1, Hsin Hua An*	passcargo	Scottish Oriental SS Co	?-28March1895-2May1895	289x37.8, 1777t	sr, 1300ihp	Blockship 1937
384	KONINGIN WILHELMINA	passferry	SMN	?-23May1895-?	320x35.75, 1964t	p, 9000ihp, 19k	Mined 1916
385	KONINGIN REGENTES	"	"	?-9July1895-?	320x35.75, 1970t	p, 9000ihp, 19k	Torpedoed, 1918
386	PRINS HENDRIK	"	"	?-22Aug1895-?	320x35.75, 1945t	p, 9000ihp, 19k	Scrapped, 1922
387	CHOW TAI, *Juan Antonio, Ana Maria, Mosquitera*		Scottish Oriental SS Co	?-17Dec1895-?	289x37.8, 1777t	sr, 211nhp	
388	WONG KOI, *Ta Chung, Yung Lung*	cargo	Scottish Oriental SS Co	?-4Dec1895-?	289x37.8, 1777t	sr, 210nhp	Scrapped, 1939
389	DUNVEGAN CASTLE	passcargo	Castle Mail S P Co	?-14April1896-?	450.5x50.9, 5958t	sr, 7000ihp, 16k	Scrapped, 1923
390	DIADEM	cruiser1stcl	Admiralty	?-21Oct1896-19July1898	435x69, 11000t	sr, 16500ihp	Scrapped, 1921
391	ST ELVIES	excursion	Liverpool & North Wales SS Co	?-13April1896-14May1896	240.6x28.3, 567t	p, 2750ihp, 18.5k	Scrapped, 1930
392	EMPRESS QUEEN	passferry	Isle of Man Stm Pkt Co	?-4March1897-8July1897	360x42, 2140t	p, 10000ihp, 20k	Wrecked, 1916
393	TINTAGEL CASTLE, *Liger*	passcargo	Castle Mail Stm Pkt Co	?-12Sep1896-17Nov1896	425x49.75, 5513t	sr, 3800ihp, 13.5k	Scrapped, 1923
394	AVONDALE CASTLE, *Garonna*	"	"	?-5Nov1896-13Jan1897	425x49.75, 5531t	sr, 3800ihp, 13.5k	"
395	GIPSY	destroyer	Admiralty	?-9March1897-27April1898	209x21, 355t	sr, 6000ihp	
396	FAIRY	"	"	?-29May1897-7July1898	209.75x21, 380t	sr, 6300ihp, 30k	Foundered, 1918
397	OSPREY	"	"	?-17April1897-1June1898	219x21, 355t	sr, 6000ihp, 30k	Scrapped, 1919
398	ARGONAUT	cruiser1stc	"	?-24Jan1898 -11July1899	435x69, 11000t	sr, 18000ihp, 19k	Scrapped, 1921
399	REGELE CAROL 1, *Korol Karl, Regele Carol 1*	passferry	Roumanian State Railways	?-22March1898-16June1898	350x42, 2369t	sr, 6564ihp, 18k	Mined, 1941
400	CARISBROOK CASTLE	passcargo	Castle Mail Stm Pkt Co	?-28Oct1897-?	500x56, 7626t	sr, 7500ihp, 17.5k	Scrapped, 1922
401	HERMES	cruiser2ndcl	Admiralty	?-7April1898-5Oct1899	350x54, 5600t	sr, 10000ihp	Torpedoed, 1914
402	HIGHFLYER	"	"	?-4June1898-7Dec1899	350x54, 5600t	sr, 10000ihp	Scrapped, 1921
403	ATMAH, *Aegean Star*	yachtsteam	Baron E de Rothschild	?-5May1898-14July1898	270x34, 1333t	sr, 3500ihp	Scrapped, 1950
404	OMRAH	passcargo	Orient Steam Navigation Co	?-3Sep1898-?	490.7x56.7, 8291t	sr, 1772nhp, 16k	Torpedoed, 1918
405	LEVEN	destroyer	Admiralty	?-28June1898-30May1899	209.75x21, 370t	sr, 6300ihp	Scrapped, 1920
407	KINFAUNS CASTLE	passcargo	Castle Mail Stm Pkt Co	?-12May1899-24Aug1899	515x59, 9664t	sr, 9800ihp, 17.5k	Scrapped, 1927
408	KILDONAN CASTLE	"	"	?-22Aug1899-24Oct1899	515x59, 9688t	sr, 9800ihp, 17.5k	Scrapped, 1931
409	CRESSY	cruiserarm'd	Admiralty	?-4Dec1899-28May1901	440x69.5, 12000t	sr, 21000ihp, 20k	Torpedoed, 1914
410	ABOUKIR	"	"	?-16May1900-3April1902	440x69.5, 12000t	sr, 21000ihp, 20k	"
411	GOOD HOPE	"	"	?-21Feb1901-8Nov1902	500x71, 14100t	sr, 30000ihp, 20k	Gunfire, 1914
412	FALCON	destroyer	"	?-29Dec1899-5Sep1900	209x21, 375t	sr, 6163ihpt	Collision 1918
413	OSTRICH	"	"	?-22March1900-5Oct1900	208x21, 375t	sr, 6250ihpt	Scrapped 1920
414	BEDFORD	cruiserarm'd	"	?-31Aug1901-11Nov1903	440x66, 9800t	sr, 22000ihp	Wrecked 1910
415	HARDINGE	troopship	Royal Indian Marine	?-11Aug1900-15Dec1900	407x50.75, 5434t	sr, 9366ihp	
416	KUMANO MARU	passcargo	Nippon Yusen KK	?-1 June1901-Aug1901	400.2x48.7, 5076	sr, 788nhp	Collision 1927
417	DONEGAL	cruiserarm'd	Admiralty	?-4Sep1902-5Nov1903	440x66, 9800t	sr, 22000ihp	Scrapped 1927
418	ORONTES	passcargo	Orient Stm Nav Co	?-10May1902-Sep1902	513.7x58.2, 9023t	sr, 10000ihp, 16.5k	Scrapped 1925
419	PANAMA, *Maine*	"	Pacific Stm Nav Co	?-8March1902-14May1902	401.2x52.3, 5981t	sr, 4000ihp	Scrapped 1948
420	VICTORIA	"	"	?-21June1902-29Aug1902	400x52, 5967t	sr, 4220ihp, 13.5t	Scrapped 1923
423	COMMONWEALTH	battleship	Admiralty	17June1902-13May03-9May1905	425x78, 16350t	sr, 18000ihp, 17.5k	Scrapped 1922
424	ARMADALE CASTLE	passcargo	Union-Castle Mail S S Co	?-11Aug1903-14Nov1903	570.1x64.5, 12973t	sr, 12500ihp, 16.5k	Scrapped 1936
425	BOSPHORUS No.47, *Tarzinevin*	passferry	Chirket-i-Hairié	?-24Jan1903-2April1903	100x19.33, 114t	sr, 345ihp	
426	BOSPHORUS No.48, *Dilnisin*	"	"	?-31Jan1903-2April1903	100x19.33, 114t	sr, 335ihp	
427	BOSPHORUS No.49, *Hale*	"	"	?-14Feb1903-25April1903	170x23, 298t	p, 961ihp	
428	BOSPHORUS No.50, *Seyyale*	"	"	?-28Feb1903-25April1903	170x23, 298	p, 949ihp	
431	FORWARD	cruiser	Admiralty	29Aug1904-26Aug1905	379x39, 2850t	sr, 16500ihp, 25k	Scrapped 1921
432	DUCHESS OF FIFE, *Duchess, Duchess of Fife*	excursion	Caledonian S P Co	9May1903-5June1903	210x25, 336t	p, 1668ihp, 15.5k	Scrapped 953
433	DURHAM CASTLE	passcargo	Union-Castle Mail S S Co	?-17Dec1903-?	475.4x56.7, 8217t	sr, 5200ihp, 14.5k	Mined 1940
434	FORESIGHT	cruiser2ndcl	Admiralty	?-8Oct1904-8Sep1905	365x38.75, 2850t	sr, 16500ihp, 23k	Scrapped 1920
435	COCHRANE	cruiserarm'd	"	?-20May1905-20Feb1907	480x73.5, 13550t	sr, 23500ihp, 23k	Stranded 1918
436	QUEEN OF SCOTS, *Dolores*	yachtsteam	W A Coats	?-19March1904-?	162x28.25, 606t	sr, 112nhp	Scrapped 1952
437	NARCISSUS, *Narcissus II, Narcissus, Grive*	yachtsteam	A E M Murphy	?-20Dec1904-4March1905	210x27.6, 661t	st, 275nhp	Mined 1940
438	FÜRST BISMARCK, *Friedrichsruhe, Amboise*	passcargo	Hamburg America Line (HAPAG)	?-22March1905-6June1905	469.2x55.1, 8332t	sr, 6800ihp, 14.5k	Scrapped 1935
439	DIEPPE, *Rosaura*	passferry	London, Brighton & S Coast Rly	?-6April1905-9June1905	273.5x34.8, 1216t	st, 950shp, 22k	Mined 1941
440	BOSPHORUS No.51, *Sureyya*	"	Chirket-i-Hairié	?-25March1905-2May1905	100x19.33, 116t	sr, 288ihp	
441	BOSPHORUS No.52, *Sihap*	"	"	?-30March1905-2May1905	100x19.33, 116t	sr, 301ihpt	
442	EMPRESS OF BRITAIN, *Montroyal*	passenger	Canadian Pacific Rlwy Co	?-11Nov1905-20April1906	550x65.5, 14188t	sr, 18000ihp, 18k	Scrapped 1930
443	EMPRESS OF IRELAND	"	"	?-27Jan1906 -12June1906	550x65.5, 14191t	sr, 18000ihp, 18k	Collision 1914
444	VIPER, *Snaefell*	passferry	G & J Burns	?-10March1906-?	315x39.6, 1713t	st, 950shp, 22k	Scrapped 1949

445	INDOMITABLE	battlecruiser	Admiralty	1Mar 1906-16Mar1907-25Jun1908	567x78.5, 17250t	st, 41000shp, 25k	Scrapped 1922
446	BOSPHORUS No.57, *Tarabya*	passferry	Chirket-i-Hairié	?-20March1906-18April1906	100x19.33, 122t	sr, 304ihp	
447	BOSPHORUS No.58, *Nimet*	"	"	?-24March1906-18April1906	100x19.33, 122t	sr, 335ihp	
448	VOLTURNO	passcargo	Volturno SS Co	?-5Sep1906-27Nov1906	340x43, 3586t	sr, 2500ihp	Burned 1913
449	HELIOPOLIS, *Royal George*	"	Egyptian Mail SS Co	?-28May1907-6Nov1907	525x60, 10897t	st, 18000shp, 19k	Scrapped 1922
450	CAIRO, *Royal Edward*	"	"	?-11July1907-9Jan1908	525x60, 10864t	st, 18000shp, 19k	Torpedoed 1915
451	HAZEL, *Mona*	"	Laird Line	?-13April1907-1June1907	260x36, 1241t	sr, 3141ihp	Scrapped 1938
452	ASSINIBOIA	"	Canadian Pacific Rlwy Co	?-25June1907-31Aug1907	336x43.5, 3880t	sr, 3300ihp, 15k	Burned 1969
453	KEEWATIN	"	"	?-6July1907-14Sep1907	336x43.5, 3856t	sr, 3300ihp	Preserved 1967
456	STANLEY	repair	Canadian Government, new boilers	Sailed 9 Nov 1907			Scrapped 1936
457	PRINCESS CHARLOTTE, *Mediterranean*	passferry	Canadian Pacific Rlwy Co	?-27June1908-16Sep1908	330x46.7, 3844t	sr ,795nhp, 19k	Scrapped 1965
458	LADY FRASER, *Mekhala*	pilotvessel	Bengal Govt	?-14March1908-17April1908	282x38, 1669t	sr, 142nhp	
459	OTWAY	passcargo	Orient Steam Nav Co	?-21Nov1908-25May1909	535.9x63.2, 12077t	sr, 14000ihp, 16k	TorpedoedU 1917
461	PRINSES JULIANA	passferry	Stoomvaart Maatschappij Zeeland	?-22May1909-4Sep1909	350x42.5, 2885t	sr, 9330ihp, 22k	Mined 1916
462	ORANJE NASSAU	"	"	?-3July1909-27Nov1909	350x42.5, 2885t	sr, 9562 ihp, 22k	Scrapped 1954
463	MECKLENBURG	"	"	?-25Oct1909-18Dec1909	350x42.5, 2885t	sr, 7921ihp, 22k	Mined 1916
464	GRASSHOPPER	destroyer	Admiralty	17Apl1909-23Nov1909-Jul 1910	271x27.83, 923t	st, 14300shp, 27k	Scrapped 1921
465	MOSQUITO	"	"	?-27Jan1910-11Aug1910	275x27.7, 945t	st, 12500shp, 27k	Scrapped1920
466	SCORPION	"	"	?-19Feb1910-30Aug1910	275x27.5, 916t	st,12500shp, 27k	Scrapped 1921
467	GLASGOW	lightcruiser	"	25Mar1909-30Sep1909-Sep1910	453x47, 4800t	st, 22000shp, 24k	Scrapped 1927
468	BALMORAL CASTLE	passcargo	Union-Castle Mail S S Co	?-13Nov1909-?	570x64.5, 13361t	sr, 16.5k	Scrapped 1939
469	PARRAMATTA	destroyer	Royal Australian Navy	17Mar1909-9Feb1910-1Sep1910	245x24.27, 750t	st, 500shp, 26k	Dismantled 1929
470	WARREGO	"	"	Dec1910-4April1911-1June1912	245x24.27, 750t	st, 10900shp, 26k	Sunk 1931
		Shipped in secctions to Cockatoo Island Dockyard, Sydney, to give Australian workers experience in ship construction.					
471	CAMELEON	destroyer	Admiralty	9Dec1909-1June1910-1Dec1910	240x25.25, 747t	st, 13500shp, 27k	Scrapped 1922
472	COMET	"	"	?-23June1910-12Jan1911	240x25.324, 747t	st, 13500shp, 27k	Torpedoed 1918
473	GOLDFINCH	"	"	?-12July1910-15Feb1911	240x25.3, 755t	st, 13500shp	Wrecked 1915
474	PRINCESS ADELAIDE, *Angelica*	passferry	Canadian Pacific Rlwy Co	?-5July1910-?	290x46.1, 3061t	sr, 4050ihp, 17k	Scrapped 1967
475	BOSPHORUS No.65, *Sarayburnu*	"	Chirket-i-Hairié	?-29June1910-9Aug1910	152x26, 434t	sr, 600ihp, 10k	
476	BOSPHORUS No.66, *Bogazici*	"	"	?-7July1910-31Aug1910	152x26, 433t	sr, 600ihp, 10k	Collision 1992
477	NEW ZEALAND	battlecruiser	Admiralty	?-1July1911-20Nov1912	555x 79.85, 18800t	st, 44000shp, 25k	Scrapped 1923
478	GLOUCESTER CASTLE	passcargo	Union Castle Mail SS Co	?-13May1911-10Aug1911	452x56, 7999t	sr, 3750ihp, 12k	Gunfire 1942
479	MAUNGANUI, *Cyrenia*	"	Union SS Co of New Zealand	?-24Aug1911-?	420.8x55.7, 7527t	sr, 1304nhp	Scrapped 1957
480	SAGUENAY, *Kiang Yong, Yangtse Phoenix*	passferry	Richelieu & Ontario Nav Co	?-22April1911-?	275.3x40, 2777t	sr, 2100ihp	Wrecked 1951
481	NORMANNIA	passferry	London & South Western Rly Co	?-9Nov1911-Feb1912	290.3x36.1, 1567t	st, 4950shp, 19.5k	Bombed 1940
482	HANTONIA	"	"	?-23Dec1911-1May1912	290.3x36.1, 1560t	st, 4750shp	Scrapped 1952
483	ENDEAVOUR	survey	Admiralty	?-30March1912-June1912	214.8x34.5, 1280t	sr, 1100ihp, 13k	Scrapped 1953
484	EMPRESS OF RUSSIA	passenger	Canadian Pacific Rlwy Co	?-28Aug1912-24March1913	570.2x68.2, 16810t	st, 22500shp, 19.5k	Scrapped 1946
485	EMPRESS OF ASIA	"	"	?-23Nov1912-?	570x68, 16909t	st, 22500shp, 19.5k	Air attack 1942
486	ERMINE	passcargo	Laird Line	15June1912-3Sep1912	300x40.2, 1845t	sr, 518nhp	Mined 1917
487	CALGARIAN	"	Allan Line S S Co	19April1913-16March1914	568.8x70.3, 17515t	st, 21000shp, 19k	Torpedoed 1918
488	FORTUNE	destroyer	Admiralty	17May1913-22Dec1913	267.5x27, 1089t	st, 24000ihp, 29k	Gunfire 1916
489	LAFOREY	"	"	22Aug1913-19Feb1914	269x26.75, 995t	st, 24500shp, 29k	Mined 1917
490	LAWFORD	"	"	30Oct1913-31March1914	269x26.75, 965t	st, 24500shp, 29k	Scrapped 1922
491	LOUIS	"	"	30Dec1913-8May1914	269x26.75, 965t	st, 24500shp, 29k	Wrecked 1915
492	LYDIARD	"	"	26Feb1914-11June1914	269x26.75, 965t	st, 24500shp, 29k	Scrapped 1921
494	LLANSTEPHAN CASTLE	passcargo	Union Castle Mail SS Co	?-29Aug1913-?	500.5x63.3, 11923t	st, 5800ihp, 14k	Scrapped 1952
495	UNDAUNTED	cruiserlight	Admiralty	?-28April1914-1Sep1914	410x39, 3512t	st, 40000shp, 28.5k	Scrapped 1923
497	VALIANT	battleship	"	31Jan1913-4Nov1914-13Jan1916	640x90, 27500t	st, 75000shp, 24k	Scrapped 1948
499	AVENGER	AMC	"	?-30June1915-21March1916	548x68.2, 14744t	st, 18k	Torpedoed 1917
500	MELAMPUS (ex Greek)	destroyer	"	?-16Dec1914-20June1915	265x26.67, 1037t	st, 25000shp, 32k	Scrapped 1922
501	MELPOMENE (ex Greek)	"	"	?-1Feb1915-Aug 1915	265x26.67, 1040t	st, 25000shp, 32k	Scrapped 1921
502	WATERWITCH, *Bosphorus No. 71, Halas*	despatch	"	?-17Oct1914-7June1915	161x26, 584t	sr, 625ihp, 13k	
503	RENOWN	battlecruiser	"	25Jan1916-4March1916-18Sep1916	794x90, 26500	st, 112000shp, 32k	Scrapped 1948
504	MANDATE	destroyer	"	27April1915-1Aug1915	265x26.67, 1037t	st, 25000shp, 34k	Scrapped 1922
505	MANNERS	"	Admiralty	15June1915-21Sep1915	265x26.67, 1040t	st, 25000shp	Scrapped 1921
507	E37	submarine	"	?-2Sep1915-17Mar1916	181x15.04, 662t	d, 1600bhp, 15k surf. e, 840hp, 10k subm..	War loss 1916
508	E38	"	"	?-13June1916-10Jul1917	181x15.1, 662t	d, 1600bhp, 15k surf. e, 840hp 10k subm.	Scrapped 1922

No.	Name	Type	Owner	Dates	Dimensions	Machinery	Fate
509	MINDFUL	destroyer	"	24Aug1915-10Nov1915	265x26.75, 1034t	st, 25000shp, 34k	Scrapped 1921
510	MISCHIEF	"	"	12Oct1915-16Dec1915	265x26.75, 1025t	st, 25000shp, 34k	Scrapped 1922
511	ONSLAUGHT	"	"	?-4Dec1915-4Feb1916	265x26.75, 1025t	st, 25000shp, 34k	"
512	ONSLOW	"	"	?-15Feb1916-15April1916	265x26.75, 1025t	st, 25000shp, 34k	Scrapped 1921
513	E47	submarine	"	?-29May1916-?	176x15.1, 662t	d, 1600bhp, 15k surf.	War loss 1917
	To Beardmore Dalmuir 29 May 1916 for completion.					e, 840hp 10k sub.	
514	E48	"	"	?-2Aug1916-Feb1917	181x15.1, 662t	d, 1600bhp, 15k	Scrapped 1928
	To Beardmore Dalmuir 2 August 1916 for completion.					e, 840hp 10k subm.	
516	OBSERVER	destroyer	"	1May1916-17Jun1916	273x26.75, 1100t	st, 25000shp, 34k	Scrapped 1922
517	OFFA	"	"	7June1916-31Jul1916	273x26.75, 1025	st, 25000shp, 34k	Scrapped 1921
518	ORCADIA	"	"	26July1916-29Sep1916	265x26.75, 1025t	st, 25000shp, 34k	Scrapped 1923
519	ORIANA	"	"	23Sep1916-23Oct1916	265x26.67, 1027t	st, 25000shp, 34k	Scrapped 1921
520	PHEASANT	"	"	23Oct1916-10Nov1916	265x26.75, 994t	st, 25000shp, 34k	Mined 1917
521	PHOEBE	"	"	20Nov1916-28Dec 1916	265x26.76, 1044t	st, 25000shp, 34k	Scrapped 1921
522	K13, *K22*	submarine	"	?-11Nov1916-5Nov1917	335x26.5, 1883t	st, 10000shp, 24k surf.	Scrapped 1926
						e, 1600hp 8k subm.	
523	K14	"	"	?-8Feb1917-22May1917	335x26.5, 1883t	st, 10000shp, 21k	Scrapped 1926
						e, 1600hp 8k subm.	
524	TEMPEST	destrmnlayr	"	?-26Jan1917-20Mar1917	276x26.75, 1065t	st, 27000shp, 36k	Scrapped 1937
525	UNDINE	destroyer	"	26Sep1916-22Mar1917-26May1917	276x26.81,1085t	st, 27000shp, 36k	Wrecked 1928
526	CARDIFF	lightcruiser	"	22Jul1916-12Apr1917-25Jun1917	425x43, 4190t	st, 40000shp, 29k	Scrapped 1946
527	RODNEY	battlecruiser	"	Laid down 9Oct1916. Cancelled March 1919 and approx, 6000tons broken up on slip.			
528	EMPRESS OF CANADA	passenger	Canadian Pacific Rly Co	?-17Aug1920-28Apr1922	627x77.9, 21516t	st, 4592nhp, 20k	TorpedoedU 1943
529	MONTROSE, *Forfar*	passcargo	"	?-14Dec1920-17Mar1922	548x70, 16402t	st, 14000shp, 16k	Torpedoed 1940
530	VENDETTA	destroyer	Admiralty	Nov1916-3Sep1917-17Oct1917	312x29.5, 1295t	st, 27000shp, 35k	Scuttled 1948
531	VENETIA	"	"	2Feb1917-29Oct1917-19Dec1917	312x29.5, 1295t	st, 27000shp, 35k	Mined 1940
532	L15	submarine	"	16Nov1916-16Jan1918-4May1918	238.6x23.5, 929t	d, 2,400bhp, 17k surf.	Scrapped 1932
						e, 1,600hp 10.5k subm.	
533	L16	submarine	"	21Nov1916-9April1918-31May1918	238.6x23.5, 929t	d, 2,400bhp 17k surf.	Scrapped 1934
						1, 600hp, 10.5k subm.	
534	WALRUS	destroyer	"	Feb 1917-27Dec1917-12Feb1918	300x26.75, 1,100t	st, 27000shp, 34k	Stranded 1938
535	WOLFHOUND	"	"	April1917-14Mar1918-27April1918	300x26.75, 1,100t	st, 27000shp, 34k	Scrapped 1948
536	WAR PATRIOT, *Adna, Dentice, Faireno, Arayat, Faireno, Dentice*	tanker	Shipping Controller	?-17Jan1918-18April1918	400x52, 5197t	sr, 2500ihp, 11k	Scrapped 1953
537	WAR PIONEER, *Brazilier, Lomas, Avon Cliff, Italo Balbo*	cargo	"	?1April1919-14May1919	400x52, 5256t	sr, 2500ihp, 11k	Torpedoed 1941
538		Machinery for Alex. Stephen & Sons Ltd.					
539	L55, *Bezbozhnik*	submarine	Admiralty	?-21Sep1918-19Dec1918	230.5x15.71, 960t	d, 2400bhp, 17k surf.	Mined 1919
						e, 1600hp, 10.5k subm.	
540	TRAIN FERRY No.3, *Daffodil*	trainferry	War Office	?-12Sep1917-2Nov1917	350x58.5, 2672t	sr, 410nhp, 12k	Sunk March 1945
541		(Cancelled)	Canadian Pacific				
542	Standard ship B type.	transferred to Harland & Wolff Govan yard					
543	Standard ship B type.	(Cancelled March 1918)					
544	Ac.1151	lighter	War Office (Director of Inland Waterways)	?-20Sep1917-?	125.75x16.33, 130t		
545	Ac.1152	"	"	?-8Oct1917 ?	125.75x16.33, 130t		
546	Ac.1153	"	"	?-13Nov1917-11Jan1918	125.75x16.33, 130t		
547	Ac.1154	"	"	?-7Feb1918-11Feb1918	125.75x16.33, 130t		
548	Ac.1155	"	"	?-21Feb1918-?	125.75x16.33, 130t		
549	Ac.1156	"	"	?-22April1918-?	125.75x16.33, 130t		
550	Ac.1157 to 561 Ac.1167	"	"	delivered in sections from 12 December 1917 to 29 October 1918. Particulars as above.			
561	Ac.1168	"	"		125.75x16.33, 130t		
562 to 567		machinery for trawlers. Admiralty					
568	L56	submarine	Admiralty	16Oct1917-29May1919-3Sep1919	235x23.5, 960t	d, 2400bhp, 20k surf.	Scrapped 1938
						e,1600hp, 12k, subm.	
569	L57	"	Admiralty (Cancelled 28 March 1919)				
570	L58	"	" (Cancelled 28 March 1919)				
571	SIKH	destroyer	Admiralty	?-7May1918-29June1918	276x26.75, 1016t	st, 27000shp, 34k	Scrapped 1927
572	SIRDAR	"	"	?-6July1918-6Sep1918	276x26.75, 1016t	st, 27000shp, 34k	Scrapped 1934
573	SOMME	"	"	Nov1917-10Sep1918-4Nov1918	276x26.75, 1016t	st,27000shp, 34k	Sceapped 1932

No.	Name	Type	Owner	Dates	Dimensions	Machinery	Fate
574	IRVINE	minesweepr	"	?-8Dec191-1Feb1918	231x28, 710t	sr, 2200ihp, 16k	Scrapped 1923
575	KENDAL	"	"	?-9Feb1918-27March1918	231x28, 710t	sr, 2200ihp, 16k	Scrapped 1928
576	KINROSS	"	"	?-4July1918-?	231x28, 710t	sr, 2200ihp, 16k	Mined 1919
577	LYDD	"	"	?-4Dec1918-?	231x28, 710t	sr, 2200ihp, 16k	Scrapped 1947
578		Seaplane carrier, ", (Cancelled)					
579		Submarine engines, ", (Cancelled)					
580	CARLISLE	lightcruiser	Admiralty	2Oct1917-9July1918-16Nov1918	451x43.75, 4290t	st, 40000shp, 29k	Scrapped 1949
581	COLOMBO	"	"	8Dec1917-18Dec1918-18June1919	451x43.75, 4290t	st, 40000shp, 29k	Scrapped 1948
582	SPEAR	destroyer	"	?-9Nov1918-28Dec1918	265x26.6, 1000t	st, 27000shp, 36k	Scrapped 1926
583	SPINDRIFT	"	"	?-30Dec1918-2April1919	265x26.6, 1000t	st, 27000shp, 36k	Scrapped 1936
584		Machinery for HMS *Reindeer*					
585		Machinery parts for HMS *Sardonyx* under construction at Alex Stephen & Sons Ltd.					
586	WANDERER	destroyer	Admiralty	7Aug1918-1May1919-18Sep1919	300x29.5, 1325t	st, 27000shp, 32k	Scrapped 1946
587	WARREN	"	", (Cancelled)				
588	EUPHRATES	lightcruiser	" (Cancelled)				
589	DESPATCH	"	Admiralty	8July1918-24Sep1919-30June1922	445x46.5, 4850t	st, 40000shp, 29k	Scrapped 1946
590	WATSON	destroyer	", (Cancelled)				
591	WAVE	"	","	Hull scrapped and engined fitted in HMS *Worcester*.			
592	WEAZEL	"	", (Cancelled)				
593	L62	submarine	", (Cancelled)				
594	WHITE BEAR	destroyer	", (Cancelled)				
595	TUSCANIA, *Nea Hellas, New York*	passcargo	Anchor Line (Henderson Bros)	?-4Oct1921-8Sep1922	550x70, 16991t	st, 12500shp, 15.75k	Scrapped 1961
596	ATHENIA	"	Anchor Donaldson Line	?-28Jan1922-27March1923	526.3x66.4, 13465t	st, 8700shp, 14.5k	Torpedoed 1939
597	LADY DENISON-PENDER	cablelayer	Eastern Telegraph Co	?-29May1920-17Sep1920	263x38, 1984t	sr, 352nhp, 12k	Scrapped 1963
598 and 599		vessels for Holder line cancelled July 1920					
600	TRANSYLVANIA	passcargo	Anchor Line (Henderson Bros)	1Dec1919-11Mar1925-3Sep1925	550x70, 16923t	st, 12500shp, 15.75k	Torpedoed 1940
601	LETITIA, *Empire Brent, Captain Cook*	passcargo	Anchor Donaldson Line	25Feb1920-14Oct1924-4April1925	536x66, 13475t	st, 8700shp, 14.5k	Scrapped 1960
602	TIGER, *Benares*	tanker	A/S Vestlandske Petroleums Co	?-26Feb1921-July1921	357x47, 3916t	sr, 347nhp	Scrapped 1955
603	AORANGI	passcarqo	Union S S Co of N Z	23Nov1922-17June1924-30Dec1924	580x72, 17486t	d, 13000bhp, 17k	Scrapped 1953
604	FLORENTINO AMEGHINO	tanker	Argentine Government	10May1920- 8April1925-1Aug1925	430x56.75, 6948t	sr, 632nhp	Explosion 1961
605 to 608		machinery contracts for Wm Doxford & Sons					
609	SANTA CRUZ	tanker	Argentine Government	?- 22June1921-Nov1921	350x46.75, 3934t	sr, 347nhp	Scrapped, 1948
610		repeat of Ship No. 604.					
611 to 614		Triple expansion engines for ships and barges building at Monmouth Shipbuildinq Co Ltd. Delivered Chepstow 2 September 1921 to November 1922.					
615	G3 battlecruiser	Admiralty (Cancelled February 1922)					
616	BERWICK	heavycruiser	Admiralty	15Sep24-30Mar26-12July1927	630x68.33, 9750t	st, 80000shp, 30.5k	Scrapped 1948
617	UPWEY GRANGE	cargorefr	Houlder Line	10July1924-23June1925-14Nov1925	430x62.25, 9130t	d, 6400bhp	TorpedoedU 1940
618	ST TUDNO	excursion	Liverpool & North Wales SS Co	20June1925-2Feb1926-15May1926	320x44, 2326t	st, 4500shp, 18.5k	Scrapped 1963
619	SHROPSHIRE, *Salopian*	passcargo	Bibby Line	26Aug1925-10June1926-8Oct1926	482x60, 10560t	d, 8000bhp, 15k	Torpedoed 1941
620	CHESHIRE	"	"	1March1925-20April1927-10July1927	482x60, 10560t	d, 8000bhp, 15k	Scrapped 1957
621		Brown Curtis turbine machinery for HMS *London* building a Portsmouth RND					
622	DUNSTER GRANGE, *Vaasa, Kinyo Maru, Yoko Maru*	cargorefrige	Houlder Line	1June1926-25Oct1927-20Jan1928	430x64.25, 9494t	d, 6400bhp, 15k	Scrapped 1974
623	Reconditioning of *Empress of Australia*		arrived Fairfield 9 Sept 1926 departed 10 June 1927				
624		Shaw Savill, passenger ship, cancelled 4 Feb 1927.					
625	TARANAKI	cargorefr	Shaw, Savill & Albion Co	26Feb1927-10Dec1927-3May1928	481x64, 8286t	d, 7450bhp, 14.5k	Scrapped 1963
626	KARAMEA	"	"	27April1927-8May1928-23Aug1928	481x64, 8281t	d, 7450bhp, 14.5	Scrapped 1960
628	NORFOLK	heavycruiser	Admiralty	8July1927-12Dec1928-1May1930	590x66, 9925t	st, 80000shp, 31k	Scrapped 1950
629	EL ARGENTINO	cargorefrige	British & Argentine S N C	19Feb1927-11Jan1928-12April1928	430x64.25, 9501t	d, 6400bhp, 15k	Air attack 1943
630	STAFFORDSHIRE, *Stafford Marchu*	passcargo	Bibby Line	16Nov1927-29Oct1928-28Jan1929	482x62, 10595t	d, 8000bhp, 15k	Scrapped 1959
631	ROSSINGTON COURT	cargo	United British SS Co	26March1928-30Aug1928-5Oct1928	420x56.17, 6922t	sr, 574nhp	Collision 1940
632	PRINCESS NORAH, *Queen of the North, Canadian Prince, Beachcomber*	passcargo	Canadian Pacific Railway Co	2April1928-27Sep1928-20Dec1928	250x48, 2731t	sr, 437nhp, 16k	Beached 1981
633		Canadian Pacific Steamships Co Ltd. New machinery for Empress of Canada. Arrived Fairfield 29 Jan 1929, left 19 August 1929.					
634	EMPRESS OF JAPAN, *Empress of Scotland, Scotland, Hanseatic*	passenger	Canadian Pacific Rlwy Co	9Aug1928-17Dec1929-8June1930	640x83.5, 26032t	st, 34000shp, 21k	Burned 1966
635	BURMA	P Henderson & Co., recconditioning. Left Fairfield 28 December 1928.					
636	ALTINKUM, *Bosphorus No. 74, Altinkum*	passferry	Chirket-i-Hairié	7Nov1928-14March1929-4 April1929	153x26, 412t	sr, 580ihp, 10k	Scrapped 2000
637		Blue Star Line. Alterations to *Arandora Star*. Arrived Fairfield 5 January 1929, left 31 May1929.					

638	PRINCESS ELIZABETH, *Pegasus, Highland Queen*	passvehcle	Canadian Pacific Railway Co	11July1929-16Jan1930-27March1930	351x52, 5251t	sr, 3600ihp, 16k	Scrapped 1976
639	PRINCESS JOAN, *Hermes*	"	"	4March1929-4Feb1930-15April1930	351x52, 5251t	sr, 3700ihp, 16k	Scrapped 1974
640	WORCESTERSHIRE, *Kannon Marchu*	passcargo	Bibby Line	2Sept1929-8Oct1930-4Feb1931	483x64.2, 10139t	d, 8000bhp, 15k	Scrapped 1961
641	SOUTHSEA	excursion	Southern Railway	28Oct1929-2April1930-5May1930	244x30, 825t	p, 213nhp	Mined 1941
642	WHIPPINGHAM	"	"	11Nov1929-1May1930-31May1930	244x30.1, 825t	p, 196nhp	Scrapped 1963
643	ST SEIRIOL	"	Liverpool & North Wales SS Co	10July1930-5March1931-12May1931	269x37, 1586t	st, 596rhp	Scrapped 1962
644	JEANIE DEANS	"	London & North Eastern Rly	9Jan1930-7April1931-18May1931	250x30, 635t	p, 2200ihp, 18k	Scrapped 1967
645	DAINTY	destroyer	Admiralty	2Feb1931-11May1932-22Dec1932	317.75x33, 1375t	st, 36000shp, 36k	Air attack 1941
646	DELIGHT	"	"	8April1931-8June1932-31Jan1933	317.75x33, 1375t	st, 36000shp, 36k	Bombed 1940
647		Machinery main and aux for HMS *Exmouth*.					
648	WOOLWICH	depotship	Admiralty	29May1933-20Sep1934-28June1935	1935575x64, 8750t	st, 6500shp, 15.5k	Scrapped 1962
649	MERCURY	passferry	London, Mi'd & Scottish Rly Co	2Aug1933-16Jan1934-22Feb1934	223x30, 621t	p, 230nhp	Mined 1940
650	KINGFISHER	corvette	Admiralty	1June1934-14Feb1935-18June1935	234x26.5, 510t	st, 3600shp, 20k	Scrapped 1947
651	GARLAND, *Marchnix*	destroyer	"	22Aug1934-24Oct1935-3March1936	312x33, 1335t	st, 34000shp, 36k	Scrapped 1964
652	GIPSY	"	"	4Sept1934-7Nov1935-22Feb1936	312x33, 1335t	st, 34000shp, 36k	Mined 1939
653	DERBYSHIRE	passcargo	Bibby Line	2Aug1934-14June1935-19Oct1935	482x66, 11660t	d, 8000bhp, 15k	Scrapped 1964
654	MARCHIONESS OF LORNE	passferry	Caledonian Stm Pkt Co	18Sep1934-19Feb1935-29Mar1935	199.5x27, 449t	p, 950ihp, 12k	Scrapped 1955
655	Caisson for Devonshire Dock	caisson	Vickers Armstrong, Barrow	Work started on 14 Jan 1935. Left Fairfield on 10 June 1935			
656	MARCHIONESS OF GRAHAM, *Hellas, Nea Hellas, Galaxias, El Greco*	passferry	Caledonian Stm Pkt Co	10Oct1935-6March1936-7April1936	221.5x30, 585t	st, 267nhp, 18k	Scrapped 1975
657	ST SILIO, *St Trillo, Thrillo*	excursion	Liverpool & North Wales S'hip Co	18Mar1935-24Mar1936-29April1936	150x27, 314t	d, 232nhp, 18.5k	Scrapped 1975
658	LIVERPOOL	cruiser	Admiralty	17Feb1936-24March1937-2Nov1938	558x63.31, 9400t	st, 82500shp, 31k	Scrapped 1958
659	GURKHA	destroyer	"	6July1936-7July1937-21Oct1938	355.5x36.5, 1870t	st, 44000shp, 36k	Bombed 1940
660	MAORI	"	"	6July1936-2Sep1937-30Nov1938	355.5x36.5, 1870t	st, 44000shp, 36k	Air attack1942
661	CIRCASSIA	passcargo	Anchor Line Ltd	25June1936-8June1937-23Oct1937	480x66, 11136t	d, 11400bhp, 16k	Scrapped 1966
662	JUPITER, *Scawfell, Jupiter*	passferry	Caledonian Stm Pkt Co	29Oct1936-9April1937-28May1937	223x30, 642t	p, 288nhp, 17.5k	Scrapped 1961
663	JUNO, *Helvellyn*	excursion	"	29Oct1936-5May1937-24June1937	223x30, 642t	p, 288nhp, 17.5k	Bombed 1941
664	CILICIA, *Jan Backx*	passcargo	Anchor Line Ltd	7Dec1936-21Oct1937-20May1938	480x66, 11136t	d, 11400bhp, 16k	Scrapped 1980
665		machinery for battleship *Queen Elizabeth* under reconstruction at Portsmouth Dockyard.					
666	PHOEBE	cruiser	Admiralty	2Sept37-25March1939-30Sep1940	485x50, 5450t	st, 62000shp, 30k	Scrapped 1956
667	JUNO	destroyer	"	5Oct37-8Dec1938-25Aug1939	339.5x35.67, 1690t	st, 40000shp, 36k	Bombed 1941
668	KELVIN	"	"	5Oct37-19Jan1939-29Nov1939	339.5x35.67, 1690	st, 40000shp, 36k	Scrapped 1949
669	HOWE	battleship	"	1June1937-9April1940-29Aug1942	745x102.83, 44000tfl	t, 110000shp, 28k	Scrapped 1958
670	DEVONSHIRE, *Devonia*	troopship	Bibby Line	26Oct1937-20Dec1938-7July1939	497x63, 11275t	d, 6300bhp, 14.5k	Scrapped 1967
671	Caisson,No.7 Dock Portsmouth	caisson	Admiralty	23Feb1939-17June1939-?	83.08x23, 600t		
672	IMPLACABLE	aircrftcar	Admiralty	21Feb1939-10Dec1942-28Aug1944	766x95, 25690t	st, 148000shp, 32.5k	Scrapped 1955
673	NAPIER	destrleadr	"	26July1939-22May1940-11Dec1940	339.5x35.67, 1773t	st, 40000shp, 36k	Scrapped 1956
674	NESTOR	destroyer	"	26July1939-9July1940-12Feb1941	339.5x35.67, 1773t	st, 40000shp, 36k	Bombed 1942
675	MUSKETEER	"	"	7Dec1939-2Dec1941-18Sep1942	362x37, 1920t	st, 48000shp, 36k	Scrapped 1955
676	MYRMIDON, *Orkan*	"	"	7Dec1939-2March1942-20Nov1942	362x37, 1920t	st, 48000shp, 36k	Sunk 1943
677 and 678		ferry	LMS Railway Co	Ferries for Holyhead/Kingston service. Contracts placed but cancelled following declaratiion of war.			
679	BELLONA	cruiser	Admiralty	30Nov1939-29Sep1942-29Oct1943	512x50.5, 5770t	st, 62000shp, 32k	Scrapped 1959
680	OFFA, *Tariq*	destroyer	"	15Jan1940-14Jan1941-5July1941	345x35, 1610t	st, 40000shp, 37k	"
681	ORIBI, *Gayret*	"	"	15Jan1940-11Mar1941-20Sep1941	345x35, 1610t	st, 40000shp, 36k	Scrapped 1965
682	PANTHER	"	"	15July1940-28May1941-12Dec1941	345x35, 1610t	st, 40000shp, 37k	Bombed 1943
683	PARTRIDGE	"	"	3June1940-5Aug1941-22Feb1942	345x35, 1610t	st, 40000shp, 37k	Torpedoed 1942
684	ONSLAUGHT, *Tughril* (orig. *Pathfinder*)	"	"	14Jan1941-9Oct1941-19June1942	345x35, 1610t	st, 40000shp	Scrapped 1977
685	REDOUBT	"	"	contract transferred to John Brown & Co., Clydebank			
686	RELENTLESS	"	"	contract transferred to John Brown & Co., Clydebank			
687 and 688		machinery	"	engines, shafting and propellers for emergency vessels x2			
689	LCM.87	landing craft	"	MLCs (small) 8 in total, order received in July 1940, completed March 1941			
690	LCT.7	"	"	?-5Dec1940-5Jan1941	135x28, 226t	pe, 500bhp, 9k	War loss 1941
691	LCT.8	"	"	?-10Dec1940-5Jan1941	135x28, 226t	pe, 500bhp, 9k	Bombed 1941
692	LCT.9	"	"	?-24Dec1940-12Jan1941	135x28, 226t	pe, 500bhp, 9k	
693	LCT.10	"	"	?-30Dec1940-12Jan1941	135x28, 226t	pe, 500bhp, 9k	Bombed1941
694	GOATHLAND	destroyer	"	20Jan1941-3Feb1942-6Nov1942	264.3x31.5, 1087t	st, 19000shp, 25k	Scrapped 1945
695	LA COMBATTANTE	"	French Navy	16Jan1941-27April1942-30Dec1942	264.21x31.5, 1087t	st, 19000shp	Mined 1945
	Laid down as HMS *Haldon*. Damaged by bombing 14 March 1941						

No.	Name	Type	Owner	Dates	Dimensions	Machinery	Fate
696		landing craft	Admiralty	October 1940			
697, 698, 699, 700			"	May 1941. Reserve sets of engines, shafting and propellers.			
701	STARLING	sloop	"	21Oct41-14Oct1942-1April1943	283x38.5, 1430t	st, 4300shp, 18.5k	Scrapped 1965
702	WOODCOCK	"	"	21Oct1941-26Nov1942-29May1943	283x38.5, 1470t	st, 4300shp, 18.5k	Scrapped 1955
703	VENUS	destroyer	"	12Jan1942-23Feb1943-28Aug1943	362.75x35.6, 1710	st, 40000shp, 36k	Scrapped 1972
704	VERULAM	"	"	26Jun1942-22Apr1943-10Dec1943	362.75x35.6, 1710	st, 40000shp, 36k	"
705	WAKEFUL, originally *Zebra*	"	"	3June1942-30June1943-17Feb1944	362.75x35.6 2505t	st, 40000shp, 36k	Scrapped 1971
706	WESSEX, orig. *Zenith*, *Jan Van Riebeeck*	"	"	20Aug1942-2Sep1943-11May1944	362.75x35.6 2505t	st, 40000shp, 36k	Target 1980
707	CAMPERDOWN	destroyer	"	30Oct1942-8Feb1944-18June1945	355x40, 2315t	st, 50000shp, 35k	Scrapped 1970
708	FINISTERRE	"	"	8Dec1942-22June1944-11Sep1945	355x40.3, 2315t	st, 50000shp, 35k	Scrapped 1967
709	THESEUS	aircrftcarrier	"	6Jan1943-6July1944-9Feb1946	630x80, 13350t	st, 40000shp, 24k	Scrapped 1962
710	BLAKE	cruiser	"	17Aug1942-20Dec1945-18Mar1961	538x64, 9550t	st, 80000shp, 29.5k	Scrapped 1982
711	CADIZ, *Khaibur*	destroyer	"	10May1943-16Sep1944-12April1946	355x40.25, 2315t	st, 50000shp, 35k	Sunk 1971
712	ST JAMES	"	"	20May1943-7June1945-12July1946	355x40, 2325t	st, 50000shp, 35k	Scrapped 1961
713	VIGO	"	"	1Sep1943-27Sep1945-9Nov1946	355x40.25, 2315t	st, 50000shp, 34k	Scrapped 1964
714	BELLE ISLE	"	"	10Nov1943-7Feb1946-Cancelled September 1945. Hull launched to clear slip.			Scrapped at Troon
715	OMDURMAN	"	"	8March1944- Scrapping began 6Dec1945, completed 11Jan1946			Scrapped on berth
716	WATERLOO	"	"	14June1945- Scrapping began 1Oct1945, completed 28Nov1945			"
717	DISDAIN, originally *Ypres*	"	"		Cancelled September 1945 and used as basis for DELIGHT		
718, 719, 720		merchant	Canadian Pacific Railway Co	Built by sister company, Lithgows Ltd, Port Glasgow			
721	MONMOUTH	aircrftcarrier	Admiralty (Cancelled)		736x120, 27,015t	st, 76,000shp, 28k	
722	AFRICA	"	" (Cancelled)		897 (oa)x115.5, 56,800t (deep) t, 200,000shp, 33k		
723	LST.3037, *Evan Gibb*	LST3	"	?-30Jan1945-31May1945	345x54, 2256t	sr, 5500ihp	Scrapped 1963
724	LST.3038, *HMS Fighter*, *Empire Grebe*	"	"	?-14March1945-27July1945	345x54, 2256t	sr, 5500ihp	Scrapped 1968
725	LST.3039, *Rio Douro*, *San Bernardo*	"	"	?-27June1945-?	345x54, 2256t	sr, 5500ihp	
726	LST.3040	"	"		345x54, 2256t	sr, 5500ihp	
727	DELIGHT	destroyer	"	5Sep1946-21Dec1950-9Oct1953	390x43, 2810t	st, 54000shp, 30.25k	Scrapped 1970
728	BEAVERCOVE, *Maplecove*, *Beavercove*, *Giovanna Costa*. This number originally allocated to a cancelled Battle Class destroyer	cargorefr	Canadian Pacific Railway Co	?-16July1946-26June1947	465x64, 9824t	te, 9675shp, 16k	Scrapped 1971
729	PRINCESS MARGUERITE	passferry	Canadian Pacific Railway Co	26May1948-5March1949-?	373x56, 5911t	te,16000shp	Scrapped 1996
730	PRINCESS PATRICIA	"	"	?-5Oct1948-?	373x56, 5911t	te, 16000shp	Scrapped 1989
731	EMPRESS OF BRITAIN, *Queen Anna Marchia*, *Carnivale*, *Fiesta Marchina*, *Olympic*, *Topaz*.	passenger	"	?-22June1955-20March1956	600x85, 25516t	st, 30000shp, 21k	Scrapped 2008
732	CALEDONIA	passcargo	Anchor Line	?-12March1947-27March1948	480x66, 11252t	d, 11400bhp, 16k	Scrapped 1971
733	WARWICKSHIRE, *Chania*, *Sirius*, *Chania*	passcgoref	Bibby Line	?-14Aug1947-19Aug1948	498x60, 8903t	st, 7000shp, 15.5k	Scrapped 1980
738	CARDIFF QUEEN	excursion	P & A Campbell	?-25Feb1947-4June1947	240x30.1, 765t	p, 2200shp, 16.5k	Scrapped 1968
739 and 740		machinery	Doxford diesel for Lithgows				
741	DORIC	cargorefr	Shaw, Savill & Albion Co	12Mar1947-21Dec1948-27Jun1949	482.33x65.8, 10674t	d, 11600 bhp, 17k	Scrapped 1969
742	RUPERTSLAND, *Victorian Marchie*, *Blanc Sablon*, *Marchine Voyager*, *Freda I*	cargo	Hudsons Bay Co	?-30Nov1948-?	183x31, 663t	d, 600bhp	
744	CASLON, *Nina G*, *Orient Transporter*	cargo	Pachesham S S Co	?-17Feb1949 -8June1949	434.5v59, 5684t	st, 3960shp, 13k	Scrapped 1971
745	LEICESTERSHIRE, *Heraklion*	passcgorefr	Bibby Line	?-29June1949-17Dec1949	471x60, 8908t	sr, 7700shp	Capsized 1966
746	ENGLISH STAR	cargorefr	Blue Star Line	?-6Dec1949-?	470x70, 9996t	d, 11600bhp, 16k	Scrapped 1973
747	SCOTTISH STAR, Kavo Yerakas	"	"	?-15May1950-?	470x70, 9996t	d, 11600bhp, 16k	Scrapped 1979
749	BRITISH REALM	tanker	British Tanker Co	?-7June1951-19April1952	619x81, 18571t	st, 13750shp, 15k	Scrapped 1970
750	PRINCESS OF NANAIMO, *Princess of Acadia*, *Princess Of Nanaimo*, *Henry Osborne*	passvehcle	Canadian Pacific Railway Co	?-14Sep1950-?	337x62, 6787t	st, 8000shp	Scrapped 1974
752	CIUDAD DE MEDELLIN, *Covadonga*	cargorefr	Flota Mercante Grancolombiana S.A	?-23Jan1951-?	395x55.3, 4219t	d, 4500bhp, 16k	Scrapped 1986
753	CIUDAD DE BARQUISIMETO, *Libertador Bolivar*	"		?-22May1951-?	396x55.3, 4214t	d, 4635bhp, 16k	Deleted 1988
754	BOMI HILLS, *Santona*	ore/oil	Liberian Nav Corp	?-30Aug1951-?	600x80, 17853	d, 8500bhp, 14.75k	Scrapped 1971
755	ENDURO, *Santurce*	"	Skibs A/S Orenor	?-28April1952-?	602.9x80.3, 17881t	d, 8000bhp, 14k	Scrapped 1975
756	ATLANTIC BARON	tanker	Atlantic Oil Carriers	?-24Sep1952-26Jan1953	545x71, 12696t	d, 8000shp, 15k	Scrapped 1973
757	ATLANTIC BARONESS	"	"	?-15April1953-16Aug1953	545x71, 12712t	d, 8000shp, 15k	"
758	SIRA, *South African Pioneer*, *S A Pioneer*	cargo	Gorissen & Klaveness	?-3Dec1952-15May1953	460x62.5, 9539t	st, 7000shp, 15k	Collision 1972
759	SIMOA, *South African Transport*, *S A Transporter*	cargo	"	?-30April1953-10Nov1953	460x62.5, 9517t	s t, 7000shp, 15k	Scrapped1972
760	SJOA, *South African Trader*, *S A Trader*	cargo	Gorrissen & Klaveness	?-8April1954-18Sept1954	460x62.5 9506t	st, 7000shp, 15k	Scrapped 1972
761	SOUTH AFRICAN MERCHANT *S A Merchant*	cargo	South African Marchine Corp	?-14Dec1954-10May1955	470.2x62.8, 9506t	st, 7700shp, 15k	Scrapped 1977
762	TABOGA, *Triton*	tanker	Soc Tropica de Carga SA	?-26Oct1953-Feb1954	559.7x72, 11913t	d,8100bhp, 13k	Scrapped 1974

763		engine for *Border Laird* built at Lithgows					
764	CHATEAUGAY, *Santander*	ore/oil	Skibs A/S Orenor	?-20Jan1954-June1954	638x80, 17999tt	st, 9250shp, 15k	Scrapped 1972
765	MOISIE BAY, *Pericles G. C.*	"	"	?-30Sep1954-Feb1955	638x85, 17999t	st, 9250shp	Scrapped 1977
766	FREE STATE, *Sanovia, Antonis*	"	Torvald Klaveness	?-16March1956-June 1956	639.7x80.5,19624t	st, 9250shp	"
767	CUYAHOGA, *Santos, Ocean Builder 1, DB Ocean Builder*	"	Skibs A/S Orenor	?-6Dec1956-?	603.7x80.4, 19600t	st, 9250shp, 15k	Scrapped 2003
771	CHICHESTER	frigate	Admiralty	26Jun1953 -21Apr55 -16May1958	340x40, 2170t	d,14400bhp 23k	Scrapped 1981
772	LINCOLN	"	"	1Jun1955 - 6Apr1959 - 7Jul1960	330x40, 2170t	d, 14400bhp 24k	Scrapped 1983
775	OXFORDSHIRE, *Fairstar, Ripa*	troopship	Bibby Line Ltd	?-15Dec1955 - 13Feb1957	609x78, 20586t	st, 18000shp 20k	Scrapped 1997
779	QUEENSLAND STAR, *Brasilia Star, Queensland Star, Brasilia Star*	cargoliner	Blue Star Line Ltd	?-15March1957 - 29Nov1957	511.5x68, 10657t	d,11600bhp	Scrapped 1979
781	IRISH HAWTHORN, *Esperis, Global Unity*	tanker	Irish Shipping Ltd	?-30Aug1957-26Feb1958	530x69.25, 12168t	st, 8250shp 14.5k	Scrapped 1979
782	BRITISH ENERGY	tanker	B P Tanker Co Ltd	?-11Dec1957 - 3Oct1958	640x86, 23124t	st, 15500shp15.5k	Scrapped 1975
783	NORDIC HAWK, *Thiresia Venizelos*	"	Nordic Tankships A/S	?-30June1958-13Jan 1959	640x86, 22920t	st, 15500shp	Scrapped 1977
784	SHELL ARAMARE, *Aramare*	"	Cia Shell de Venezuela Ltda	?-8Sep1959-Feb1960	640x85, 23525t	st, 11000shp 15k	Scrapped 1981
785	SHELL NAIGUATA, *Naiguata*	"	"	?-12May1960-Sept1960	640x85, 23533	st, 11000shp	Scrapped 1979
786	IRISH BLACKTHORN, *Profitis Elias, Golden Piper, Captain Elias*	"	Irish Shipping Ltd	?-27Jan1959-3June1959	555.5x 69.58, 12168t	st, 8250shp 14.5k	Scrapped 1969
788	SHROPSHIRE, *Argiro, Naftilos*	cargoliner	Bibby Line Ltd	?-14Nov1958-2April1959	451.4x63.6, 8775t	d, 8100bhp 15.5k	Scrapped 1985
789	Engine for Rankin & Blackmore Ltd.						
790	EXETER	Type 61 frigate	Admiralty, (Cancelled 18 March 1957)				
792	YORKSHIRE, *Eastern Princess, Yorkshire, Bordabekoa, Sea Reliance*	cargoliner	Bibby Line Ltd	?-27April1960-?	491x63.9 7218t	d, 8100bhp 16k	Scrapped 1984
793 and 794		Sulzer engines for Henry Robb					
796	BRITISH GUARDSMAN	tanker	B P Tanker Co Ltd	?-12Dec1962-14June1963	760x97, 54611tdwt	st, 16000 shp 15.5k	Scrapped 1975
797	Stork diesel engine						
801	ALERT, *C.S.Alert, Al*	cablelayer	H M Postmaster General	?-7Nov1960-May 1961	417x54, 6413t	d, 7440bhp	Scrapped1995
802	KUZGUNCUK	passferry	Denizcilik Bankasi TAO	?-21Dec1960-2June1961	210x36, 781t	sr, 1600ihp 13k	Scrapped 2000
803	KANLICA	"	"	?-21Dec1960-2June1961	210x36, 781t	sr, 1600ihp 13k	
804	PENDIK	"	"	?-6Feb1961-30July1961	210x36, 781t	sr, 1600ihp 13k	Burned 1992
805	A KAVAGI, *Tunc*	"	"	?-6Feb1961-15Aug1961	210x36, 781t	sr, 1600ihp 13k	Burned 1985
806	ATAKOY	"	"	?-6April1961-15Aug1961	210x36, 781t	sr, 1600ihp 13k	Wrecked 2004
807	INKILAP	"	"	?-19April1961-1Sep1961	210x36, 781t	sr, 1600ihp 13k	
808	HARBIYE	"	"	?-19April1961-19Sep1961	210x36, 781t	sr, 1600ihp 13k	Scrapped 1995
809	IHSAN KALMAZ, *Tegmen Ali Ihsan Kalmz*	"	"	?-3July1961-27Oct1961	210x36, 781t	sr, 1600ihp 13k	Scrapped 2009
810	TURAN EMEKSIZ	"	"	?-3July1961-27Oct1961	210x36, 781t	sr, 1600ihp 13k	
811	LEECLIFFE HALL	bulk	Hall Corporation of Canada Ltd	?-18May1961-19Aug1961	729x75.42, 18071t	st, 8200shp	Collision 1964
812	GLENOGLE, *Harvest, Yang Cheng*	cargoliner	Glen Line Ltd	?-3May1962-26Sep1962	543.7x74.5, 11918t	d, 18000bhp 20k	Deleted 2002
813	GLENFALLOCH, *Qing He Cheng*	"	"	?-3July1962-26Jan1963	543.7x74.5, 11918t	d, 18000bhp 20k	"
814		Sulzer diesel for Burntisland Shipbuilding Co.					
815		Dock entrance gate for Inchgreen Dry Dock					
817	FIFE, *Blanco Encalada*	destroyer	MOD (Navy)	1June1962-9July1964-22June1966	521x53.5, 6200t	COGAS 30000shp 30k	Scrapped 2007
818	SIR LANCELOT, *Lowland Lancer, Perseverance, Glenn Braveheart, Ark*	logistic	Min of Transport	?-25June1963-16Jan1964	413x58, 6390tfl	d, 9520bhp 17k	Scrapped 2008
819	LANCASHIRE, *Safina-E-Haider*	cargo	Bibby Line Ltd	?-7May1963-29Aug1963	436x65, 8919t	d, 9600bhp	Scrapped 1993
820	cancelled ship for Bibby Line						
821 to 824		Sulzer engines for Lithgows Ltd.					
825	NILI, *Jamaica Queen, Nili, Arion*	passvehcle	Nili-Somerfin Car Ferries	?-11Aug1964-12June1965	414.6x62, 7851t	d, 11000bhp 20k	Scrapped 1983
826	AUSTRALIAN CITY, *Wilkawa, Chikuma, Marchina, Miranda*	bulk	Reardon Smith Line Ltd	?-14July1964 - 4Nov1964	601x80.2, 18621t	d, 9600bhp 15k	Scrapped 1986
827	EASTERN CITY, *Chiyoda, President Aguinaldo, Lucky Twenty Six*	"	"	?-5March1965-June1965	601x80.2, 18621t	d, 9600bhp 15k	Scrapped 1986
828	DUHALLOW, *Jana Vijay*	"	Charter Shipping Co Ltd	?-26Oct1965-14March1966	641.5x90, 25503t	d, 14700bhp 15.5k	Scrapped 1985
829	CLUDEN, *Garthnewydd, Suzanne, Suzanne H, Golden Day*	bulk	Matheson Shipping Co Ltd	?-25Sep1964 -?	626.2x85.2, 22341t	d, 13800bhp 15k	"
830	WAHINE	passvehcle	Union SS Co of New Zealand Ltd	?-14July1965-19June1966	440x 8948t	st, 18000shp 22k	Wrecked 1968

Fairfield (Glasgow) Ltd

No.	Name(s)	Type	Owner	Dates	Particulars		Fate
795	BRITISH COMMODORE	tanker	BP Tanker Co	?-18Aug1966-23Feb1967	819.6x108.3, 67862t	d, 18000bhp, k	Scrapped 1982
831	ATLANTIC CITY, *Atlantic, Olmeca, Orient Coral*	bulk	Reardon Smith Line	?-10Feb1967-20April1967	631.4x95.1, 45780t	d, 13800bhp, 15.2k	Scrapped 1985
832	FERNIE,	"	Trident Tankers	Cancelled c1965			
834	INDIAN CITY, *Eastport, Alkmini*	bulk	Reardon Smith Line	?-11Jul1967-28Sept1967	631.4x95, 45910t	d, 13800bhp, 15.2k	Scrapped 1987

Upper Clyde Shipbuilders Ltd (Govan Division)

No.	Name(s)	Type	Owner	Dates	Particulars		Fate
833	ANTRIM, *Cochrane*	destroyer	MOD (Navy)	?-19Oct1967-23Nov1970	505x53.5, 5587t	COGAS, 30000shp, 28k	Chile. Scrap'd 2010
835	CHAUVENET, *Texas Clipper II, Pacific Collector*	survey	United States Navy	?-13May1968-12Nov1970	394x52, 3035t	d, 3600bhp, 15k	
836	HARKNESS	"	"	?-12Jun1968-29Jan1971	394x52, 3035t	d, 3600bhp, 15k	
837	WELSH CITY, *Jade Star, Long Phu*	cargo	Reardon Smith Line	?-28Mar1968-6Jul1968	476.5x72, 16223t	d, 9000bhp, 16k	Scrapped 1992
838	CORNISH CITY, *Neptune Star, Neptune Star 1, Long Khan*	"	"	?-17Apr1969-17Jul1969	476.5x72, 16501t	d, 9000bhp, 16k	Scrapped 1997
839	JERVIS BAY	container	Shaw, Savill	?-3April1969-May 1970	721x100.2, 29262t	st, 32450bhp, 24k	Wrecked 24Jan1984.
840	VANCOUVER CITY, *Stasia, Male, Male II*	bulk	Reardon Smith Line	?-28Oct1969-18Mar1970	535x83.5, 26288t	d, 11600bhp, 15k	Scrapped 1997
841	PRINCE RUPERT CITY, *Yaqui, Singa Sky, Zhong Nong Xin, Herun, He Yuan, Hua Kai, Run Tong 6*	"	"	?-24Mar1970-17Jul1970	535x83.5, 16639t	d, 11600bhp, 15k	
842	PACIFIQUE	dredge	D.O.S. Dredging Co	?-26Nov1969-?	421.1x63, 1x7349t	d, 4400bhp, 13k	Scrapped 2011
844	FRESNO CITY, *Olmeca, Singa Sailor*	bulk	Reardon Smith Line	?-3Jul1970-22Dec1970	536x83.5, 25874t	d, 11600bhp, 15.25k	Scrapped 1994
101	TEMPLE BAR, *Lake Nipigon, Laketon, Lake Nipigon, Algonorth*	"	Lambert Bros Shipping	?-1Sep1970-8Apr1971	495x75, 22157t	d, 12000bhp, 15k	Scrapped 2012
108	DELTA BAY, *Goryoho, King Richard X*	dredge	Costain Blankevoort (UK) Dred. Co	?-26Jan1971-14Jul1971	410x64.5, 16714t	d, 10264bhp, 13.5k	Sank 2020
113	NORSE PILOT, *Homeric Dawn, Global Seahorse, Seahorse G, Star of Dubai*	bulk	Cardigan Shipping Co	?-26Mar1971-14Sep1971	535x83.5, 17187t	d, 11600bhp	Scrapped 1996
114	NORSE MARSHAL, *Astro Carrier, Carrie, Glory Sky, Fotini, Paris*	"	"	?-14May1971-8Feb1972	535x83.5, 17187t	d, 11600bhp, 15.25k	Wrecked 1995
115	NEW WESTMINSTER CITY, *Kassia, Fox, Singa Wilranger, Norman Ranger, Singa Ranger, Myo Hyang, Manpok*	"	Reardon Smith Nav Co	?-14Jul1971-1Feb1972	535x83.5, 25832t	d, 10600bhp, 15.25k	Scrapped 1998
116	TACOMA CITY, *Bougainville, Rosebud, Rosaleen M, Petalouda, Macforce, Force, Gokar, Gok*	"	"	?-5Nov1971-14Mar1972	536x83.5, 25842t	d, 10600bhp, 15.5k	Scrapped 1998
130	PORT ALBERNI CITY, *Northern Valley, Vitaluck, Abdul Rahman S, Ha Ni, An Da*	"	"	?-16Mar1972-14Sep1972	535x83.5, 25795t	d, 10600bhp, 15.25k	Scrapped 2000

Govan Shipbuilders Ltd

No.	Name(s)	Type	Owner	Dates	Particulars		Fate
109	GUANABARA, *Estado Da Guanabara, Draga Copacabana*	Suction D	Brazilian Government	?-15May1972-10May1973	341.3x59, 6655t	d, 6700bhp, 13k	Scrapped 2016
126	IRISH PINE, *Christinaki*	Bulk	Irish Shipping Co.	?-29June1972-?	535x83.5, 26091t	d, 10600bhp, 15k	Foundered 1994
127	IRISH MAPLE, *Lepetane, Evina, Zachary Arabella*	"	"	?-24Oct1972-21March1973	535x83.5, 25679t	d, 10600bhp, 15k	
131	IRISH OAK, *Alev, Ales K, Tahir Kiran*	"	"	?-19Jan1973-17May1973	535x83.5, 25679t	d, 10600bhp, 15k	Burned 1997
132	IRISH LARCH	"	"	?-3April1973-11July1973	535x83.5, 25707t	d, 10600bhp, 15k	Scrapped 1984
201	GOLDEN ANNE, *Zeynep K*	Bulk	World-Wide Shpg. Agency	?-1June1973-17Oct1973	535x83.5, 26369t	d, 10600bhp, 15k	Scrapped 1998
202	GOLDEN ORIOLE, *Star Oriole, Golden Oriole Nazli K, Glory Moon*	"	"	?-29Aug1973-?	535x83.5, 26369t	d, 10600bhp, 15.5k	Scrapped 2010
203	NORSE TRADER, *An Da Hai, Ji Li, Sheng*	"	Cardigan Shipping Co	?-14Nov1973-12March1974	535x83.5, 26000t	d, 10600bhp, 15k	Scrapped 2009?
207	AL MUBARAKIAH, *Rickmers Tianjin, Trident Amber, Al Mubarakiah*	Cargo	Kuwait Shipping Co	?-20Feb1974-5July1974	550.1x76.6, 23451t	d, 12500bhp, 16k	Scrapped 2000
208	AL SALIMIAH, *Trident Arrow*	"	"	23May1974-11Oct1974	551.5x76.8, 23792t	d, 12500bhp, 16k	Scrapped 2001
209	IBN BATTOTAH, *Trident Arc, Ibn Battotah*	"	"	?-3Sept1974-?	551.5x76.8, 23841t	d, 12500bhp, 16k	"
210	IBN RUSHD, *Trident Bond, Ibn Rushd*	"	"	?-29Nov1974-7March1975	551.5x76.8, 23841t	d, 12500bhp, 16k	Scrapped 2000
211	IBN HAYYAN, *Trident Beauty*	"	"	?-12Feb1975-?	551.5x76.8, 23841t	d, 12500bhp, 16k	"
212	IBN TUFAIL, *Trident Baltic, Ibn Tufail*	"	"	29April1975-24June1975	551.5x76.6, 23846t	d, 12500bhp, 16k	Scrapped 1999
213	CAMARA, *Elpidoforos, Toxon II, Yun Feng Ling, Wintec*	Bulk	A/S Det Ostasiatiske Komp.	?-24June1975-Sept1975	535x83, 26119t	d, 10600bhp, 15k	Scrapped 2002
214	CINCHONA, *Anastasios, Nini, Ontario, Tari Tari 1, Caleta, Fotini, Tini*	Bulk	A/S Det Ostasiatiske Komp	9Sept1975-27Nov1975-?	535x83.6, 26119t	d, 10600bhp, 15k	Scrapped 1998

215	*CAPE ORTEGAL, Kilmun, Esperanza V, Delena 1, Vizcaya, Andaxios, Flag Maru*	Bulk	Lyle Shipping Co	20Nov1975-2Feb1976-?	535x83.5, 26930t	d, 11600bhp, 15.5k	Scrapped 2001
216	*CAPE RODNEY, Rodney, CHL Innovator*	"	"	?-23Jan1976-17March1976-?	535x83.7, 26931t	d, 10600bhp, 15.5k	Scrapped
217	BARON NAPIER, *Anthippe L, Hira IV, Kale 1, Silver Fish, Suisen*	"	H Hogarth & Sons	?-15April1976-24June1976	535x83.5, 26931t	d, 10600bhp, 15k	Scrapped 2003
218	BARON PENTLAND, *Evangelos L, Hua Zhen*	"	"	?-29April1976-?	535x83.5, 26814t	d, 10600bhp, 15k	Scrapped
222	DONA HORTENCIA 2#, *Golden Empire, Well Speeder*	Bulk	Aklan Bulk Carriers Inc	?-14July1976-16Sept1976	535x83.5, 26587t	d, 10600bhp, 15k	Foundered 1999
223	DONA SALVADOR 3#, *Singa Sun, Star Glory*	"	Coron Bulk Carriers Inc	?-23Sept1976-8Nov1976	535x83.5, 26586t	d, 10600bhp, 15k	Scrapped 2001
224	DONA MAGDALENA, *Singa Sea*	"	Ecija Bulk Carriers	?-11Nov1976-15Dec1976	535x83.5, 26586t	d, 10600bhp, 15k	Foundered 1988
225	DONA PAZ, *Dona Paz II, Singa Saga*	"	Fuga Bulk Carriers	?-18Jan1977-?	535x83.5, 26586t	d, 11600bhp, 15k	Scrapped 1998
227	IBN SINA, *Conman II, Cosman II*	Cargo	Kuwait Shipping Co	23March1977-20May1977	535x83.5, 23,829t	d, 12500bhp 15k	Scrapped 2001
228	IBN JUBAYR, *Conman I, Cosman I*	"	"	?-19Aug1977-?	535x83.5, 23830t	d, 12500bhp, 16k	"
229	IBN BAJJAH, *Trade Glory, Yi Rong*	"	United Arab Shipping Co	?-19May1977-5July1977	550x76.6, 23000t	d, 125000bhp, 16k	
230	AL YAMAMAH	"	"	?-11Oct1977-7Dec1977	550x76, 23741t	d, 11400bhp, 16k	Scrapped 2003
232	HIJAZ, *Trident Emerald, Rickmers Qingdao, Hijaz*	"	"	?-12Jan1978-24March1978	550x76.6, 23740t	d, 125000bhp, 16k	"
233	Al RAYYAN, *Rickmers Dalian, Trident Epic, Al Rayyan*	"	"	?-10March1978-11May1978	550x76.6, 23740t	d, 114000bhp, 16k	Scrapped 2002
235	SALAH ALDEEN, *Trident Energy, Sea Emperor Salah Aldeen*	"	"	?-24May1978-12July1978	550x76.6, 23740t	d, 125000bhp, 16k	Scrapped 2003
236	BOLESLAWIEC, *Amura, Kibele, Arena*	"	Polska Zegluga Morska	?-19Oct1978-31May1979	288.7x47.9, 4390t	d, 3000bhp, 12k	Scrapped 2011
244	MIELEC, *Pyramis*	"	"	?-3Oct1979-18Mar1980	288.7x47.9, 4456t	d, 3000bhp, 13.5k	"
246	MALBORK 2#, *Malbork, Fettah, Zeyno* hull only completed at Robb Caledon.	"	"	?-5Oct1979-		d, 3000bhp, 12.25k	"
247	KOPALNIA JASTRZEBIE, *Wigry*	Bulk	Anglo Polish Ship. Venture	?-10Jan1979-?12Nov1979	492.1x72.2, 16650t	d, 8100bhp, 14.5k	Scrapped 2015
248	KOPALNIA SIEMIANOWICE, *Mamry, Mamry II*	"	"	?-13Mar1979-?Dec1979	492.1x72.2, 16650t	d, 8100bhp, 14.5k	Scrapped 2012
249	KOPALNIA SIERSZA, *Wadag, Wadag II*	"	"	?-22June1979-20Feb1980	492.1x72.2, 16750t	d, 8100bhp, 14.5k	Scrapped 2013
250	KOPALNIA MIECHOWICE, *Gradno*	"	"	?-3Oct1979-18April1980	492.1x72.2, 16750t	d, 8100bhp, 14.5k	Scrapped 2015
251	LORD BYRON, *Bolnes, Wani Bird, Balsa, Ioannis AP, John P, Harutec, Jin Hai Yuan, Jina*	"	Bishopsgate Marine HP	?-12Nov1980-23Feb1981	574.5x83.75, 26354t	d, 11600bhp	Scrapped 2014
252	LORD CURZON, *Binsnes, Wani Falcon, Banak, Bamia, Pantelis P, Huyutec, Swift Splash, Haina Fountain*	"	"	?-5Feb1981-14April1981	574.5x83.75, 26354t	d, 11600bhp, 15.5k	Scrapped 2012
253	PACIFIC PEACE, *Atlantic Huron, CSL Innovator, Christoffer Oldendorf, General Priar*	"	Movex Ltd/Lombard Discount	?-15Oct1981-10March1982	744x105, 66844t	d, 15400bhp 14.5k	Scrapped 2015
254	PACIFIC PATRIOT, *D-Francisco de Almeida, Massimiliano, Red Cherry, Sea Pride*	"	Safetran/Midlaw (No9) Lon.	?-29Jan1982-Jan1983	744x105, 66980t	d, 15400bhp	Scrapped 2012
255	SIR JOHN FISHER, *Helen, Samsun Spirit, Bright Zhejiang, Heng Tong 5*	"	James Fisher & Sons	?-8June1982-10Jan1983	744x105.8, 67368t	d, 15400bhp, 14.5k	Scrapped 2013
256	SELKIRK SETTLER, *Federal St Louis, Federal Fraser, Spruceglen*	"	Misener Transportation	?-28Jan1983-25April1983	711.2x75.7, 35315t	d, 10880bhp, 12k	
257	CANADA MARQUIS, *Federal Richelieu, Federal Mackenzie, Birchglen*		Misener Shipping	?-15April1983-July1983	711.2x75.7, 35315t	d, 10880bhp, 12k	Scrapped 2015
258	SASKATCHEWAN PIONEER, *Lady Hamilton, Voyageur Pioneer, Kaministiqua*	"	Pioneer Shipping	?-13July1983-Sept 1983	711.2x75.7, 35315t	d, 10880bhp, 12k	
259	EL CONQUEROR, *Reliance 1, Jovanna, Arnes, Ernest, Emi S*	"	Eshley Cia Nav S A	?-16Sept1982-18March1983	574.2x93.3, 34913t	d, 10400bhp	Scrapped 2011
260	LAKENES, *Faethon, Antonis, Marsi*	"	Jebsens (UK) Ltd	?-9Nov1983 - 12March1984	600x101.8, 45090t	d, 9800bhp 14k	Scrapped 2015
261	LOFTNES, *General Lukban, Wani River, Havtjeld, China Progress, Huaxinghai, Saloos, Chios Sun, Sea Wave*	"	"	?-20March1984 - 14July1984	600x101.8,, 45090t	d, 9800bhp 14k	Scrapped 2012
262	SIR CHARIES PARSONS, *Barko, Princess Ayat, Daytona Prime*	"	Bishopsgate Colliers Ltd	?-7March1985 - 12July1985	485.6x80.4, 22530t	d, 5920bhp 12.5k	Scrapped 2020
263	LORD CITRINE, *Atlantic Lord, Sider Bay, Jin Bin*	"	"	?-5June1985 - 10Feb1986	485.6x80.4, 22447t	d, 5920bhp 12.5k	Scrapped 2012
264	LORD HINTON, *Umay*	"	"	?-18Sept1985 - 28Feb1986	485.6x80.4, 22447t	d, 5920bhp 12.5k	Scrapped 2015
265	NORSEA, *Pride of York, GNV Aries*	Ferry	P&O Steam Nav Co	?-9Sept1986- 7May1987	556.1x82.32, 6419	d, 26100bhp 18.5k	

Scotstoun Marine Ltd. A subsidiary of Govan Shipbuilders Ltd

No.	Name(s)	Type	Owner	Dates	Particulars		Fate
128	HARFLEET, *Chi Grace, Atlas, Antacus*	Bulk	J&C Harrison Ltd	?-6March1973-27June1973	574x83,-25,633t	d, 11,600bhp 15.5k	Foundered 1984
129	HARFLEUR, *Chi Trust, Andromeda, Arion, Arhon*	"	"	?-14Sept1973-21Dec1973	545.8x83.7,-26,016t	d, 11,699bhp 15.5k	Scrapped 1997
204	NORSE HERALD, *Norse Carrier, Kefalonia Hope, Singa Wilmara, Norman Mariner, Mariner, Mari, Maria, Roksolana, Deborah*	Bulk	Cardigan Shipping Co Ltd	?-21May1974-Sept1974	535x83.5, 26204t	d, 10600bhp, 15.25k	Scrapped 1998
205	LOCH LOMOND, *Al-Zahra, Fastelle, Geesthaven, Star Haven, Pacific Koru, Mary K, Patagonian Pride*	Reefer	Glenfarg Shiping Co.	?- 16Dec1974-July1975	157.3x22.5, 11,167t	d, 18,800 bhp, 21.5k	*Scrapped 2000*
206	LOCH MAREE, *Al-Salama, Langelle, Geestcape, Eastcape, Pacific Gala, Claire S, Rio Negro Valley, Ice Music*	Reefer	Charles Connell	?-27May1975-Jan1976	157.3x22.5, 11,168t	d, 18,800 bhp, 21.5k	Scrapped 2006
219	IBN ABDOUN, *Noor*	Cargo	Kuwait Shipping Co.	?-16Jan1976-July1976	175x23m , 15040t,	d, 12500bhp 16k	Scrapped 2003
220	IBN-AL HAITHAM	"	"	?-8July1976-Oct1976	175x23m , 15040t,	d, 12500bhp 16k	"
221	IBN HAZM, *Trident Dusk, Sea Dragon, Ibn Hazm,*	"	"	?-22Dec1976-Feb1977	175x23m , 15040t,	d, 12500bhp 16k	Scrapped 2002
226	IBN ZUHR, *Trade Bright, Yi Ming, Wan Tong*	"	"	?-18May1977-July1977	175x23m , 15040t,	d, 12500bhp 16k	
231	ARAFAT	"	"	?-30Nov1977--Feb1978	175x23m , 15040t,	d, 12500bhp 16k	"
234	AL MUHARRAQ, *Magdalena Oldendorff, Haris, Togo Beauty, Arak*	Cargo	United Arab Shipping Co.	?-25April1978-?	175x23m , 15040t,	d, 12500bhp 16k	Scrapped 2000
237	GNIEZNO II Sadent	Bulk	Anglo Polish Ship. Venture	?-17Nov1978-15June1979	95x14.6m, 2997t,	d, 3000bhp	Scrapped 2009
238	SIERADZ *TKB Takano, JB Rubin, Orion Nebula*	"	"	?-30Nov1978-May1979	95x14.6m, 2997t,	d, 3000bhp	Scrapped 2014
239	WYSZKOW *Sider Shine, Gloriall*	"	"	?-1Mar1979-8Oct1979	95x14.6m, 2997t,	d, 3000bhp	Scrapped 2011
240	CHORZO *Selena, Sunshine, Vanino, Ser-2011 Macro*	"	"	?-29Mar1979-Jan1980	95x14.6m, 2997t,	d, 3000bhp	Scrapped 2012
241	LOMZO *Flipper, Nazmi Abi*	"	"	?-17Nov1978-15June1979	95x14.6m, 2997t,	d, 3000bhp	Scrapped 2011
242	WIELUN *Bereket, Ahmet Yusuf*	"	"	?-11Jul1979-May1980	95x14.6m, 2997t,	d, 3000bhp	"
243	WARKA *Joy 2 Milano*	"	"	?-21Sept1979-6June1980	95x14.6m, 2997t,	d, 3000bhp	Scrapped 2012

Kvaerner Govan Ltd

No.	Name(s)	Type	Owner	Dates	Particulars		Fate/Notes
266	TAIHE	container	COSCO	?-25Jan1989-Sept1989	734.94x105.65, 46136t	d, 22770bhp, 19k	Scrapped 2014
267	PUHE	"	"	?-9Jun1989-7Mar1990	734.94x105.65, 46136t	d, 22770bhp, 19k	"
301	HELICE, *Maharishi Vamadeva*	LPGtanker	Spey Marine	?-7Sept1990-April1991	635.2x105.65, 48700t	d, 1550bhp,16k	
302	HAVFROST, *BW Havfrost*	"	"	?-16Jul1991-Dec1991	635.2x105.65, 48730t	d, 1550bhp,15.7k	
303	HELIOS, *BW Helios*	"	"	?-3April1992-Aug1992	635.2x105.65, 48730t	d, 1550bhp,16k	
304	HAVIS, *BW Havis*	"	Havgas Partners	?-28Nov1992-April1993	635.2x105.65, 48730t	d, 1550bhp,15.75k	
305	JO SELJE, *Selje*	chem. tanker	Jo Selje Shipping CV	?-10Jun1993-Nov1993	598.5x 105.3, 36220t	d, 14150bhp, 15.5k	
306	BOW FLOWER	"	Slaney Shipping Co	?-25Feb1994-July1994	580.4x105.65, 36630t	d, 14190bhp, 16k	
307	NCC RIYAD, *Bow Riyad*	"	Nat. Chemical Carriers Riyadh	?-9Sept1994-Feb1995	580.4x105.65, 37252t	d, 141,00bhp	
308	BOW CLIPPER	"	West Coast Maritime	?-22April1995-15Jul1995	580.4x105.65, 37221t	d, 14190bhp, 16k	
309	OCEAN, *Atlantico*	helocarrier (LPH)	MOD (Navy)	30May1994-11Oct1995-?	623.4x93.5, 20500t	d, 18360bhp, 18k	Arr Barrow 27.11.96.
310	STIRLING FORTH, *Inverforth, Havila Fortress*	supply vessel	Stirling Marine	?-3May1996-23Aug1996	248.7x62.34, 4693t	d, 6690bhp, 14.5k	
311	EMILY P G	tankercoastal	Emily PG	?-29July1996-22Nov1996	6249x314.48, 3627t	d, 300bhp, 12.5k	
312	SEA LAUNCH COMMANDER	command ship	Swale Shipping	?-12Dec1996-22Dec1997	599.1x105.85, 50023t	d, 21200bhp, 20k	
		Rocket handling & control eqpment installed at St Petersburg Jun 1998.					
314	JO SYPRESS	chem. tanker	J O Odfjell Chem. Carriers	?-1997- March 1998	577.78x104.99, 36752t,	d, 14193bhp, 15.5k	
315	TOISA INTREPID, *GSP Altair*	supply vessel	Sealion Shipping	?-28Feb1998-24Jun1998	248.7x62.34, 4693t,	d, 6690bhp, 14k	
316	TOISA INVINCIBLE	"	"	?-1May1998-30July1998	248.7x62.34, 4689t,	d, 6690bhp, 14k	
317	FAR STRIDER	"	Farstad Supply AS	?-12Oct1998-12Jan1999	248.7x62.34, 4605t,	d, 6690bhp, 14k	
318	FAR SUPPLIER	"	"	?-20Nov1998-12Mar1999	248.7x62.34, 4546t,	d, 6690bhp, 14k	
319	CRYSTAL OCEAN	oilrecovery vessel	Crystal Ocean AS	?-16April1999-23Nov1999	100x21, 8346t, 10376t	d, 14.5k	Delivered by BAE

BAE Systems Ltd

No.	Name(s)	Type	Owner	Dates	Particulars	Notes
231	L.9732	LCU Mk10 Landing Craft	MOD (Navy)	?-Jan2002-28Jun 2002	29.80x7.40m, 112grt, 1172bhp, 9.5k	Transferred from Ailsa Nov 2000
232	L.9733	"	"	?-Feb2002-28Jun2002	29.80x7.40m, 112grt, 1172bhp, 9.5k	"
233	L.9734	LCU Mk10 Landing Craft	MOD (Navy)	?-March2002-20Aug 2002	29.80x7.40m, 112grt, 1172bhp, 9.5k	"
234	L.9735	"	"	?-24Sep2002	29.80x7.40m, 112grt, 1172bhp, 9.5k	"
235	L.9736	"	"	?-2002-7Oct2002	29.80x7.40m, 112grt, 1172bhp, 9.5k	"
236	L.9737	"	"	2002	29.80x7.40m, 112grt, 1172bhp, 9.5k	"
237	L.9738	"	"	2002	29.80x7.40m, 112grt, 1172bhp, 9.5k	"
238	L.9739	"	"	2002	29.80x7.40m, 112grt, 1172bhp, 9.5k	"
320	BALDER VIKING, *Jean Goodwill*	supply tug	Viking Offshore A/S	?-28Oct1999-14Oct2000	83.7x18.05m 3382gt, d,18272bhp, 16k	Fit-out by Havyard Leirvik, Norway.
321	WAVE RULER	oiler	MOD (Navy)	?-9Feb2001-27April2003	196.5x28.25, 31500t, d, 27200bhp, 18k	Contracted by Kvaerner
322	STIRLING ISLAY, *Islay*	supply tug	Stirling Shipmangt.	19Oct2001-11March2002	75x16m, 2544grt, d, 15000bhp, 15k	
323	MOUNTS BAY	Bay Class LSDock	MOD (Navy)	25Aug2002-9Apr2004-15Dec2005	176.6x26.4, 16160t, 16000bhp,17k	
324	CARDIGAN BAY	"	"	13Oct2003-9April2005-3Aug2006	176.6x26.4, 16160t, 16000bhp,17k	
325	LYME BAY	"	Hull built by Swan Hunter. Towed to BAE Systems, Govan, for completion arriving 22July2006. Commissioned 26Nov2007			
1062	DAUNTLESS	Type 45 destroyer	MOD (Navy)	28Aug2004-23Jan2007-3Dec2009	152.40x21.20, 8,400t, 59000shp, 30k	Fit-out at Scotstoun
1063	DIAMOND	"	"	25Feb2005-27Nov2007-22Sept2010	152.40x21.20, 8,400t, 59000shp, 30k	Fit-out at Scotstoun
1064	DRAGON	"	"	19Dec2005-17Nov2008-31Aug2011	152.40x21.20, 8,400t, 59000shp, 30k	Fit-out at Scotstoun
1065	DEFENDER	"	"	1July2006-21Oct2009-25July2012	152.40x21.20, 8,400t, 59000shp, 30k	Fit-out at Scotstoun
1066	DUNCAN	"	"	26Jan2007-11Oct2010-30March2013	152.40x21.20, 8,400t, 59000shp, 30k	Fit-out at Scotstoun
341081	QUEEN ELIZABETH	QE Class aircrftcarrier	MOD (Navy)	7July2009-17July2014-7Dec2017	(Lower Blocks 3 and 4)	
805006	FORTH	River Class OPV, batch 2	"	10Oct2014-20Aug2016-9March2017	90.5x13m, 2000t, 25k	Taken by barge to KGV dock and floated off. Fit-out at Scotstoun thereafter
805007	MEDWAY	River Class OPV, batch 2	"	8June 2015-23Aug2017-19Sept2019	90.5x13m, 2000t, 25k	"
805008	TRENT	"	"	7Oct2015-20March2018-3Aug1010	90.5x13m, 2000t, 25k	"
341081	PRINCE OF WALES	QE Class aircraftcarrier	"	16May2011-21Dec2017-10Dec2019	(Lower Blocks 3 and 4)	
805009	TAMAR	River Class OPV, batch 2	"	8Dec2016-10Oct2018-17Dec2020	90.5x13m, 2000t, 25k	as Forth
805010	SPEY	"	"	21April2017-19Jun2019-18Jun2021	90.5x13m, 2000t, 25k	"
350011	GLASGOW	Type 26 frigate	"	20July2017-3Dec2022-?	149.9x20.8m, 8,000t, 26k	Taken by barge to Loch Long (Glenmallan) and floated off for fit-out at Scotstoun
350012	CARDIFF	"	"	14Aug2019 (first steel cut)		
350013	BELFAST	"	"	29June2021 (first steel cut)		
350014	BIRMINGHAM	Type 26 frigate, batch 2	"	4April2023 (first steel cut)		
350015	SHEFFIELD	"	"			
350016	NEWCASTLE	"	"			
350017	LONDON	"	"			
350018	EDINBURGH	"	"			

CHAPTER NOTES

Notes to Era One 1830–1884

[1] This title is taken from page 63 of David Bremner's *Industries of Scotland*, published in 1869. Bremner offers a comprehensive description of the early development of the marine steam engine and shipbuilding.

[2] The principal engine builders were John Robertson at Tradeston, James Cook, Tradeston and Duncan M'Arthur, Camlachie and later Washington Street. They were followed by David Napier and Robert Napier at Camlachie. The 73 foot long *Albion* from 1816 was a typical product. The hull was built by John Wood & Co at Port Glasgow at a cost of £1000, the engines and paddle wheels by James Cook at Tradeston at a cost of £1600 and the furnishings and other fittings for a further £850, making a total of £3450 or £278,000 at 2023 prices. (Bank of England inflation calculator). They were followed by David Napier and Robert Napier at Camlachie.

[3] Duncan McArthur and James Cook also manufactured several steam engines for the vessels that would be built soon after *Comet*'s introduction.

[4] Camlachie was a district in the east end of Glasgow. Camlachie Foundry was established in 1814/15.

[5] *Memoirs and Portraits of One Hundred Glasgow Men*, James Maclehose & Sons, Glasgow 1886.

[6] John Elder's youngest brother Alexander was a partner in the Elder, Dempster Line formed in 1868.

[7] *Memoirs and Portraits of One Hundred Glasgow Men*, James Maclehose & Sons, Glasgow 1886.

[8] Charles Randolph obituary. *Glasgow Herald*, 23 November 1878.

[9] See *A Memoir of John Elder*, Macquorn Rankin.

[10] Information from *A Memoir of John Elder* by Macquorn Rankine.

[11] *Mitchell's Steam Shipping Journal*, as reprinted in *Paisley Herald and Renfrewshire Advertiser*, 2 June 1860.

[12] *Glasgow Herald*, 23 November 1878.

[13] *Glasgow Morning Journal*, 30 October 1863 and 3 November 1863.

[14] Royal dockyards which included shipbuilding and where the majority of Royal Navy warships were built at that time were significantly larger than privately owned shipyards.

[15] *Engineer*, 5 January 1866.

[16] *The Liverpool Albion* 20 June 1864.

[17] UCS2/7/16. Financial statement, Mitchell Library, Glasgow.

[18] The wreck of *Condor* is now on the US National Register of Historic Places and a North Carolina Heritage Dive Site as one of the best preserved of all the sunken Civil War blockade runners.

[19] *Glasgow Morning Journal*, 17 January 1865.

[20] Syndicated article, *The Dundee, Perth and Cupar Advertiser*, 2 February 1864.

[21] George Reith was the great uncle of John Charles Reith, the first Director General of the British Broadcasting Corporation.

[22] TCN16/9/1, Records of the Clyde Navigation Trust held at the Mitchell Library, Glasgow. All correspondence in this section is based on these records.

[23] The Amalgamated Society of Engineers, the Boilermakers and Iron Shipbuilders' Society, the Operative Blacksmiths' Society, the Shipwrights' Society, the Ship Joiners' Society.

[24] *Glasgow Evening Citizen*, 19 June 1866.

[25] UCS2/6/1, Minute of agreement, Charles Randolph, John Elder and Richard Stedman Cunliff, 13 and 14 April 1868.

[26] Andrew Duncan and, from March 1869 onward, John Deas.

[27] *Journal of the Royal United Service Institution*, Volume XII, 1869. Published by W Mitchell and Co., London

[28] *A Memoir of John Elder, Engineer and Shipbuilder*, Glasgow. W J Macquorn Rankine, LLD, Professor of Civil Engineering at the University of Glasgow.

[29] Later Sir James Bain, Lord Provost of Glasgow 1874-77 and Deputy Chairman of the Clyde Navigation Trust.

[30] *From Ticket Boy to Director, 1871-1929*, George Strachan. Privately published.

[31] A sister ship, HMS *Northampton,* was under construction at Napier's yard further up the River Clyde.

[32] Daughter of Sir Michael Robert Shaw Stewart, Lord Lieutenant of Renfrewshire. Also present at the launch were Nathaniel Barnaby, Director of Naval Construction and Admiral Houston Stewart, Third Sea Lord and Controller.

[33] *Glasgow Herald*, 6 November 1876.

[34] *On Some Results of Recent Improvements in Naval Architecture*, a paper read by William Pearce, 19 January 1881.

[35] *The Scotsman*, 8 July 1880.

[36] *Evening Citizen*, 22 September 1883.

[37] These works, still in existence, were A-Listed in 1987.

[38] *The Glasgow Herald*, 31 March 1882. A lengthy article quoting A C Kirk's paper on triple expansion engines to the Institution of Naval Architects.

[39] *Evening News and Star* 28 May 1884.

[40] *Glasgow Evening Post*, 29 November 1884.

Notes to ERA Two 1885–1965

[1] *The Engineer*, 21 December 1888, Sir William Pearce obituary.

[2] *Ibid*, 6 January 1888.

[3] Pearce, whose Glasgow address was at 10 Park Terrace, left a personal estate of £1,069,669 15s 1d, (over £142m at 2021 rates) which went to his wife, Lady Pearce.

[4] *Centenary of John Elder* by Dr Percy A Hillhouse, read on 20 November 1924. Published in the *Old Govan Club Transactions*, Volume 3, 1924.

[5] *Lloyds List,* 13 July 1889.

[6] Bank of England Inflation calculator.

[7] Minutes for March and April 1892.

[8] *Dundee Evening Telegraph,* 6 August 1892.

[9] *Dundee Courier*, 23 July 1982.

[10] Minutes, 9 December 1892

[11] Minutes, 14 June 1893.

[12] Richard Saxton White subsequently returned to Walker shipyard

as General Manager. The yard was by this time known as Armstrong Whitworth & Co Ltd. He retired from that position in 1920 and died two years later.
[13] UCS2/31/2 Private Ledger, 30 June 1894.
[14] *Warshipbuilding on the Clyde*, Volume 2, Hugh Peebles.
[15] Lloyds Banking Group Archives. GB1830 BOS/10/62/9/2 Procedure Book: Glasgow. Vol 2 1884-1893.
[16] In 1903 William Rhodes was made a director of Fairfield and, at the same time, manager of the London office. After four years, in October 1907, he resigned from the Fairfield Board and at the same time ceased to be manager of the London office and New Palace Steamers. In December 1907, George Strachan, Fairfield's Secretary, was made a director of New Palace Steamers.
[17] UCS2/. letter of 22 August 1895 to Sir Donald Currie.
[18] Scotland's People, National Probate Index.
[19] Based on the Bank of England's inflation calculator.
[20] Alexander Stewart's nephew, Iain Maxwell Stewart, appears later in this history as the driving force behind the Fairfield Experiment of 1966/7.
[21] Equivalent to £2.7m at 2023 prices. Bank of England Inflation Calculator.
[22] University of Liverpool, Cunard Archive. D42/B1/4 - Minutes: Board of Directors, Vol. 2 - 13 May 1886-10 Apr 1896.
[23] University of Liverpool, Cunard Archive. D42/B1/4 - Minutes: Board of Directors, Vol. 2 - 13 May 1886-10 Apr 1896. Fairfield Board Minutes
[24] Dr Elgar who was offered a fee of £2000 per annum for three years, subsequently negotiated a fee of £3500.
This agreement was terminated at the end of 1906.
[25] Minutes, 12 October 1904, held at the London office.
[26] Minutes, 16 November 1905.
[27] Minutes, 10 September 1909.
[28] *Steel, Ships and Men: Cammell Laird 1824-1993*, Kenneth Warren, Liverpool University Press 1999.
[29] One of the ships was SS *Mount Lebanon*.
[30] This deal also included the small steamer *Volturno* whose owner had also defaulted. This vessel was sold to MacKenzie Mann & Co for £55,000.
[31] Articles in *The Scotsman* 14 Jan 1910, 29 July 1910 and 9 Nov 1910.
[32] UCS1/1/13 Minutes of John Brown & Co Ltd, 29 September 1910, held by Glasgow University Archives. In August 1910, John Brown & Co had written to the Deputy Minister of the Canadian Naval Service, wishing to submit a joint tender for building, in Canada, warships then thought likely to be built for the new Canadian Navy. In 2014, a number of drawings were found in the plant files of BAE Systems Govan yard (formerly the Fairfield shipyard). These included several from 1910 and one dated April 1911. They depicted various shipyard schemes intended for Canada. See *Warship 2017* for an extended article on this topic.
[33] *Glasgow Herald Trade Review 1913* from which the statistics that follow were taken.
[34] Taken from Thomas Bell's speech made on 17 April 1946.
[35] Glasgow University Archives. UCS1/1/1, Minutes of John Brown & Co., 31 August 1916.
[36] Ships Cover, *Renown* and *Repulse*, National Maritime Museum, Greenwich London.
[37] Minutes, 15 December 1915 at 9 Victoria Street London.
[38] Minutes, 30 October 1917.
[39] The Admiralty memo regarding the cancellation read '. . . having regard to all the circumstances, the Board agreed that the construction of HM ships, *Anson*, *Howe* and *Rodney* should forthwith be cancelled and the slips set free for merchant construction, and that in communicating this decision to the War Cabinet, it should be made clear that the question of building additional battlecruisers will be reconsidered at the earliest possible moment after the terms of peace are finally settled: as unless further battlecruisers are built in the near future, we shall before long fall behind the United States Navy in ships of that class . . .' Extract from a confidential Admiralty Board meeting dated 27 February 1919, circulated to various Admiralty departments. From Ships Cover, HMS *Hood*, National Maritime Museum.
[40] *History of the Ministry of Munitions* Volume IV, Part II, Chapter II page 49, The Coopersmiths' Strike.
[41] *History of the Ministry of Munitions*, Volume IV, Part 5 Wages and Welfare, page 34.
[42] 1,000 in semi-skilled or skilled occupations and 800 in general labouring.
[43] *History of the Ministry of Munitions*, Volume IV, Part II, page 56
[44] Christopher Addison was Parliamentary Secretary to the Ministry of Munitions as quoted by I S McLean in his thesis *The Labour Movement in Clydeside Politics*, 1914-1922. University of Oxford 1971.
[45] *The Accident to K13*, by Percy A Hillhouse, DSc., M.I. N.A., 4 March 1919, an address to the Greenock Association of Engineers and Shipbuilders.
[46] This brochure, which was issued by the Company in 1919, reviewed war contracts and future prospects and also lists a contract for an Orient Line ship. This was most likely Fairfield's bid for *Orama*, which was subsequently built by Vickers at Barrow.
[47] Minutes, 25 April 1919
[48] Sir John Esplen resigned from the Board in August 1922. Robert A Workman and Edward McKay Edgar resigned from the Board in October 1923. William Orr Workman resigned in October 1924.
[49] Minutes, various dates referring to the indebtedness of Chepstow to Fairfield including the minutes of 6 August 1924.
[50] Referred to by their design prefix G3, the orders were placed with Wm. Beardmore, Fairfield, John Brown and Swan Hunter.
[51] The cancellation of the four 48,000-ton battlecruisers was in some small measure compensated by the building of two 35,000-ton battleships, *Nelson* and *Rodney*, permissible under the terms of the Washington Treaty.
[52] *Sunday Post*, 10 September 1922.
[53] Minutes, 29 October 1923.
[54] Shipyard Diary.
[55] Hansard, National Shipbuilders Securities Limited, 20 February 1934.
[56] Minutes, 24 November 1933.
[57] *Ibid*, 13 November 1935
[58] Minutes, 27 November 1935. This overdraft was with the Bank of Scotland. For an account of the financial predicament that Fairfield was in, see *Planning and Profits*, by Christopher W Miller.
[59] Shipyard Diary for Friday 14 June 1935.
[60] *Anson* (later renamed *Duke of York*) at John Browns; *Beatty* (later

renamed *Howe*) at Fairfield and *Jellicoe* (later renamed *Anson*) at Swan Hunter & Wigham Richardson Ltd.

[61] Minutes of the Fairfield Shipbuilding & Engineering Co Ltd, 11 April 1939. This licence would have been a major distraction from core activities, although Chairman Sir James Lithgow thought it would be good for technical staff to see how modern aircraft were manufactured.

[62] Approximately £151m in 2022 values.

[63] UCS2/27/2

[64] *Merchantman Rearmed*, David W Bone, Chatto & Windus, London 1949.

[65] The Fairfield Minutes for 12 September 1939 state that the contract for *Thunderer* was accepted, although this most likely means that the tender was accepted as a ship number was not allocated and the ship suspended soon after.

[66] Interview by Ian Johnston with John Bowen at his home in Hunstanton, Norfolk, 7 August 2014. John, who came from Rye, was sent to Fairfield in 1935 to start an apprenticeship, 'as the Clyde was the place to be,' remaing there until 1947. He was a well-known modeller of miniature ships and wrote various articles and books on the subject. Many of his models were of Fairfield ships.

[67] On 29 November 1941, a tower crane collapsed bringing down with it four steel lattice derricks. Two of these crashed through the roof of the platers' shed. No one appears to have been injured.

[68] TNA BT28/319. Report by the Machine Tool Controller, Cecil Bentham, 1942.

[69] Memo dated 12 May 1944 conjectures that *Africa* would not be laid down until *Monmouth* was launched which was likely to be August 1946, with a completion date for *Africa* of sometime in 1951. *Malta* Class Cover held at the National Maritime Museum.

[70] *Ibid.*

[71] For further information see *British Aircraft Carriers* by David Hobbs.

[72] With a deep load displacement close to 60,000 tons and a length of 900 feet, the *Malta* class were approaching the size of the present day *Queen Elizabeth* class.

[73] *Daily Mail*, 24 October 1947.

[74] Ian Johnston, *Ships for all Nations: John Brown & Co Ltd, 1847-1971*, page 266.

[75] *Dumbarton Herald*, 7 June 1947.

[76] Interview with Professor Hugh Murphy, University of Glasgow, by the author. 15 October 2023.

[77] Minutes, 20 August 1962.

[78] STV news item 1963.

[79] Shipbuilding Enquiry Committee 1965-66 Report.

[80] Minutes, 29 March 1965.

[81] Interview with Ian Morrison by the author on 22 April 2014 at Portsmouth.

[82] It was subsequently decided to wait for completion of the Severn Bridge contract.

[83] Minutes for 3 May 1965.

[84] The Fairfield Shipbuilding & Engineering Co Ltd was an English registered company, which meant that receivership rather than liquidation was available to it, as receivership was not available in Scotland at that time. Interview and explanation from Sir Robert Courtney Smith, 10 April 2013.

[85] In March 1966, at a creditors meeting, it was revealed that one of the reasons why the 'old' Fairfield had gone down was because it had lost £1m on the contract to build the ferry *Nili* alone. The meeting also noted that creditors would get only 5/1d in the pound.

[86] *The Scotsman*, 5 November 1965.

[87] *Scott Lithgow: Déja Vu all over again! The Rise and Fall of a Shipbuilding Company*. Lewis Johnman and Hugh Murphy, International Maritime Economic History Association, St John's Newfoundland 2005, page 155.

[88] National Archives, EW27/118 meeting on 3 November 1965, Papers from the Department of Economic Affairs covering the period 28 October to 15 December 1965. Papers passed to the author by Professor Hugh Murphy.

[89] *Ibid*, EW27/118, paper written 2 November 1965 for the First Secretary, George Brown, with the latter's hand written comment added on 4 November.

[90] *Ibid*, EW27/118, letter written on 19 November by Iain Stewart and read by George Brown on 22nd.

[91] *Ibid*, EW27/118, note of meeting held at 7.45pm on 23 November 1965.

[92] *Ibid*, EW27/118, memo by First Secretary, George Brown, on 25 November 1965. The memo ran to over five pages.

[93] except the ETU.

[94] Derek Palmar was an accountant, appointed to the Department of Economic Affairs under George Brown as an industrial advisor. He was heavily involved and proactive throughout all Government deliberations associated with keeping Fairfields going.

[95] National Archives, EW27/118, from the report by Derek Palmar written on 6 December 1965.

[96] Hansard, Volume 722, 9 December 1965.

Page notes Era Three 1966–1999

[1] *The Scotsman*, 28 December 1965.

[2] These included Lord Thomson, Sir Isaac Woolfson, Hugh Stenhouse and Christian Salveson.

[3] These were the Amalgamated Engineering Union, Amalgamated Society of Woodworkers, the Clerical and Administrative Workers' Union, the National Union of General and Municipal Workers and the Electrical Trades Union.

[4] Of these, 300 men who were responsible for machinery installations onboard ship were taken on by Fairfield (Glasgow) Ltd.

[5] *The Town of Chepstow*, part 2, by Ivor Waters, published by the Chepstow Society.

[6] One dictionary defines this as: 'a method of assessing wages in which a daily production target is set and a daily wage agreed on this basis. If the target is reached, the worker receives the agreed daily wage: if it is not reached or is exceeded, the worker is paid pro rata.'
Oliver Blanford had previously been a consultant with P A Management Consultants Ltd and later, Manager of the Productivity Department at Unilever (Port Sunlight) where he had introduced Work Study.

[7] The unions were: The Amalgamated Engineering Union; The Amalgamated Society of Woodworkers; The Electrical Trades Union; The National Union of General and Municipal Workers; The National Union of Sheet Metal Workers and coppersmiths;

The Plumbing Trades Union; The Amalgamated Association of Painters and Decorators; The Transport and General Workers Union; The United Paternmakers' Association; The Amalgamated Society of Boilermakers, Shipwrights, Blacksmiths and Structural Workers; The Association of Supervisory Staffs, Executives and Technicians; The Clerical and Administrative Workers' Union; The Draughtsmen's and Allied Technicians Association.

[8] *Fairfield News*, January 1967, Fairfield Heritage.

[9] *Ibid*, Volume 1, N0.12, February 1967. Fairfield Heritage.

[10] *Ibid*

[11] *Fairfield News*, Volume 2, No.4, June 1967. Fairfield Heritage.

[12] *Ibid*, Volume 1, No.5, August 1967. Fairfield Heritage.

[13] Hansard 9 March 1967, Debate on the Shipbuilding Industry Bill.

[14] *Financial Times*, November 1967

[15] *Fairfield News*, Volume 2, No.7, October 1967. Fairfield Heritage.

[16] Sir Iain M Stewart's private papers. A 13 page report on Fairfields (Glasgow) Ltd dated 17 August 1967.

[17] *Ibid*, He was referring to the books *Fairfields. A Study in Industrial Change* by Ken Alexander and C Jenkins (1970) and *Whatever Happened at Fairfields* by Sydney Paulden and Bill Hawkins (1969).

[18] The remit was 'to coordinate the study and evaluation of shipbuilding assets and facilities on the Upper Clyde and to make proposals resulting in the formation of a group for the consideration of the firms concerned in accordance with the recommendations of the Geddes Report'

[19] A copy of this report is in the papers of Iain Maxwell Stewart, held by the Stewart family.

[20] The report included Barclay Curle's recently modernised Clydeholm yard but this yard was later excluded.

[21] Quote from the minutes of Fairfields (Glasgow) Ltd Board meeting 6 February 1968. Papers belonging to the estate of Sir Iain Maxwell Stewart. The Board of Fairfields (Glasgow) Ltd remained in existence until April 1968.

[22] Sir Iain Stewart's son Max in conversation with the author, recalls that his father was infuriated that Fairfield (Glasgow) Ltd was not allowed to run its course rather than be cut short and incorporated into Upper Clyde Shipbuilders. Stewart felt certain that had George Brown been in a position to influence events that Fairfield would have continued to a conclusion particularly as it was on its way to profitability. He was equally infuriated and saddened that Jim Houston, the architect of many of the new methods introduced at Fairfield was not given a management position in Upper Clyde Shipbuilders.

[23] UCS2/7/9 Report of visit to Swedish yards 14-19 October 1968.

[24] UCS1 5/1/2, Minutes of Upper Clyde Shipbuilders 3 February 1970.

[25] Report dated March 1968. Papers belonging to the estate of Sir Iain Maxwell Stewart.

[26] *The Scotsman* 19 December 1969.

[27] *Ibid*, 6 February 1970

[28] *Ibid*, 2 September and 22 October 1970

[29] *Ibid*, 2 September 1970.

[30] *Upper Clyde News*, Volume 2 No 3 July 1970.

[31] *Ibid*, Volume 3 No 1 March 1971.

[32] An expression used by the Chancellor of the Exchequer (Anthony Barber) and the Secretary of State for Trade and Industry, John Davies, on 4 November 1970.

[33] UCS1 5/1/2, Minutes of Upper Clyde Shipbuilders, 10 June 1971.

[34] Issues at Clydebank prolonged the Work-in at that yard until early October.

[35] Sir Robert Courtney Smith was Interviewed by the author at Fairfield Heritage on 10 April 2013.

[36] Master Plan reports by Maynard PRC, February 1973 held at Fairfield Heritage.

[37] *The Daily Record*, 29 March 1974

[38] *The Scotsman*, 13 February 1975

[39] *Ibid*, 30 April 1976.

[40] *The Scotsman*, 6 August 1976.

[41] The Shipbuilding Intervention Fund of £65m was established in 1977.

[42] *Glasgow Herald*, 24 May 1978

[43] Press release 'News from British Shipbuilders' 14 November 1978. Fairfield Heritage.

[44] Of the fourteen Polish ships allocated to Govan Shipbuilders, seven were built at Govan and eight at Scotstoun. *Malbork 2* was launched at Govan but completed by Robb Caledon at Leith.

[45] Between March 1979 and March 1982, a total of 2390 workers at Govan Shipbuilders and Scotstoun Marine had been made redundant.

[46] Jack Webster, *Daily Express*, 12 December 1979.

[47] Three months after the Gilchrist debacle, Michael Casey was replaced by Robert Atkinson.

[48] Liberty Maritime: Default of Charter Hire. A paper for British Shipbuilders Executive Committee Meeting 7 September 1982 by T J Parker. Fairfield Heritage.

[49] *Shipbuilding News*, August 1980.

[50] Author's meeting with Bill Gallagher on 29 November 2022 and emails thereafter. Bill was Deputy Managing Director and Production Director 1980-88 and Operation Director at Kvaerner Govan 1988-90.

[51] British Shipbuilders Report and Accounts 1980-81, Fairfield Heritage.

[52] Hansard British Shipbuilders, volume 64, 25 July 1984.

[53] *Ibid*, various dates.

[54] Two ferries were ordered, the second one originally going to Dutch shipbuilders. When they were unable to proceed, the order was given to Nippon Kokan, Yokohama.

[55] *Clydesider*, Spring 1986

[56] The French Government was subsequently fined by the EEC.

[57] See - https://www.company-histories.com/Kvaerner-ASA-Company-History.html

[58] Kvaerner Govan Board Meeting, 6 December 1988 held at Fairfield Heritage.

[59] Comment made to the author by Steinar Draegebo.

[60] Steinar Draegebo was awarded the OBE for service to the British shipbuilding industry in 1993.

[61] Study of Hull Erection Methods at Kvaerner Govan Ltd. Phase IIA 'Ship Production Strategy' by Total Transportation Systems (International) AS, 23 May 1989. Fairfield Heritage.

[62] *Ibid*, 23 May 1989.

[63] Online interview with Steinar Draegebo, Wednesday 18 May 2022

[64] Email correspondence between the author and Steinar Draegebo, 9 October 2023.
[65] GD1326/8/8/1 Kvaerner Govan Board Meeting, 28 September 1989.
[66] KGL's price amounted to £28.8m, while Japanese and Korean bids were £25.8m. However, Chinese, Brazilian and Spanish yards tendered at £22m.
[67] KGL's price of £29m was £6m above the Japanese competition. The Minutes note that the high price of the yen and our finance package had closed the gap.
[68] Kvaerner Govan Board Meeting, 6 December 1988 and 0211.T Board Meeting 1 November 1988. Fairfield Heritage.
[69] *Clydesider*, March 1990.
[70] Email correspondence with David Millar, formerly Quality Control Manager, Kvaerner Govan Ltd., on 21 December 2021.
[71] *Clydesider* February 1991.
[72] Email correspondence between Bobby Gordon, Personnel, Employee Relations & Communications
Director and the author on 21 and 22 April 2022. Bobby describes Steiner Draegebo as the best boss he ever worked for.
[73] Email correspondence between the author and Steinar Draegebo, 9 October 2023.
[74] *Ibid*, 9 October 2023.
[75] it would eventually be reduced to just over 3%.
[76] *Clydesider*, July 1991.
[77] *Ibid*, November 1991.
[78] *Clydesider*, Issue 23, May 1991. The Govan team won 6-2.
[79] *Ibid*, December 1992.
[80] Jo Tankers was owned by Odfjell from 1915 until 2016.
[81] *Clydesider*, Issue 36, December 1993.
[82] GD1326/8/8/1, KGL management meeting.
[83] *Clydesider*, Issue 41, April 1995.
[84] *The Scotsman*, 15 February 1995.
[85] Landing Platform Helicopter.
[86] *Clydesider*, Issue 42, April 1996
[87] GD1326/8/8/1, KGL management meeting 6 July 1992.
[88] Report by the Comptroller and Auditor General. Ministry of Defence: The Award of the Contract for the Landing Platform for Helicopters. National Audit Office. HMSO 29 July 1993. Section 3.33
[89] GD1326/8/8/1. Production report to week 36 (1/9/95). Everything was expressed in man-hours. As Bobby Gordon said, man-hours were king!
[90] GD1326/8/8/1, KGL management meeting 14 February 1994.
[91] *Ibid*, 16 May 1994
[92] *Ibid*, 16 May 1994.
[93] Although the contract was signed, following design changes requested, the contract subsequently expired. Ship No. 313 had been allocated to this contract
[94] GD1326/8/8/1, KGL management meeting 16 May 1994 and 6 November 1995.
[95] GD1326/8/8/1, KGL management meeting 27 September 1995.
[96] *Clydesider*, Issue 43 July 1996.
[97] letter from Sigbjørn Ellingsen to all employees dated 20 March 1997 by courtesy of Michael Conway.
[98] Two survey ships were ordered from Appledore Shipbuilders in June 2000.
[99] *The Scotsman*, 22 September 1997.
[100] *Ibid*, 12 Nov 1997.
[101] *Ibid*,18 Nov 1997 and 19 Nov 1997.
[102] Phone call with Paul Friedberg, 20 July 2023.
[103] Email correspondence with David Millar on 21 December 2021.
[104] *Back from the Brink*, Jamie Webster, p20.
[105] https://www.navylookout.com/in-focus-the-wave-class-tankers-of-the-royal-fleet-auxiliary/
[106] https://www.heraldscotland.com/news/12212076.bae-systems-will-take-20-year-lease-on-govan-site-after-225m-purchase-kvaerner-full-of-praise-for-clydeports-sale-role/
[107] The order for the six RoRo ferries, announced in the Strategic Defence Review of 1998, was ultimately given to the German Flensburger yard (4) and Harland & Wolff (2), the latter yard also in difficulties.

Page notes Era Four 1999–2024

[1] The Strategic Defence Review, 1997-1998 published on 3 September 1998, House of Commons Library.
[2] These vessels would later emerge as Type 45 air defence destroyers, Type 26 anti-submarine frigates and Type 31 general purpose frigates
[3] Hansard, 11 July 2000, Type 45 destroyers.
[4] Harland & Wolff completed its last ship *Anvil Point* in 2003.
[5] BAeSEMA was a joint venture between British Aerospace and the French Sema Group formed in 1991. BAeSEMA was dissolved in 1998 when British Aerospace acquired Sema's share of the business.
[6] Information about the 'Bay' class contracts from The Landing Ship Dock (Auxiliary) Project, published by the National Audit Office in November 2007.
[7] Hansard, 14 July 2006 Landing Ship Dock (Auxiliary).
[8] On completion in May 2004, *Woolston* was the first vessel to be built in Portsmouth for 37 years. The Woolston yard closed in the Autumn of 2003.
[9] See *World Naval Review 2014*, published by Scaforth, 'From Daring to Duncan: Building the Type 45 Destroyers', Ian Johnston and Paul Sweeney.
[10] Interviewed by the author on 28 September 2012
[11] BAE Systems retained Maritime Services and high-end naval equipment and combat systems business there where it employed 3,200 people.
[12] *Naval Technology*, 25 May 2014.
[13] https://www.navaltoday.com/2021/01/13/final-river-class-batch-2-opv-joins-royal-navy/
[14] National Security Strategy and Strategic Defence and Security Review 2015.
[15] Reported in *Glasgow Live*, July 2022
[16] Email to the author, 23 August 2023.
[17] In June 2018, BAE Systems pulled off a major sales success when it announced that the Royal Australian Navy had selected the Type 26 design as the basis for its *Hunter* class frigates. A year later the Royal Canadian Navy announced that it too had selected the Type 26 as the basis design for its future frigate programme. While these vessels would be built in local shipyards, all design work associated with them will be conducted in the UK.

SOURCES

Glasgow City Archives
Most of the records of the Fairfield shipyard are held by Glasgow City Archives based at the Mitchell Library, Glasgow. The first deposits were made following the failure of Upper Clyde Shipbuilders and thereafter following the eclipse of Govan Shipbuilders and Kvaerner Govan. Glasgow City Archives also hold the records of the Clyde Navigation Trust who played at times a less than constructive role in the development of the shipyard.

In putting this book together, the main source of information has been:

UCS2/1/1 Minutes of the Fairfield Shipbuilding & Engineering Co Ltd, from 1888 to 1965. The minute books from January 1895 to July 1899 and November 1951 to October 1959 are missing.
UCS2/27 Shipyard Diary.
UCS2/6/1-15 Various documents concerning John Elder & Co.
UCS2/118 Ships' plans.
UCS2/132 Photographs.
UCS2/6/28 West Yard construction.
GD1326, Kvaerner Govan Ltd. The minutes of Management Meetings from August 1989 to January 1996, held at the Mitchell Library in Glasgow. The minutes from February 1996 onwards appear not to have been preserved.

Clyde Navigation Trust
TCN16/9/1 correspondence between George Reith and John Elder 1863–1873.
TCN16/9/2 basic yard plan with details of bridge at mouth of fitting-out basin 1870–1890

University of Liverpool
Special Collections & Archives. The Cunard Archives:
D42/S14/8/4 to 8 Minutes 1886–1890
D42/B1/4 to 5 Minutes 1896–1903

Ballast Trust
Thanks to Kiara King, for access to papers relating to the Govan yard prior to BAE Systems ownership.

Lloyds Banking Group Archives, Sighthill, Edinburgh
Thanks to Sian Yates for access to John Elder & Co, banking records

The British Newspaper Archive, various dates.

Photographs are by courtesy of:
Glasgow City Archives
The National Records of Scotland
Imperial War Museum
Fairfield Heritage
BAE Systems
The author's collection
or as otherwise indicated.

BIBLIOGRAPHY

Books

A History of Shipbuilding on the Clyde, Various authors, BAE Systems, Naval Ships, 2006
Alexander, K J W and Jenkins, C L. *Fairfields. A Study of Industrial Change*, Allen Lane The Penguin Press 1970
Blanford, Oliver, *The Fairfields Experiment*, The Industrial Society, London 1969
Bone, David W, *Merchantman Rearmed*, Chatto & Windus, London 1949
Bremner, David, *The Industries of Scotland. Their Rise, Progress, and Present Condition*, Edinburgh 1869
Brotchie, T C F, *The History of Govan Burgh and Parish*, Cossar, Govan 1905.
David Napier Engineer 1790–1869, an Autobiographical Sketch with Notes (David Bell), James Maclehose & Sons, Glasgow 1912.
Fletcher, R A, *Steamships: The Story of their Development to the Present Day*, J B Lippincott Co, Philadelphia 1910
History of the Ministry of Munitions, HMSO, London, 1922
Hume, J R, *Govan Shipbuilding*, Friends of Govan Old, 2016
Johnman, Lewis and Murphy, Hugh, *British Shipbuilding and the State since 1918*, University of Exeter Press 2002
Johnman, Lewis and Murphy, Hugh, *Scott Lithgow: Déja Vu all over again! The Rise and Fall of a Shipbuilding Company*, International Maritime Economic History Association, St John's Newfoundland 2005
Johnston, Ian, *Ships for all Nations:, John Brown & Co Ltd, 1847–1971*, Seaforth Publishing 2014
Macquorn Rankine, William J, *A Memoir of John Elder Engineer and Shipbuilder*, Wm Blackwood & Sons, Edinburgh 1871
MacLehose, James, *Memoirs and Portraits of 100 Glasgow Men*, Glasgow 1886.
Meek, Marshall, *There Go the Ships*, The Memoir Club, Spenymoor, County Durham 2003
Miller, Christopher W, *Planning and Profits*, Liverpool University Press 2018
Mortimer, J E, *History of the Boilermakers' Society*, George Allen & Unwin 1973
Napier, James, *The Life of Robert Napier*, Blackwood, Edinburgh 1904
Paulden, Sydney and Hawkins, Bill, *Whatever Happened at Fairfields?*, Gower Press, London 1969.
Peebles, Hugh B, *Warshipbuilding on the Clyde*, John Donald, Edinburgh 1987
Riddell, John, *Clyde Navigation*, John Donald, Edinburgh 1979
Shields, John, *Clyde Built*, MacLennan, Glasgow 1949
The Fairfield Shipbuilding and Engineering Works, Engineering, London 1909
Thornley, F C, *Past and Present Steamers of North Wales*, T Stephenson & Sons Ltd 1952
Warren, Mark D, *Campania and Lucania*, reprinted from *Engineering 1893*, Patrick Stephens 1993

Warren, Kenneth, *Steel, Ships and Men: Cammell Laird 1824–1993*, Liverpool University Press 1999

Webster, Jamie and Walker, Russell, *Back from the Brink*, Brown Son & Ferguson Ltd, Glasgow 2008

Fairfield 1860–1960, Barbour MacLaren, Glasgow 1960

Papers, Reports and Articles

Diaper, S, 'The Sperling Combine and the Shipbuilding Industry: Merchant banking and industrial finance in the 1920s', in J J Van Helten and Y Cassis (eds), *Capitalism in a Mature Economy*, Edward Elgar Publishing 1990

'The Late David Elder', *Engineering*, 16 February 1866

Elder, John, 'Circular Ships of War with Immersed Motive Power,' A paper read at the Royal United Service Institution, London 1869

Financial Exports, Capital Exports and British Industry, 1870–1939, Aldershot 1990, pp71–94

Goulaeff, E E, 'The Russian Imperial Yacht *Livadia*', *Journal of the Franklin Institute*, August 1880

Hillhouse, Percy A, DSc, MINA, 'The Accident to *K13*', an address to the Greenock Association of Engineers and Shipbuilders, 4 March 1919

Historical and Descriptive Account of the Works of the Fairfield Shipbuilding and Engineering Co Ltd. A 56-page booklet published privately in 1896

The Iron and Allied Industries of Scotland VII. Messrs John Elder and Cos works, Compound Engines, Iron, 2 May 1874

author's

Johnston, Ian and Sweeney, Paul, 'From *Daring* to *Duncan*, Building the Type 45 Destroyers', *World Naval Review 2014*, Seaforth Publishing 2013

Millar, W J C F, 'On Early Clyde Built Steamers', *Transactions of the Institution of Engineers and Shipbuilders in Scotland*, 1881

Murphy, Hugh, 'An Anatomy of Speculative Failure: Wm Doxford & Sons Ltd, Sunderland, and the Northumberland Shipbuilding Company of Howdon on Tyne, 1919–1945', *The Mariner's Mirror*, 104:1 2018, pp58–72

Hillhouse, Percy, 'A Centenary of John Elder', *Old Govan Club Transactions*, read on 20 November 1924

Pearce, William, 'On Some Results of Recent Improvements in Naval Architecture', a paper read at the Naval and Marine Engineering Exhibition on 19 January 1881. The Exhibition ran from December 1880 to January 1881 in Glasgow

Shipbuilding Enquiry Committee 1965–66 (The Geddes Report), London, HMSO 1966

Skelton, Vice Admiral Sir Reginald W, 'The Work of Andrew Laing, Engineer', *North East Coast Institution of Engineers and Shipbuilders*, 1932

Strachan, George, *From Ticket Boy to Director, 1871–1929*, Privately published

Waters, Ivor, *The Town of Chepstow, part 2*, the Chepstow Society

Journals

The Engineer
Engineering
The Marine Engineer
Marine Week

Shipyard House Journals:
Fairfield News
Shipbuilding News
Upper Clyde News
Clydesider

INDEX

Shipyard cranes

Fitting-out cranes

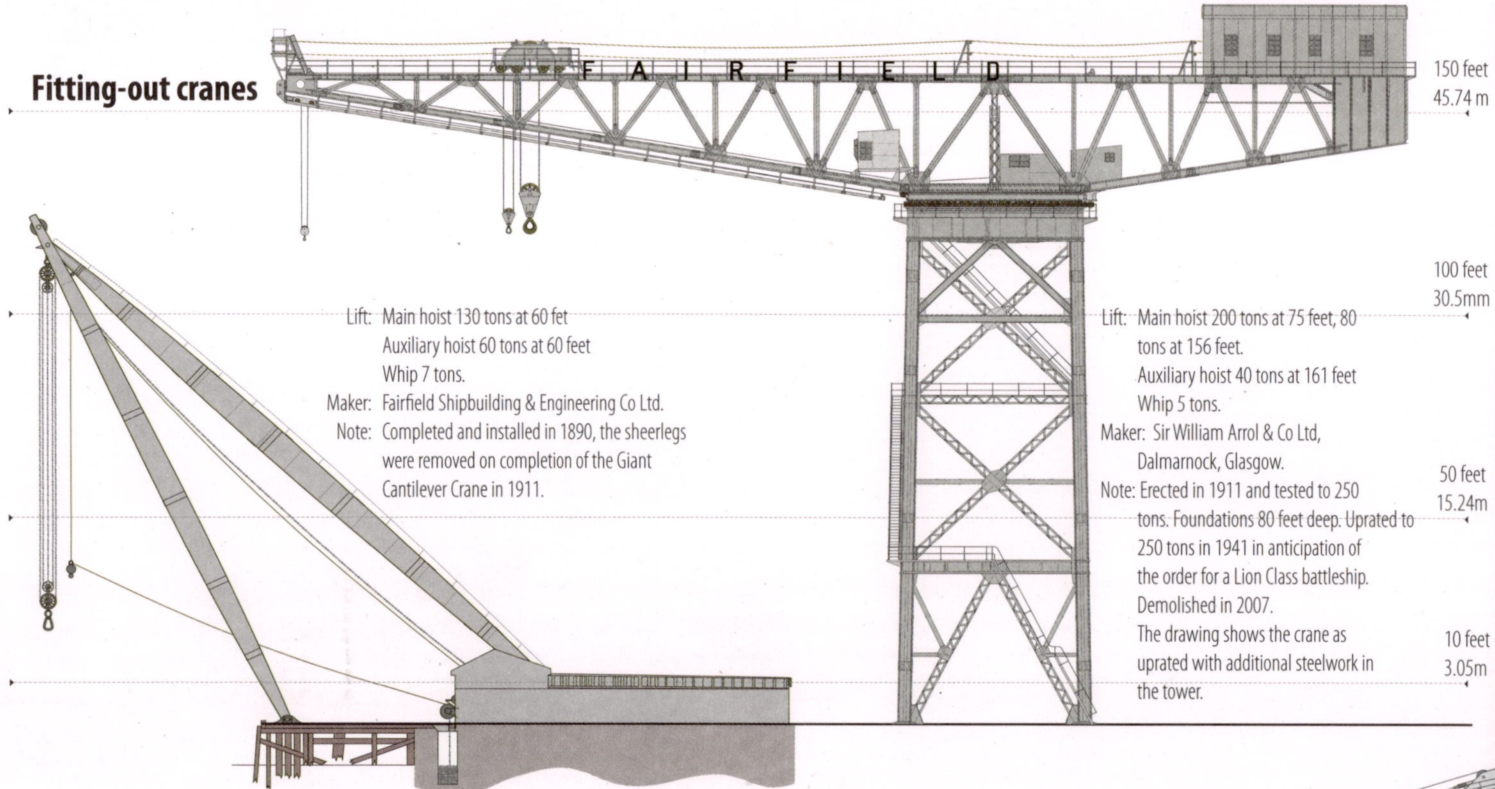

Lift: Main hoist 130 tons at 60 fet
Auxiliary hoist 60 tons at 60 feet
Whip 7 tons.
Maker: Fairfield Shipbuilding & Engineering Co Ltd.
Note: Completed and installed in 1890, the sheerlegs were removed on completion of the Giant Cantilever Crane in 1911.

Lift: Main hoist 200 tons at 75 feet, 80 tons at 156 feet.
Auxiliary hoist 40 tons at 161 feet
Whip 5 tons.
Maker: Sir William Arrol & Co Ltd, Dalmarnock, Glasgow.
Note: Erected in 1911 and tested to 250 tons. Foundations 80 feet deep. Uprated to 250 tons in 1941 in anticipation of the order for a Lion Class battleship. Demolished in 2007.
The drawing shows the crane as uprated with additional steelwork in the tower.

Berth cranes

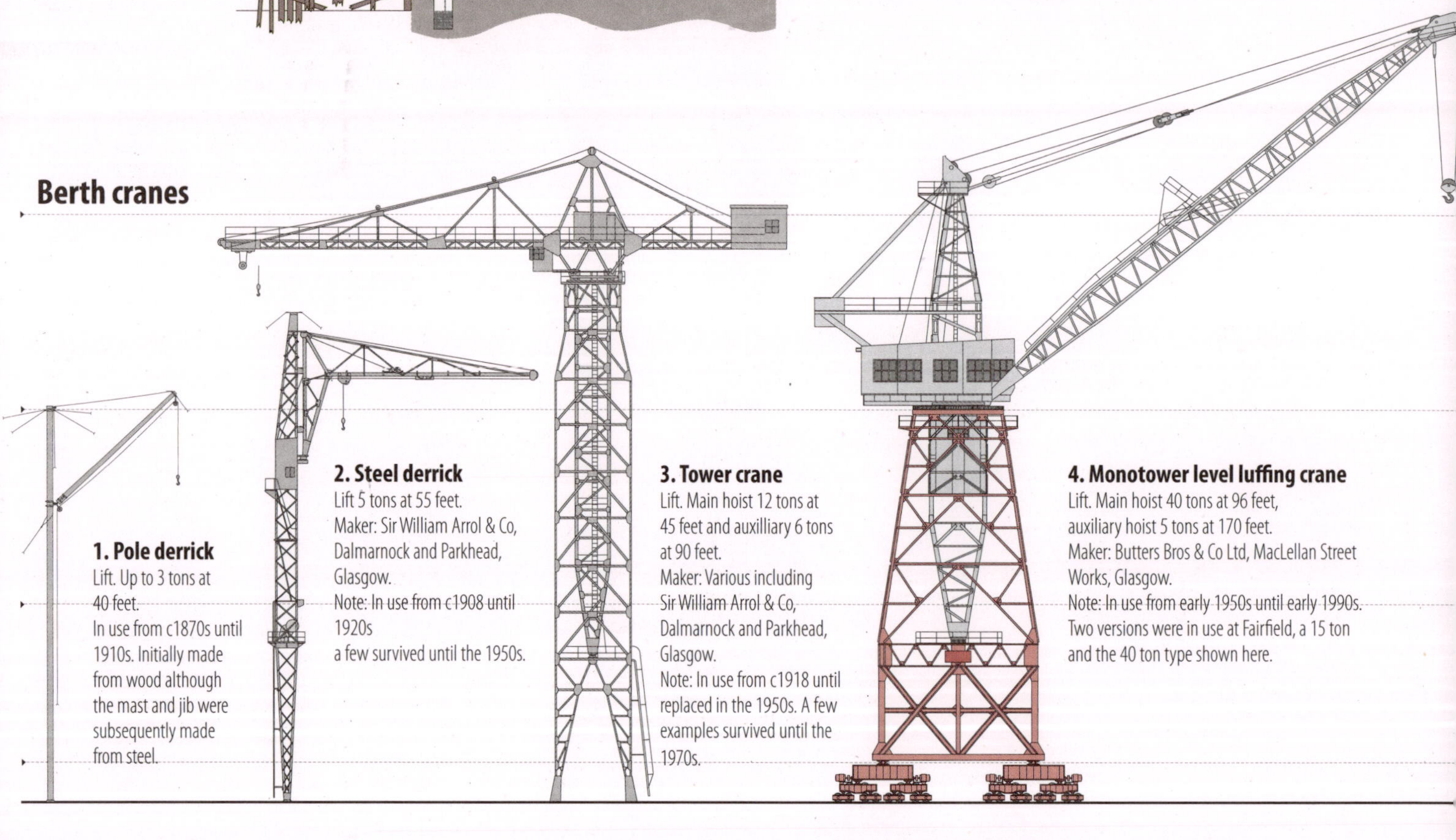

1. Pole derrick

Lift. Up to 3 tons at 40 feet.
In use from c1870s until 1910s. Initially made from wood although the mast and jib were subsequently made from steel.

2. Steel derrick

Lift 5 tons at 55 feet.
Maker: Sir William Arrol & Co, Dalmarnock and Parkhead, Glasgow.
Note: In use from c1908 until 1920s
a few survived until the 1950s.

3. Tower crane

Lift. Main hoist 12 tons at 45 feet and auxilliary 6 tons at 90 feet.
Maker: Various including Sir William Arrol & Co, Dalmarnock and Parkhead, Glasgow.
Note: In use from c1918 until replaced in the 1950s. A few examples survived until the 1970s.

4. Monotower level luffing crane

Lift. Main hoist 40 tons at 96 feet, auxiliary hoist 5 tons at 170 feet.
Maker: Butters Bros & Co Ltd, MacLellan Street Works, Glasgow.
Note: In use from early 1950s until early 1990s. Two versions were in use at Fairfield, a 15 ton and the 40 ton type shown here.